Conceptual Thermodynamics

Denis Sunko

Conceptual Thermodynamics

A Contextual Introduction to Statistical Physics

Springer

Denis Sunko
Department of Physics
University of Zagreb
Zagreb, Croatia

ISBN 978-3-032-20876-7 ISBN 978-3-032-20877-4 (eBook)
https://doi.org/10.1007/978-3-032-20877-4

This Springer imprint is published by the registered company Springer Nature Switzerland AG
The registered company address is: Gewerbestrasse 11, 6330 Cham, Switzerland

This book is dedicated to all who are motivated intrinsically.

Preface

This book is intended primarily for future researchers in physics, meeting statistical physics as a separate subject for the first time. It is also intended for all who search for a framework to think about the world they see. In the past, such frameworks were provided by a set of concentric social circles, one's family, friends, local community, nation, religion.

The rise of digital communication has destabilized this hierarchy, providing an intuitive experience of macroscopic thermodynamic phenomena to a large global population. With its amazing plasticity, the human mind has already started to make sense of this new reality in common truisms such as "if you are not the customer, you are the product." The rise in artificial intelligence has only increased the need for such simple and robust insights.

The ambition of this book is to make as many as possible aware that thermodynamics, together with its statistical rationalization in terms of constituent atoms, is a framework already in place which makes it possible to have a coherent discussion of these developments. Coherent means that the same words have the same meaning to all participants. The thermodynamic framework is a language, which one can learn to speak understandably. Learning to speak it at the level of an introductory textbook in statistical physics is the easiest way to learn it.

Thus, I make no apology for the hard physics, which makes the foundation and bulk of this book. I have tried to make it accessible to any student who has been exposed to mathematical analysis beyond high school, such as is taught in first-year technical studies or economics. I believe that that accessibility will help students of physics as well, at least countervailing the tendency of beginners to get embroiled in technical detail. Still, the book should be sufficient to study for an examination at the level of a third-year course in statistical physics in most physics departments today, if paired with a problem set adapted to that course's particular requirements.

With the above wider purpose in mind, I have tried to enrich the physics core of the book with as much context as possible within reasonable bounds. Examples are taken from biology, economics, sociology, and history. They are meant to emphasize that observations in these fields have a commonality based on underlying stabilization mechanisms, in other words: It is possible to compare them in a *literal*, not *literary*, sense.

Zagreb, Croatia Denis Sunko

Competing Interests The author has no competing interests to declare that are relevant to the content of this manuscript.

Ethics Approval The author declares that this manuscript does not include primary studies involving animals.

How to Read This Book

With this book's multiple purposes in mind, it has been arranged to facilitate various approaches to reading. In particular, chapters may be read in almost any order. Other variations include:

1. Look at the figures only. Imagine other captions for the same figure, or other figures for the same caption.
2. Skip the subsections. All the forgettable technicalities are in subsections, even if some subsections are not technical.
3. Read the Literature chapter only. Use it to avoid textbooks.
4. Re-read just one chapter until you know it by heart.
5. Open the book at random and read a few pages. Repeat at will.
6. Read systematically from first to last page, in that order. Pass the exam.

Contents

1 Introduction 1
- 1.1 Stability and Existence 1
- 1.2 The Ontological Framework 2
- 1.3 The Gnoseological Framework 5
- 1.4 Basic Thermodynamic Concepts 6

2 Thermodynamics 11
- 2.1 Absolute Zero 12
- 2.2 The Second Law of Thermodynamics 13
- 2.3 The First Law of Thermodynamics 14
- 2.4 Entropy 16
  - 2.4.1 Calibrating the Thermometer 16
  - 2.4.2 Reversible and Irreversible Motion 17
  - 2.4.3 Analysis of a Carnot Cycle 18
  - 2.4.4 The Efficiency of a Reversible Engine 18
  - 2.4.5 Clausius' Theorem. Maximization of Entropy 20
- 2.5 Mole, Molecule, Temperature 21
- 2.6 Thermodynamic Potentials. Equations of State 22
  - 2.6.1 Minimization of Internal Energy 24
  - 2.6.2 Enthalpy 25
  - 2.6.3 Legendre Transforms. The Gibbs-Duhem Relation 26
  - 2.6.4 Massieu Potentials 28
  - 2.6.5 Practical Calculations. Susceptibilities 29
  - 2.6.6 Stability. Le Chatelier's Principle 31
- 2.7 Useful Engines. Feedbacks 33
  - 2.7.1 Invisible Control 35
  - 2.7.2 Spontaneous Symmetry Breaking. Purposeful Design 36
- 2.8 Purposeful Engines 38

3 Statistical Foundations 41
- 3.1 Random Equalization of Concentrations 42
- 3.2 Thermalization in Classical Mechanics 44

3.2.1 Compatibility of Classical Mechanics and Statistics 45
3.2.2 Classical Mechanics of a System in a Container 47
3.2.3 The Thermalization Time. Choice of Statistical Distribution 50
3.3 Statistical Independence and Additive Invariants 52
3.4 Statistical Entropy 53
3.4.1 Microscopic Entropy 54
3.4.2 Relevant Entropy 54
3.5 Connection with Thermodynamics 56
3.6 The Ensemble: A Universal Random Model 58
3.6.1 Thermodynamic Equations of State 59
3.6.2 Statistical Derivation of the First Law of Thermodynamics 62
3.7 The Probability of a Fluctuation 64
3.7.1 Volume Fluctuations 65
3.7.2 Nonlinear Responses 66
3.7.3 Purposeful Governance 66
3.7.4 Mathematical and Theoretical Physics 67
3.8 The Denominator of the Fluctuation Probability 69

4 Ensembles 71
4.1 Classical Ideal Gas in the Microcanonical Ensemble 72
4.2 Indistinguishability of Identical Particles 76
4.2.1 Gibbs' Explanation 77
4.2.2 Extensivity in Society 78
4.2.3 Macroscopic Manipulation of Microscopic Distinctions 80
4.2.4 The Sackur-Tetrode Formula Revisited. Non-extensivity 81
4.3 Fixed Temperature: Canonical Ensemble 83
4.3.1 Classical Ideal Gas at Fixed Temperature 85
4.3.2 Heat Capacity and Latent Heat 86
4.3.3 Entropy and Heat 87
4.4 Fixed Chemical Potential: Grand Canonical Ensemble 88
4.4.1 Classical Ideal Gas at Fixed Chemical Potential 89
4.5 The Maxwell-Boltzmann Distribution 91
4.5.1 Maximization of Boltzmann's Entropy. Quasiparticles ... 92
4.6 Equipartition of Energy 95

5 Quantum Statistical Physics 97
5.1 Indistinguishability 98
5.1.1 Quantum Correlations 99
5.1.2 Connection Between Spin and Statistics 100
5.2 Parametric and Laboratory Space 101
5.3 Volume of Phase Space 102
5.4 The Gauge Degree of Freedom 102

5.4.1 Example: The Meissner Effect 104
5.5 Simple Quantum Systems 106
5.5.1 Arbitrary Number of States of the Same Energy 106
5.5.2 Two Particles Without Interaction 107
5.5.3 Two States of Different Energies 109
5.6 Sums over States for Free Particles 110
5.6.1 Particle Number and Occupation Probability 111
5.6.2 Classical Ideal Gas 112
5.7 Time in Quantum Mechanics 114
5.8 Coherence and Dissipation 115
5.8.1 Bell's Theorem 117
5.8.2 Bell's Inequalities 119
5.8.3 The Logic of Bell's Inequality 122
5.9 Logical Entropy 123
5.9.1 Counting Distinctions 125
5.9.2 Subsets and Partitions 126
5.9.3 Logical Entropy of Quantum States. Density Matrices ... 127
5.9.4 Entangled States and Bell's Inequality 129
5.10 Coherence in Society 132
5.10.1 The Rug-Seller's Story 132
5.10.2 Prisoner's Dilemma 133
5.10.3 Ostracism 134
5.10.4 Heresy of the Second Kind 135
5.11 Quantum Mechanics and Free Will 135
5.12 Timeless Strangeness 136

6 Fermions 139
6.1 Ground State. Fermi Level 139
6.2 Fermions at Low Temperatures 140
6.3 Level Density 143
6.3.1 The Case of Free Particles 144
6.4 Number Fluctuation 145
6.5 Entropy and Heat Capacity 147
6.6 Practical Determination of the Fermi Energy 149
6.6.1 Fermi Energy at $T = 0$ 149
6.6.2 Fermi Energy at $T > 0$. Sommerfeld Expansion 150
6.6.3 Behavior of the Chemical Potential 152
6.7 Landau Fermi Liquid 153
6.7.1 Evolution of One-Particle States with the Interaction 153
6.7.2 Effective Mass 155
6.8 Superconductivity 156
6.8.1 Superfluidity in ^{3}He 157

7 Bosons 159
7.1 Bose-Einstein Condensation 160
7.2 Chemical Potential and Transition Temperature 161

7.2.1 The Meaning of the Chemical Potential ... 162
7.3 Entropy and Heat Capacity ... 162
7.4 Continuous Phase Transitions ... 163
7.5 Superfluidity ... 165
7.5.1 BEC in a Superfluid ... 166
7.5.2 Long-Range Order in Space ... 168

8 Light ... 169
8.1 Stefan-Boltzmann Equation of State ... 170
8.2 Comparison with the Classical Approach ... 171
8.3 Planck's Insight, Modern Version ... 172
8.4 Planck's Insight, Original Version ... 174
8.5 Photons ... 176
8.6 Atoms ... 177
8.7 Fermi's Golden Rule ... 179
8.8 Bosons ... 180
8.9 Vacuum Energy ... 181

9 Sound ... 183
9.1 Heat Capacity of Crystals ... 184
9.1.1 Sound and Light ... 185
9.1.2 Debye Model ... 186
9.1.3 Optical Modes: Einstein's Model ... 187
9.1.4 Real Crystals ... 189
9.2 Zero Sound ... 191
9.3 Second Sound ... 192

10 Applications ... 197
10.1 Barometric Formula ... 197
10.2 Diatomic Molecules ... 199
10.2.1 Molecules of Identical Atoms ... 201
10.3 Van der Waals Model of Gas Liquefaction ... 203
10.3.1 The Law of Corresponding States ... 204
10.3.2 Maxwell's Construction ... 205
10.4 Paramagnetism ... 207
10.4.1 Equations of State ... 208
10.4.2 Negative Temperatures ... 211
10.4.3 Heat Capacity and Classical Limit ... 211
10.4.4 Susceptibility and Entropy ... 213
10.5 Ferromagnetism ... 214
10.5.1 Weiss's Phenomenological Model ... 215
10.5.2 Susceptibility and Magnetization ... 216
10.6 The Origin of Magnetism ... 218
10.6.1 Hund's Rule ... 218
10.7 Landau's Description of Phase Transitions ... 219
10.7.1 Superconductivity ... 221

10.8 Particle-hole Excitations ... 223
10.8.1 Scattering of Electrons on Nuclei ... 224
10.8.2 Intuitive Calculation ... 225
10.8.3 Precise Calculation ... 227
10.8.4 Thermodynamics of Atomic Nuclei ... 228

11 Fluctuations and Nonequilibrium Processes ... 231
11.1 Brownian Motion ... 232
11.1.1 One-dimensional Random Walk ... 233
11.1.2 Continuum Limit ... 234
11.1.3 Multi-dimensional Walker ... 235
11.1.4 The Policeman's Lecture on One-dimensional Fluctuations ... 237
11.1.5 Massive Walker ... 238
11.1.6 Motion of a Brownian Particle ... 239
11.1.7 Einstein's Relation and the Fluctuation-dissipation Theorem ... 241
11.1.8 Comparison of Brownian and Thermodynamic Fluctuations ... 242
11.2 Wiener-Khintchine Theorem ... 243
11.2.1 Derivation of the Theorem ... 244
11.2.2 Examples ... 245
11.3 Nyquist Theorem ... 246
11.3.1 Macroscopic Derivation of the Theorem ... 246
11.3.2 Comparison with Einstein's Relation ... 248
11.3.3 Output Noise and the Generalized Nyquist Theorem ... 248
11.3.4 Galvanometer Fluctuations ... 250
11.3.5 Limits of Measurement Accuracy ... 251
11.3.6 Evolution of Clocks ... 253
11.4 Return to Equilibrium as an Irreversible Process ... 254
11.4.1 Dissipation ... 254
11.4.2 Susceptibilities and Correlations Out of Equilibrium ... 255
11.5 Time Evolution of Fluctuations and Onsager Relations ... 257
11.5.1 Thermoelectric Effect ... 258
11.6 Relevant Time Scales in Society ... 260
11.6.1 The Playing Cards ... 261
11.6.2 The Jubilee ... 262
11.6.3 IKEA ... 262
11.6.4 The State as an Engine ... 263

12 Literature ... 265
References ... 271

Index ... 283

List of Figures

Fig. 2.1 Absolute zero 12
Fig. 2.2 Air spring 14
Fig. 2.3 Clausius' theorem 20
Fig. 2.4 At equilibrium 23
Fig. 2.5 Restoring force 23
Fig. 2.6 A different air spring 26
Fig. 2.7 Signed volume 32
Fig. 2.8 Not good enough 32
Fig. 2.9 Irreversible cycle 34
Fig. 2.10 Spontaneous symmetry breaking 37
Fig. 2.11 Universal purposeful engine 39
Fig. 3.1 Top: Binomial distribution in relative scale, for $N = 11$, 101, and 10,000, in that order. Bottom: The black rectangle is the distribution in absolute scale for $N = 100$ 43
Fig. 3.2 Random gravity. For an arbitrary sequence of numbers, there exists an initial condition such that the intervals $t_{i+1} - t_i$ of the transit times of the small sphere between the two large ones, orbiting each other in the horizontal plane, are that sequence 49
Fig. 3.3 Anharmonicity and instability 66
Fig. 4.1 Classical limit 75
Fig. 4.2 Negative chemical potential 76
Fig. 4.3 Quantum limit 76
Fig. 4.4 Line, parabola, entropy of mixing 79
Fig. 5.1 Two bosons in two states 99
Fig. 5.2 Rotation of a spin $1/2$ 100
Fig. 5.3 States of two particles 108

Fig. 5.4 Fermi-Dirac distribution (5.35), Maxwell-Boltzmann distribution (4.71), and Bose-Einstein distribution (5.38), with chemical potential in the classical regime $\mu = -2.5kT$. The shaded surface for physical values $\varepsilon > 0$ is the concentration, according to (4.72) 113

Fig. 5.5 The logic of Bell's inequality. The dark surface corresponds to the left-hand side of the inequality (5.75), while the dark and light surfaces together correspond to the right-hand side 122

Fig. 5.6 Bits and dits. Left: Subsets ordered by inclusion. Lines indicate branches that always contain the element a. Right: Partitions ordered by refinement. Lines indicate branches along which the element a is resolved last 127

Fig. 6.1 Fermi-Dirac distribution with chemical potential in the quantum regime, $\mu = 25\,kT$. The curve is centrally symmetric around the point $(\mu, 1/2)$. The arrows show the step width estimate $\mu \pm 2kT$. The thin solid line is the tangent through the inflection point 141

Fig. 6.2 Graphical representation of the dependence $\langle N \rangle\,(\mu)$, for fifteen unequally spaced levels, at low temperature. Thick line: $\Delta\varepsilon < kT \ll \Delta$. Thin line: $kT < \Delta\varepsilon \ll \Delta$. Dashed line: Solution of the equation of state for given $N = 7$, $\mu^\star(7) = (\varepsilon_8 - \varepsilon_7)/2$. The right-hand side of the equation of state (6.4) is integer-valued at the points of horizontal inflection, which are at the energy midpoints due to the central symmetry of the Fermi distribution 142

Fig. 7.1 Bose-Einstein distribution with chemical potential in the quantum regime, $\mu = -kT/100$. The closed circle indicates the expected number of particles at rest, $N_0 \approx -kT/\mu = 100$ 161

Fig. 9.1 Heat capacity of simple crystals, fitted with Debye's and Einstein's models. [Roger Balian, From Microphysics to Macrophysics, Vol. II. Springer, Berlin 1982. Figure 11.20. Reproduced with permission.] 187

Fig. 9.2 Acoustic and optical vibrations 190

Fig. 10.1 Isotherms of the reduced van der Waals equations (10.20). The temperature values are, in order from bottom to top, $T/T_c = 0.85, 0.9, 1, 1.1$, with $v = V/N$. The inflection point (p_c, v_c) is marked with an open circle 204

Fig. 10.2 Simultaneous depiction of the isotherm $T = 0.9T_c$ from Fig. 10.1 (*thin full line*) and the reciprocal of the corresponding compressibility, $1/\kappa_T$ (*dashed line*, arbitrary units). The thick full line is Maxwell's construction 206

Fig. 10.3 Paramagnetic heat capacitance (10.41), for various spin values, with scaling (10.44) 212
Fig. 10.4 Solving the mean field Eq. (10.56) for $J = 1/2$. *Solid lines* are Brillouin's functions for two temperatures, and the *dashed line* is B_m/B_0. The *filled circle* indicates a solution with spontaneous magnetization 216
Fig. 10.5 Landau's description of phase transitions according to Eq. (10.67) 220
Fig. 10.6 Scattering of electrons on nuclei. *Full lines* are the calculated expressions (10.91), shifted by the binding energy of the nucleus. The discrepancy at high energy is due to pion production, which opens an additional dissipation channel [John W. Negele and Henri Orland, Quantum Many-Particle Systems. CRC Press Taylor and Francis Group, 2019. Fig. 5.19. Reprinted with permission.] 224
Fig. 10.7 The locus of allowed states for particle-hole excitations is depicted as *vertical solid lines*. *Dashed lines* are the locus of given energy. Allowed states are outside the first Fermi sphere and inside the second. Left: $q > 2k_F$, the locus is a disk. Right: $q < 2k_F$ and $\omega < 2q - q^2$, the locus is a circular annulus 226
Fig. 11.1 Evolution of a single one-dimensional Brownian walker (left), and of samples from an ensemble (right). The figure at right shows the sum of spin orientations of a thousand randomly oriented spins $J = 1/2$, for a thousand randomly chosen members of the ensemble 242
Fig. 11.2 Time scales in society 261

Chapter 1
Introduction

> *No one has ever seen the inside of a brick. Every time you break the brick, you only see the surface.*
>
> *Richard Feynman*

Thermodynamics is a language developed to describe physical observations. Just as mathematics is not any language, but is logic added to language, so thermodynamics is stability added to language. Both languages are constructed to avoid discussing things which do not exist. The nature of this mistake is different in the two cases. Mathematics precludes discussion of objects having contradictory properties, which cannot exist at all. Thermodynamics precludes discussion of objects which do not last long enough in the time frame of interest. In the narrowest sense, it limits itself to those which would persist forever, if the circumstances in which they were created did not change. Such objects are said to be stable under those circumstances.

1.1 Stability and Existence

While existence is just a name for persistence in time, stability implies a physical reason for this persistence. Here, a distinction is made between things that are made of other things and those that are not. The latter are the elementary particles and fields, whose existence must be taken as given. Stability is merely a word for them, because no physical mechanism is known to explain their permanence. The former are the true subject of thermodynamics: Nuclei, atoms, and all matter consisting of them. Statistical physics is based on the complete physical understanding of the mechanism, called thermalization, by which they achieve stability. The cornerstone of this understanding is precisely that they are made of other things, which can themselves be made of other things. Surprisingly enough, that idea alone—the so-called "molecular hypothesis"—has been sufficiently fertile to explain both the stability and the behavior of all aggregate matter, even if the main components of the explanation were settled only after nearly a century of development from Avogadro,

D. Sunko, *Conceptual Thermodynamics*,
https://doi.org/10.1007/978-3-032-20877-4_1

who first enunciated the hypothesis in 1811, to Einstein, who found a direct way to measure Avogadro's number in 1905. Like embryos repeating the evolution of the species in their own development, the readers will be occasionally invited to revisit this history, to try to see the scientific enterprise through the eyes of those who made lasting progress in specific situations.

1.2 The Ontological Framework

Ontology is the philosophy of existence. Its older version, introduced by Aristotle, studies the existence of things. The newer one, appearing in the seventeenth century, studies existence as such. The two are not independent. A fundamental exact result of many-body theory is that elementary particles do not interact directly by elementary fields, but only by collective excitations of the medium between them, or sound in the broadest meaning. Thus, any kind of existence can be inferred only by the mediation of actual things. This result means that the vacuum is also a thing.

Thermodynamics is an ontology in Aristotle's sense, however limited to the existence of aggregate matter. This limitation is less restrictive than it might seem, because the language of thermodynamics provides a framework to understand the existence of systems far removed from its original purview of gases and chemical reactions. These include living organisms and evolutionary and economic equilibria. In these cases the precise mathematical apparatus of thermodynamics is difficult to apply quantitatively, but it still allows qualitative insight into the relations between observations. A number of such examples are scattered throughout the book. The ontological framework is introduced in general terms now.

Take Helmholtz's well-known expression for the free energy,

$$F = U - TS. \tag{1.1}$$

(It will be properly derived in due course, only its structure is of concern here.) In its original context, it means that not all the change in the internal energy ΔU of a gas during isothermal expansion is available for visible work, because a part $T\Delta S$ will be invisibly lost as heat. The remainder ΔF is the energy actually free to do work, or free energy.

The same statement can be read much more generally. Call F "what is" (the observation), U "what ought to be" (the laws of motion) and S "random events." Then one can say "what ought to happen will not actually happen because of random events." The influence of random events is tuned by the parameter T, called the temperature. If T is zero, everything is deterministic: $F = U$, so the observations conform to the laws of motion.

Understanding the existence of a thing is then a matter of discernment: How much of its stability is due to some cause, and how much is random? This basic ontological conundrum is neatly captured by the English saying "one cannot get an ought from an is." Clearly, the temperature T deserves some additional consideration. As we shall see later, it turns out that in a stable state the right-hand side of Eq. (1.1) contains two independent unknowns, not three as it seems. One can take either T or S to be given arbitrarily, but not both.

As a simple illustration, imagine playing a game of cards. One knows one's own hand (U), but the other players' hands are distributed at random (S). Can one predict the outcome of the game (F)? The answer hinges on how one's assessment of one's own hand changes when the others reveal theirs. But that is precisely the meaning of temperature:

$$T = \frac{\partial U}{\partial S}. \tag{1.2}$$

If one has a very strong hand, it does not matter what the others hold, so U does not depend on S. Then T is zero, and one can predict the outcome with certainty: One wins deterministically. Conversely, the value of a weak hand (U) depends strongly on the card distribution (S), so the temperature is high, and the influence of the known part (U) on the outcome (F) is negligible.

As another illustration, it is commonly understood that good decisions lead to favorable outcomes. However, that is a deterministic statement. Random events play a much greater role in the lives of the poor than of the rich. There is a specific record of a man becoming homeless for six years because he could not afford to repair his car, which made him lose his job, and consequently his home. Evidently, a rich person needs to suffer much more misfortune to become homeless. One can say that the two live at different temperatures.

Given that observations present us with a combination of two unknowns, one might despair of ever being able to understand what is really happening. There are two ways to meet this challenge, mystical and physical. Mystically, one can either say that $S = 0$ always (mysticism of the *first kind*), or $U = 0$ always (*second kind*). Either everything is really deterministic, only we humans do not see how, or everything is really random, only we humans project some sense on it for our own reasons. In both cases, the mystical attitude is that observations are fundamentally marred by human weakness, so they are not truly objective. Strong or enlightened individuals live in a different reality, because they either know everything, or have no reason to make sense of anything. The physical attitude is that random events really do happen, but that is not the whole story, and it is possible to take them into account in a way which preserves both the notion of objective reality and understandable causal relations.

The key to the physical approach is to distinguish between correlation and response. One can disentangle the laws of motion from random events by testing causal relations. Technically, a statistical correlation function is the difference between two expectations, that of events A and B occurring at two instances in time anyhow, and of their occurring at those instances independently:

$$\left\langle A(t)B(t')\right\rangle - \langle A(t)\rangle \left\langle B(t')\right\rangle, \tag{1.3}$$

where expectation, or average value, is denoted by angular brackets. On the other hand, a response function takes into account which of the two occurred first, i.e., it orders them in time:

$$\left[\left\langle A(t)B(t')\right\rangle - \langle A(t)\rangle \left\langle B(t')\right\rangle\right]\theta(t-t'). \tag{1.4}$$

Here, θ is a step function, also known as Heaviside's theta-function:

$$\theta(t) = \begin{cases} 1 \text{ for } t \geq 0, \\ 0 \text{ for } t < 0. \end{cases} \tag{1.5}$$

Clearly, if A causes B, the response function will be zero, because A must precede B, but the step function tests the other direction. One can say that a response is a correlation armed with causation. It avoids the error known as post hoc, ergo propter hoc, which the correlation function does not: The mere fact that one event occurs after another does not mean that it was caused by it.

The above discussion allows a precise technical definition of magical thinking. Magic is the idea that reality is determined by appearances (F), without being aware of the distinction between predictable (U) and random (S) events.[1] A well-known example are cargo cults, some of whose followers built life-size models of airfields in the hope that an airplane would land with supplies. There is also the story of a man who worried that there might be a bomb on his flight, so he decided to bring one himself, reasoning that it would be quite improbable to have two bombs on the same plane. Technically, magic drops the theta-function, which equates responses and correlations, confusing cause and effect. Notably, at equilibrium there are only correlations, because it appears after all responses have died out, which hides the distinction between U and S.

The need to consider responses in addition to correlations means that one cannot understand physical existence purely by passive observation of an equilibrium state. One must probe a thing somehow, disturb the stable state to observe its recovery. Physics of the stable state is revealed by the forces raised when it is disturbed. These forces are due to the unstable states near equilibrium, so the two must be linked in a coherent narrative for the inanimate object under study to be understood. That narrative, however, is the function of an animate object, the human mind that struggles to "connect the dots." The ontological framework requires a gnoseological one in order to distinguish physical reality from constructions of the human mind, the building from the scaffolding.

[1] Unlike mysticism, which acknowledges the distinction, but claims that it is illusory.

1.3 The Gnoseological Framework

Gnoseology is the philosophy of knowledge. The ontology of thermodynamics was finally settled only at the beginning of the twentieth century, with Einstein's realization that Brownian motion is thermalization made visible. By contrast, a gnoseological framework sufficient for scientific research was put in place already in the thirteenth century by scholastic philosophers, in particular St. Albertus Magnus, Roger Bacon, and St. Thomas Aquinas. Their effort to reconcile faith and reason culminated in the work of Aquinas, whose philosophical apparatus established for that purpose is fully general, reflecting, it seems, some innate features of human cognition.

The central gnoseological question is how to know the truth. What makes one think that something is true? In order to understand Aquinas' answer, one must understand the foundation on which it is built. It is simply the idea that truth is objective and knowable. The converse ideas that truth is a social construct, or is unknowable, undermine not only his philosophy, but also the foundations of the Christian faith as they have been understood at least since St. Augustine. However, if that were the case, scientific research could hardly claim any lasting progress in understanding reality either.

Aquinas admits two ways to reach the truth.[2] One he calls "philosophical truth," which is perceived as proven by argument. It is certain, *but* depends on the argument being correct. The other he calls "revealed truth," which is perceived as given. It is also certain, *but* depends on an act of will by the observer to accept it. These two "buts" of Aquinas' gnoseology frame all scientific research and discovery. There needs to be both correct argument and accepted observation.

Theory is clearly in the nature of philosophical truth. Notably, a correct argument is compulsive. For example, one is forced to agree that all prime numbers greater than two are odd. To understand that experiment is in the nature of revealed truth, one must understand that the interpretation of a measurement is never compulsive. Accepting the *meaning* of an observation is an act of free will. (It may appear less so to experienced researchers, who have previously invested a considerable effort of will to reach their position.)

Anthropological studies indicate that, indeed, even the most obviously obvious is not compulsively obvious. For example, people not previously exposed to photography have been recorded not recognizing themselves and their friends in a photograph. They readily described it in terms of lines and surfaces, without reporting any figurative meaning. Clearly, the idea that a piece of paper can be connected with a living human being needs to be acted upon willingly for the connection to be made. Light rays reflected from the paper into the eyes cannot force anyone to make the inference, like some screwdriver adjusting the brain. Nevertheless, the scene really did depict specific people, not some others, or, say,

[2] Not that there are two truths, which would mean that reason and faith are incompatible.

ships and trees. This objective truth was knowable and could have been revealed to an observer freely willing to accept it from observation of the photograph.

The distinction between philosophical and revealed truth explains why discussion with flat-earthers, creationists, anti-vaxxers, and similar heretics,[3] is most often futile: It is a contest of will, not of argument. They are capable of following the arguments, but they will a different narrative. Conversely, a researcher used to the technique of nuclear magnetic resonance wills to distinguish signals of COOH and CH_3 groups in a molecule, say, just as easily as observers used to photography will to distinguish signals of their friends in a photograph. They could also will to hold it upside down and claim that the sky is the sea. Science makes progress by pondering what to accept as given. One rarely doubts the measured values as such, but what do they mean?

Scholastic gnoseology reflects the lived experience of individual researchers. Some things we accept by our will, some we work out by our reason. Both approaches are fallible, but neither is fundamentally flawed, nor provisory. The lasting value of this framework is in the realization that neither can be subsumed in the other. On the contrary, it is in the nature of truth itself that both are required. One can organize them in a dynamical system, like walking on two legs, which reaches the truth in many small exchanges between reasoning and observation, contemplation and action. On the communal level, that system is the scientific method. Individually, it is the personal evolution of a researcher, whose scientific attitude reflects the actual balance achieved between the two approaches over many years.

1.4 Basic Thermodynamic Concepts

Thermodynamics being a universal ontological language, there is almost no limit to the scope of its concepts, once properly understood. However, to achieve such general insights, one should thoroughly understand the original special context of the study of heat and gases in the seventeenth century. Modern thermodynamics was founded as the record of immediate laboratory experience with macroscopic measuring devices, such as pressure gauges and thermometers. Notably, pressure and temperature can be measured without knowing what they are, just like electric and magnetic forces. The distinction already mentioned, that the former are the effect of many particles and the latter are elementary, lay far in the future.

Thermodynamics is concerned with empirical observations only, without attempting to rationalize them. In particular, it does not try to explain why various materials have different properties. Its practical strength is that very few properties need to be measured to characterize any material for a given purpose, such as building a bridge. These measured properties are a severe test for any attempt to

[3] Of the *first kind*, we shall meet the second when we learn enough quantum mechanics.

explain them, which also gives thermodynamics a position of strength in theory, as a gatekeeper for admissible ideas: There is no argument against facts. Statistical physics is a generic approach to providing such explanations in terms of the structure of matter, but its results cannot be understood without understanding thermodynamics on its own first.

The basic observation of thermodynamics is that heat passes from a warmer object to a colder one. Or, if two bodies touch, they will eventually have the same temperature. While we wait, the process is in the domain of non-equilibrium thermodynamics; once all macroscopic changes have ceased, we speak of thermodynamic equilibrium. This equilibrium has an interesting equivalence property, sometimes called the zeroth law of thermodynamics: If two bodies are in equilibrium with a third, they are also in equilibrium with each other. This simple observation has the far-reaching consequence that any body can serve as a thermometer, and all thermometers agree which of two bodies is warmer.

As an autonomous discipline, thermodynamics has its own conceptual apparatus, based on immediate experience, and its own mathematical apparatus with predictive power. The first such concept is that of *thermodynamic limit*, which reflects the common experience that a change in quantity of a substance does not produce a change in quality: Breaking a brick in two gives two smaller pieces of brick. Like all physical concepts, it comes with observable limitations. For example, the properties of lens coatings do not change when their surface doubles, but do when their thickness doubles: Thin films are in the thermodynamic limit in two dimensions, but not the third. As one thickens the film, its properties stop changing at some point, whence we speak of bulk material.

This example suggests that the thermodynamic limit may be reached by making things bigger, which is generally a good intuition, but not without exception. Stars are not in the thermodynamic limit, because a much bigger star than the Sun has quite different properties. A material aggregate in the thermodynamic limit is called *extensive*.

Limiting ourselves to extensive objects, we give the one under observation a name: "The system." Whatever is not the system is "the surroundings." The determinative property of the system is that it can be studied separately from its surroundings. The conceptual repercussions of this intuitively obvious separation took several centuries to unfold, which we shall follow in detail later. For now, note that the separation implies that the interactions of the system with the surroundings are weak, in some sense which needs to be elucidated.

Having, then, a system under observation, it turns out that its behavior can be recorded by only a few macroscopic parameters, such as pressure and temperature, for the observation to be reproduced. For example, "the boiling point of water is 100 degrees Centigrade at standard pressure." These parameters are called "thermodynamic variables," implying that they can be set at will by the experimentalist. The next fundamental observation is that all available parameters *cannot* be set at will, but a subset can be. When equilibrium is reached after some are fixed, the remainder will be set by the system itself, and can be measured as outcomes. For example, if a fixed pressure is applied, the volume starts changing and then stops at an equilibrium value. This experience is generically described by an expression like

$$f(p, V, T) = 0, \tag{1.6}$$

which is called an *equation of state*. In this example, it means that fixing any two of the parameters, pressure p, volume V, and temperature T, causes the system to move to an equilibrium state which determines the third. The function f for a system can be determined by such measurements.

The existence of the function f encoding the interdependence of thermodynamic parameters indicates that the theory of functions of several variables is the native mathematical apparatus of thermodynamics. In order to establish it, only one further generic observation needs to be invoked: When one of the parameters changes a little, the others also change a little. In other words, the function f is *smooth*. Mathematically speaking, the equation of state defines a piecewise smooth surface in the space of parameters. Piecewise, because we must allow for the observation of phase transitions, at which either f or its derivative change abruptly. The former describes first-order, the latter second-order transitions.

For a system at equilibrium, the solutions of Eq. (1.6) are always unique. This reflects the experience that it does not matter what we have done to fix two of the variables, the system will always return the same value of the third. It is sometimes said that a system at equilibrium has forgotten its past. The mathematical expression of uniqueness is that the total change in any thermodynamic parameter A following a closed loop in the surface of equilibrium states must be zero:

$$\oint dA = 0. \tag{1.7}$$

Being unique, all thermodynamic parameters defining the equation of state necessarily have this property. To distinguish them from other possible observables, they are called "state variables." Here we reap the first mathematical fruit of the observations: Eq. (1.7) is the mathematical definition of an *exact (total) differential*. So we can say, if A is physically a state variable, then dA is mathematically an exact differential, and vice versa.

The predictive power of thermodynamics relies on smooth surfaces having mathematically nice properties. For example, a local energy minimum in, say, two dimensions can typically be fitted to a paraboloid—that is just Taylor's theorem—so, anything that can happen in the vicinity of such a point is determined by just two numbers, curvatures along its symmetry axes. The equation of state (1.6) is just the extremal equation for some such potential. The curvatures characterizing the extremum are physically susceptibilities: They are the constants of elasticity of the equivalent harmonic oscillator, like "k" in the expression $kx^2/2$ for the elastic energy. If the parameters are volume and temperature, the susceptibilities are the compressibility and the heat capacity. Measuring the two enables one to predict the *linear response* of the system when its equilibrium is perturbed.

Some years ago, there was a study on the social network Facebook, in which some users were exposed to more bad news in their feeds, and others to more good news. It was found that their posts were similarly biased, i.e., polarizing the news feed influenced their mood in the same direction by a measurable amount. In thermodynamic terms, someone had measured the susceptibility of the population. Just as a hardware store can sell springs of certified elasticity, social networks can sell calibrated changes in their members' moods.

This example shows the scope of thermodynamic phenomenology. The same equations have the same solutions, more specifically, physical stable points are usually quadratic, and a parabola is mathematically a simple object. The underlying questions are why such simplicity prevails in the real world, and is the reason for that—the microscopic mechanism—the same for gases and Facebook users. The answers to these questions lie outside of thermodynamics. They depend on the nature of thermalization, which is in the purview of statistical physics. To anticipate, it turns out that thermalization is microscopically forgetfulness: A particle of pollen in water moves randomly because it keeps forgetting its initial conditions. So, a social network is like a gas to the extent to which its members are like Brownian particles, namely: At any moment, they move in the direction of the instantaneous force on them, forgetting where they came from the moment before. This insight raises the question, what is a moment, a second, day, year? The response of a system to perturbation reveals characteristic time scales, determined by its internal dynamics, which open a window to the microscopic world of its constituents. That window into the molecular hypothesis is the capstone of thermodynamics, put in place by Einstein, to be discussed in due course. We embark on the foundations now, beginning with Boyle.

Chapter 2
Thermodynamics

[...]; we cast our eyes upon the longer leg of the Glass, on which was likewise pasted a list of Paper carefully divided into Inches and parts, and we observed, not without delight and satisfaction, that the Quicksilver in that longer part of the Tube was 29. Inches higher than the other.

Robert Boyle

The secretary of the Royal Society Robert Hooke was an exception among his learned colleagues in that he had to work for a living. He variously supplemented his income as professional surveyor and as maker of bespoke scientific instruments. He constructed a vacuum pump for the immeasurably rich Robert Boyle, perhaps the first high-technology scientific instrument in the modern sense of the word. High technologies develop by integrating autonomous skills into a new whole, while basic research stimulates this development through demanding measurements.[1] In this case, the requirement was for a wide range of pressures without loss of precision, gas leaks, or deforming the pressure chamber. Hooke's solution was for the sturdy chamber of the pump to be made by glassmakers, the smooth cylinder by gunsmiths, and the precise mechanism by watchmakers.

Boyle was thus able to confirm convincingly previous authors' reports that the product of pressure p and volume V of a gas is constant, $pV \sim const.$ Several years later, Mariotte showed that the volume increases linearly with the temperature, $V/T \sim const.$ In 1811, Avogadro observed that volumes of gases that participate in chemical reactions appear in proportion to small natural numbers. His interpretation, that it was a consequence of atoms combining into molecules, became known as the *molecular hypothesis*. The linear dependence of the volume on the number of particles in the gas, $V/N \sim const.$, is called Avogadro's law. All these observations can be subsumed into a single equation of state for an ideal gas,

$$pV = nRT = NkT, \tag{2.1}$$

[1] Unlike this systematic stimulus, results of basic research influence technology only accidentally.

D. Sunko, *Conceptual Thermodynamics*,
https://doi.org/10.1007/978-3-032-20877-4_2

where n is the number of moles of the gas, R the universal gas constant, N the number of gas molecules, and k the Boltzmann constant. The ratio $R/k =: N_A$, the number of molecules in a mole, is called Avogadro's number. In SI units, it is a fixed constant, $N_A = 6.02214076 \times 10^{23}$.

2.1 Absolute Zero

The most interesting property of the equation of state (2.1) is that there is a temperature at which the pressure is equal to zero for any quantity of gas and any volume of the container. The expression (2.1) is written so that this temperature is the origin, so it appears obvious that $p = 0$ at $T = 0$. The phenomenon itself is puzzling, however: Measure the pressure p at several near-room temperatures, and notice that the corresponding points lie on a line in the plane $p(T)$. Change the gas concentration N/V by pumping, measure again, obtain a line with a different slope, and so on up to any number of lines, to observe at last that they all intersect at a single point *far below* the room temperatures at which measurements were made, without regard which temperature scale or thermometer was used. How can all these lines agree where that far-away point is, called *absolute zero*? Stranger still: If we try to approach it by cooling, the agreement becomes worse. Various gases liquify at various temperatures, Eq. (2.1) is no longer valid, so it is not even clear that the pressure will be zero at $T = 0$. (It is not.) One needs to go in the opposite direction, to higher temperatures, where differences between gases disappear, and they all agree where absolute zero is all the better, the farther away from it they are (Fig. 2.1).

For most physical quantities, it is easy to see whether they have an absolute zero. Mass does, because it cannot be negative. Velocity does not, because motion is relative. It is not so easy to imagine the absolute zero of temperature. It is a direct consequence of the fact that every many-particle system has only one state of motion with minimal possible energy, its so-called ground state. Because there is only one, there can be no random events at that energy. Thus, motion is certainly deterministic, which means that the temperature is zero. Although this energy can vary among different systems, any system in its ground state can serve as a thermal reservoir at zero temperature, i.e., a universal standard for the origin of temperature.

Such an explanation defies classical intuition, which holds it impossible for an unimaginably large number of particles to have only one state of motion all together. The observation of an absolute zero of temperature is only the first in a line of thermodynamic observations which can only be understood as quantum effects.

Fig. 2.1 Absolute zero

Macroscopic thermodynamics, which describes the behavior of steam engines, is determined directly by the microscopic reality of atoms, described by quantum mechanics. We shall warn of that on occasion in this chapter, even if it is formally concerned with thermodynamics at the level of nineteenth-century heritage.

2.2 The Second Law of Thermodynamics

Heat passes from a warmer body to a colder one. The reverse does not happen: No matter how long we wait, two bodies in thermodynamic equilibrium will not abandon that equilibrium and develop different temperatures. Similarly, a boat in a bathtub can heat the water with its propeller, but the propeller of a freely floating boat will never start turning of itself while the water in the tub gets colder.

There are two classic formulations of the second law of thermodynamics, which describes the experience of *irreversibility* of thermodynamical observations in general. In order of the examples above, Clausius' formulation says that there cannot be a change in state of a system whose *only* effect would be to shift some quantity of heat from a colder reservoir to a warmer one. Kelvin's formulation says that there cannot be a change of state of a system whose *only* effect would be to convert a certain quantity of heat wholly into work.

Neither means that heat cannot be converted into work at all. The word "only" is essential in both cases. A feather may rise in the warm air above a stove, changing its heat into work. That is not the only effect, however, because the room is getting warmer as well. Similarly, isothermal expansion of a gas converts a certain quantity of heat wholly into work, but that is not the only effect, because the volume of the gas has increased as well.

At the phenomenological level of describing thermodynamics by functions of several variables, the second law is a restriction on movement in the space of parameters. Certain trajectories in this space are forbidden, because they describe sequences of measurements that cannot be realized in a laboratory. Thermodynamics is not equivalent to the theory of functions of several variables, but is a specialization of that theory as constrained by the second law. Mathematicians always find such a state of affairs interesting, and there are attempts in the literature to replace the second law with an axiom, which would separate thermodynamics from physics and turn it into an autonomous mathematical discipline. One could have hoped that such an axiom would also clarify the physical nature of the second law: What is heat at all, and how it knows which way to go. However, even when they were found, these axioms did not help physical understanding. They could not have, because the second law is a consequence of the granular microscopic structure of matter, which is beyond the scope of the theory of functions of several macroscopic variables.

Macroscopically, irreversibility is an asymmetry in the requirements on experimental arrangements. It is easy for a swimmer to jump into a pool, which will heat the water (slightly) when it settles. If we recorded that on film, and ran it backwards, we would see a coordinated motion of an enormous number of water molecules, coherent over large distances, all conspiring to eject the swimmer out of the pool by a cascading effect of multiple mutual collisions, appearing as a sudden pressure surge. Every single molecular collision in that cascade conforms to the microscopic laws of motion, and not a single one gives a hint which way the film is running. Irreversibility means that we cannot arrange such a coordinated motion at will, and it is not probable that it will occur spontaneously. An echo is less dramatic: We know how to make a wall speak, yet it would still be a miracle if it spoke by itself.

Microscopically, irreversibility requires a subtler discussion. We can wonder whether the information about the swimmer's jump is retained in the microscopic motion of water molecules, only we do not know how to get it back, or it disappears physically as if the swimmer never existed. The latter is true, but we shall defer that discussion to Chap. 3.

2.3 The First Law of Thermodynamics

It was clear already to Boyle that air can act like a spring. But is there any difference between the two? For a spring holding up a weight at some height h, the equilibrium value of h can be found from the so-called principle of virtual work, namely that the work ΔW involved to change h by a small amount Δh does not change the total energy U to first order in Δh:

$$\Delta W = \Delta U = \Delta \left[mgh + \frac{k}{2}(h - h_0)^2 \right] = mg\Delta h + k(h - h_0)\Delta h = 0, \qquad (2.2)$$

whence $h - h_0 = -mg/k$ (the spring is compressed). The analogous calculation with a gas in place of the spring is

$$\Delta W = mg\Delta h - p\Delta V = (mg - pA)\Delta h = 0, \qquad (2.3)$$

whence the equilibrium pressure is $p = mg/A$, with A the surface of the piston. The superficial difference between the two is that the mechanical work of the gas is $-p\Delta V$, in place of $k(h - h_0)\Delta h$ for the spring (Fig. 2.2). The fundamental difference is that $\Delta W \neq \Delta U$ for a gas, in other words, one cannot equate the work done on a gas with the change in its *internal energy* U, which is generally the sum of all kinetic and potential (interaction) energies of the molecules that comprise it.

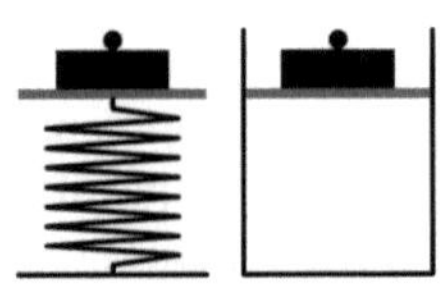

Fig. 2.2 Air spring

In order to balance the change in internal energy with the visible work done on the system, an invisible term is needed, called *heat*, denoted Q:

$$dU = \eth Q + \eth W, \tag{2.4}$$

where we have written the energy balance in differential form to emphasize that the internal energy is a state variable, so dU is an exact differential. By contrast, the symbol $\eth$ indicates an *inexact differential*, meaning that its integral over a closed path need not vanish. Equation (2.4) is called the first law of thermodynamics. It is true for the spring as well, but the heat contribution is negligible there, if the metallurgy is good and the weight not excessive.

The division between heat and work is phenomenological: Work is the visible and heat the invisible part of a change in internal energy. Throughout this book, the work part will be written explicitly as

$$\eth W = -pdV + \mu dN, \tag{2.5}$$

including mechanical and chemical work, one by change of volume, the other by change of particle number. The *chemical potential* μ plays the same role to the particle number as the pressure to the volume: It is the force felt by one who tries to change the state variable. The pair (force, variable) are called *conjugate variables*. Because the internal energy is *extensive*, i.e., proportional to the size of the system, one of the conjugate variables is always *intensive*, i.e., does not change with the size of the system, while the other is extensive.

The physical meaning of the invisible part $\eth Q$ was a mystery for a long time, and remains a stumbling block for beginners. Like children skipping impatiently to the end of a detective novel, let us expose the culprit now. After some discussion, deferred to Chap. 3 again, it turns out that the observed internal energy is microscopically a statistical expectation value, or average:

$$U = \langle E \rangle = \sum_i p_i \cdot E_i, \tag{2.6}$$

where i denotes any possible microscopic state of motion of the system, E_i its energy, and $p_i := p(E_i)$ the probability of encountering the system in a state of that energy. Upon differentiating this expression,

$$dU = \sum_i E_i \cdot dp_i + \sum_i p_i \cdot dE_i, \tag{2.7}$$

we see that the first term is a change in the probability distribution, which is naturally invisible; that is heat. On the other hand, work is a change in the energies of the various states of motion, which depend on the macroscopic physical conditions, such as volume and particle number.

In order to accept this explanation as physical, one must accept the reality of random events. Otherwise, the temptation to resolve the statistical distribution into some imagined deterministic motion cannot be put to rest. Thus, the main obstacle to accepting physical reality is psychological. The untrained intuition is prone to mystical or magical interpretations of the visible world. Operatively, the underlying error in reasoning may be avoided by a simple prescription: Do not say "one could in principle ..." when one could not, not even in principle.

2.4 Entropy

The expression $\eth W = -pdV$ suggests an interesting way to make exact differentials out of inexact ones:

$$dV = \frac{\eth W}{-p}. \tag{2.8}$$

Mathematicians call this operation *dividing by an integrating factor*, and guarantee that smooth functions always provide such a factor. Different kinds of work each have their own integrating factor, like the pressure, or chemical potential. The idea is to find the integrating factor for heat now. Importantly, it has to be universal, like heat itself. The only universal intensive variable that we know is the temperature, so the proposal is that

$$dS := \frac{\eth Q}{T} \tag{2.9}$$

is the exact differential of some state variable S that we name the *entropy*. No matter what we call it, the question is: Can we prove that

$$\oint \frac{\eth Q}{T} = 0, \tag{2.10}$$

based on the observed behavior of heat and temperature? The answer is yes, and more than that: It is hard to find an example of simpler reasoning leading to more far-reaching conclusions.

2.4.1 *Calibrating the Thermometer*

The temperature in the denominator of Eq. (2.9) is measured by a thermometer. What if we change it, using alcohol instead of mercury, say? The change is encoded by replacing T with some function $f(T)$, the *calibration curve* connecting measurements by the new thermometer with the old ones. The mathematical theory says that nothing depends on the choice of f as long as it has a unique inverse. Can we guarantee that from observations? Yes, because all thermometers agree which body is warmer, so $f(T)$ is strictly increasing, so it is invertible. Even more simply, f is invertible because the old and new thermometers can exchange roles.

Having understood that, why put T in the denominator and not T^3, for example? It turns out that the choice T is consistent with the ideal-gas equation of state (2.1). Thus, our thermometer is calibrated to the ideal gas as a working fluid.

2.4.2 Reversible and Irreversible Motion

Imagine pulling a sled on a flat surface. If one does not pull hard enough, the sled does not move. Tug a bit more or less, the sled stays in place. Evidently, the sled reacts instantaneously to the external force according to Newton's third law, which means that static friction holds it in place *reversibly*. When the sled slides, dynamic friction heats the surface, and releasing the rope does not return the sled to its initial position. Such motion is *irreversible*. As the example shows, reversible motion is microscopic, and only irreversible motion is useful, in the sense that it has some required macroscopic effect at the price of not being able to return to the initial state. The second law of thermodynamics says that one must always pay such a price, otherwise one has an eternal-motion machine, *perpetuum mobile*, the Holy Grail of self-taught inventors.

Unlike our sled, the standard model of an engine in thermodynamics follows a closed curve in the space of parameters, like a steam engine. Because $\eth Q$ and $\eth W$ are not exact differentials, one can obtain a net exchange of heat and work in such a cycle. It is at each other's expense, because the internal energy is a state variable. The simplest engine takes heat from a reservoir at one temperature, and delivers it to another at some other temperature. It is called a Carnot engine. It is reversible by construction because it can turn both ways, either delivering work at the expense of heat from the warmer reservoir, or requiring work to transfer heat from the colder to the warmer one. It can simulate a useful engine if we ignore the effort required to keep the reservoirs at fixed temperatures while it works. If we do not, the Carnot engine is reversible only for infinitely slow, or vanishingly small, paths in the surface of equilibrium states.

Mathematically, a Carnot engine explores the local geometry of the surface of equilibrium states, defined by equations of state. It reminds one of a little bug trying to find out the curvature of the world by going around a triangle and measuring the angles. (If they add up to 180°, the world is flat.) Notably, Carnot invented it before the equivalent mathematical apparatus grew out of the theory of functions of several variables and eventually developed into differential geometry, which rendered his engine obsolete as a formalism. Crucially, Carnot's engine respects the second law, so we can use it to prove that entropy is a state variable.

2.4.3 Analysis of a Carnot Cycle

For simplicity, we replace differentials with finite differences now. The total change in internal energy around a closed cycle is zero:

$$\Delta U = \Delta Q + \Delta W = Q_> - Q_< + \Delta W = 0, \tag{2.11}$$

where we have stipulated the heat taken from (delivered to) the warmer (colder) reservoir at temperature $T_>$ ($T_<$) with opposite signs. With this convention, $Q_< > 0$, otherwise we would be taking heat from both reservoirs and could use some of the work to give back the heat taken from the warmer one, leading to a net conversion of heat into work from the colder reservoir alone. Similarly $Q_> > Q_<$, or a net amount of heat could flow from cold to hot without other effects. With $Q_> > Q_< > 0$, $\Delta W < 0$, whichever way the machine is turning. The *efficiency* of the engine is

$$\eta = \frac{|\Delta W|}{Q_>} = 1 - \frac{Q_<}{Q_>} < 1 \quad \text{for} \quad Q_> > Q_< > 0. \tag{2.12}$$

The key insight now is that η is maximal for a reversible engine. It means that this maximum does not depend on the construction of the engine, because any reversible engine must have this maximal efficiency. The consequence is that the efficiency η of a reversible engine is some universal thermodynamic function of the reservoir temperatures alone, so we can write

$$1 - \eta = \frac{Q_<}{Q_>} =: g(T_<, T_>), \tag{2.13}$$

where g is such a universal function.

The proof is by contradiction with the second law. Suppose that there is an engine X with efficiency $\eta' > \eta$. Adjust it so that it takes the same amount of heat $Q_>$ from the hot reservoir as a reversible engine R. Then it must deliver less to the cold one, $Q'_< < Q_<$, or it would not be more efficient. Now run R in reverse, in parallel with X, to give back the heat $Q_>$ to the hot reservoir, taken from it by X. The net effect is that a quantity $Q_< - Q'_< > 0$ of heat is wholly converted into work at the expense of the cold reservoir alone, as forbidden by Kelvin. Taking X to be reversible as well, it follows that no reversible engine can be more efficient than another.

2.4.4 The Efficiency of a Reversible Engine

We can find the universal function $g(T_>, T_<)$ simply by considering two reversible engines in series, connecting three reservoirs at temperatures $T_1 > T_2 > T_3$. The first connects reservoirs 1 and 2 with efficiency η_{12}, while the second connects 2 and 3 with efficiency η_{23}. We can also imagine this arrangement as one engine, connecting 1 and 3. The work calculated both ways must be the same:

$$\eta_{12}Q_1 + \eta_{23}Q_2 = \eta_{13}Q_1. \tag{2.14}$$

Expanding, we see that this statement depends on the obvious relation

$$\frac{Q_3}{Q_2} \cdot \frac{Q_2}{Q_1} = \frac{Q_3}{Q_1}, \tag{2.15}$$

which is merely a consistency check on the definition of η, because the third engine is only in our imagination. But if the engines are reversible, the same relation becomes a functional equation for the universal function g:

$$g(T_3, T_2) \cdot g(T_2, T_1) = g(T_3, T_1). \tag{2.16}$$

It has the same structure as reducing the fractions above, so it can be solved by a similar fraction in terms of some universal function f of a single variable:

$$\frac{Q(T_2)}{Q(T_1)} = g(T_2, T_1) =: \frac{f(T_2)}{f(T_1)}. \tag{2.17}$$

We have elevated the index in $Q_{1,2}$ to a functional dependence on the temperatures $T_{1,2}$, because here we mean specifically the heat exchanged by a reversible engine. The two ends of this expression can be rearranged to read

$$\frac{Q(T_1)}{f(T_1)} = \frac{Q(T_2)}{f(T_2)}. \tag{2.18}$$

Now the indices 1 and 2 are separated on the two sides, which means that the ratio

$$\frac{Q(T)}{f(T)} = const., \tag{2.19}$$

independent of temperature. In other words, we have discovered that this quantity of heat is given by a universal function $f(T)$, multiplied by a universal constant.

To specify $f(T)$, we only need to recognize it as the calibration curve of some thermometer. Choosing the ideal-gas thermometer, $f(T) = T$, we find

$$\boxed{Q(T) = S_Q T,} \tag{2.20}$$

where the constant S_Q is extensive and independent of temperature. Then the efficiency of a reversible engine is given by

$$\eta = 1 - \frac{T_<}{T_>}, \tag{2.21}$$

with temperatures measured from absolute zero by any linear scale. The standard choice is the Kelvin scale K. Then S_Q has the dimension [*energy*/K].

2.4.5 Clausius' Theorem. Maximization of Entropy

We are ready to prove the proposition (2.10). Break the circular process down into a finite number N of reversible steps:

$$\oint \frac{\eth Q}{T} = \sum_{i=0}^{N} \left(\frac{Q(T_i^>)}{T_i^>} - \frac{Q(T_i^<)}{T_i^<} \right), \tag{2.22}$$

where i counts the engines connecting the reservoirs, and $i = 0$ refers to the same as $i = N$. The heat taken from the warmer reservoir, respectively delivered to the colder one, appears with opposite signs, cf. Eq. (2.11). All these fractions are equal to S_Q for reversible engines, cf. Eq. (2.20). Thus, each term in the sum is $S_Q - S_Q = 0$ (Fig. 2.3), so

$$\oint dS = \oint \frac{\eth Q}{T} = 0. \tag{2.23}$$

We have proven that the entropy S is a state variable at equilibrium, because reversibility means that the closed path lies on the surface of equilibrium states.

The argument can be extended out of equilibrium. Replacing one of the reversible engines can only lead to lower efficiency:

$$\eta' \leq \eta \implies \frac{Q'(T_i^<)}{Q'(T_i^>)} \geq \frac{Q(T_i^<)}{Q(T_i^>)} = \frac{T_i^<}{T_i^>}, \tag{2.24}$$

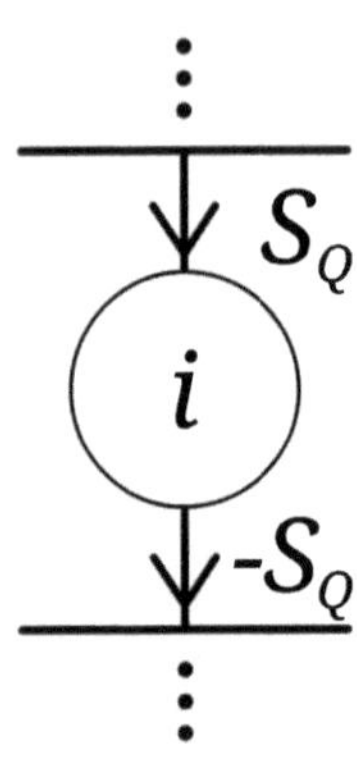

Fig. 2.3 Clausius' theorem

where the last equality holds for the replaced reversible engine. Then the inequality may be resolved to read

$$\frac{Q'(T_i^<)}{T_i^<} \geq \frac{Q'(T_i^>)}{T_i^>}, \tag{2.25}$$

so the corresponding contribution to the sum in Eq. (2.22) may be negative now. The condition on the path integral is thus relaxed:

$$\oint dS \leq 0, \tag{2.26}$$

where equality holds for a reversible process, i.e., at equilibrium, when the entropy is a state variable. Equation (2.26) is known as Clausius' theorem: The entropy is maximal at equilibrium, because it can only decrease out of equilibrium. It is a mathematical consequence of the second law of thermodynamics, a precise statement of the constraint that the second law imposes on the theory of functions of several variables. We have derived it without any idea what entropy could be physically.

2.5 Mole, Molecule, Temperature

To define a standard entropy S_Q in Eq. (2.20) is to choose some standard size of the reservoir, in other words, to define a universal measure of the quantity of matter. Setting $S_Q = R$, the gas constant, defines *one mole*. Setting $S_Q = k$, the Boltzmann constant, defines *one molecule*. The first to suggest a direct measurement of the constant k was Einstein in 1905, which is why he wrote "...then a precise determination of the true size of atoms is possible"[2] in the introduction to his first article on Brownian motion.

The relation (2.20) has three important interpretations. First, it should be read from right to left. It defines the temperature by the quantity of heat delivered or absorbed *reversibly* in contact with a reservoir of given size. This interpretation conforms to the way temperature is measured, namely by an irreversible transfer of heat to (or from) a thermometer, which stops once macroscopic equilibrium is reached, with heat exchanged reversibly after that. Only then can a reading be made.

Second, following from the first: $Q(0) = 0$, meaning that the heat capacity of all substances vanishes at absolute zero on general thermodynamic grounds. That is unsettling, because the molecular hypothesis, which has made irreversibility at least plausible so far, is suddenly turning against us. Just as were getting used to the idea that heat is some random motion of molecules, a mole of molecules no

[2] ...es ist dann eine exakte Bestimmung der wahren Atomgröße möglich.

longer take up heat! Why would even a single molecule refuse to be moved at zero temperature, let alone every molecule of every body? For the same reason that there is an absolute zero in the first place. Our intuition that absorbing heat changes states of internal motion is correct, but at absolute zero the whole system has only one state of motion, the minimal-energy ground state, so it cannot be changed, so the heat capacity is zero. We fix a desired temperature by reversible transfer of heat. We cannot reach absolute zero because there cannot be heat transfer in both directions with a state of minimal energy, so reversible fluctuations stop.

Third and perhaps most important, it describes thermal equilibrium quantitatively: Two bodies exchange heat reversibly in quantities of kT per molecule, all the time, giving rise to microscopic fluctuations in the molecules' energies. That is what it means for their common temperature T to be defined by reversible heat exchange. Being reversible, these fluctuations cannot do useful work, much like our sled stayed still while we tugged it. Thermodynamics models them by infinitesimal Carnot engines, but cannot explain them physically, so their rate—the nominal power of the heat exchange—remains mysterious. To repeat, this power is only nominal, not related to any work, because a reversible exchange of size RT per mole happens in many exchanges of size kT per molecule with both signs, adding to RT in absolute value, but invisible macroscopically. These heat exchanges are to temperature what momentum exchanges of molecules with walls are to pressure.

Notably, the thermodynamic definition of a molecule is macroscopic, just like the definition of a mole, so the Boltzmann constant could in principle be as arbitrary as the gas constant. It becomes subject to measurement only if the universal measure of the quantity of matter, expressed by the constant S_Q, is really the number of physical molecules. As the second interpretation above shows, that idea has its own problems, which were finally solved by quantum mechanics in the twentieth century.

It is surprising how much could be gleaned about the microscopic behavior of matter from macroscopic observations. In particular, the relation (2.20) has been derived by elementary arguments, not surpassing high-school algebra. The only way to make sense of such dizzying success is to realize that we live in a quantum world at all temperatures. There is no classical physics in physical reality, because Planck's constant h is never zero. What we call "the classical limit" is actually the limit of high temperatures, where quantum effects simplify to some extent, leaving us puzzled when these simplifications fail at lower temperatures, and even more puzzled when they succeed beyond all reasonable expectations, such as the prediction, made at steam-engine temperatures, of a vanishing heat capacity at zero temperature.

2.6 Thermodynamic Potentials. Equations of State

Clausius' theorem allows us to write the heat term analogously to the work term, $\eth Q = TdS$. Mathematically, that is all we need to turn the first law of thermodynamics into a practical description of systems at equilibrium, even if we cannot imagine entropy. The first law reads now

$$dU = TdS - pdV + \mu dN. \tag{2.27}$$

It states that the internal energy U is a function $U(S, V, N)$ of three independent variables, entropy, volume, and particle number. "Independent" means that each can be fixed arbitrarily, without affecting the others. The corresponding laboratory arrangement is an *isolated system*: A thermally insulated hard box, sealed hermetically. The state variables are fixed by preventing external influences that affect them.

How do we know that Eq. (2.27) refers to a stable system? We do not, at this point. Stable equilibrium means physically that the system is at the extremum of some potential, so that forces returning it to equilibrium are generated when it is disturbed. All we know is Clausius' theorem, which says that entropy is such a potential, whose mysterious tendency to increase spontaneously gives rise to these restoring forces. To make use of that insight, we rewrite the first law:

$$dS = \frac{1}{T}dU + \frac{p}{T}dV - \frac{\mu}{T}dN. \tag{2.28}$$

At equilibrium, $dS = 0$ and $d^2S < 0$, the negative sign indicating that the entropy is maximal (Fig. 2.4). The thermal insulation means $dU = d^2U = \ldots \equiv 0$, all differentials are identically zero because the internal energy is a fixed number. Then

$$\frac{p}{T}dV - \frac{\mu}{T}dN = -\frac{\eth W}{T} = 0, \tag{2.29}$$

known as the equation of mechanical balance, or principle of virtual work. That name implies a more interesting application than just setting $dV, dN \equiv 0$ because they are also fixed. Instead, we make use of the fact that no work is required to maintain a stable equilibrium. Small changes in the variables generate conjugate forces which must be balanced at the stable point (Fig. 2.5). The force-balance equations in thermodynamics are called equations of state, and they follow directly

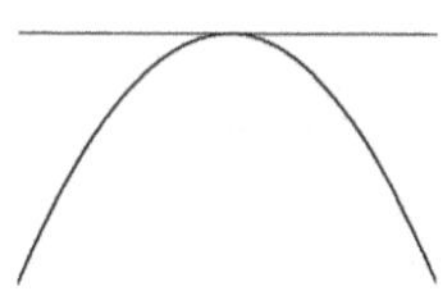

Fig. 2.4 At equilibrium

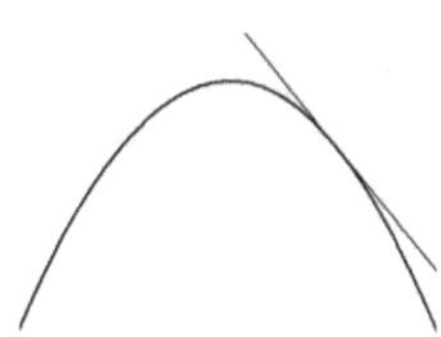

Fig. 2.5 Restoring force

from Eq. (2.28):

$$\frac{1}{T} = \frac{\partial S}{\partial U}, \quad \frac{p}{T} = \frac{\partial S}{\partial V}, \quad \frac{\mu}{T} = -\frac{\partial S}{\partial N}. \tag{2.30}$$

The partial-derivative notation means, as usual, that the other variables are fixed when taking the derivative, so their differentials in Eq. (2.28) vanish. Identifying these forces in the equation of mechanical balance, we can use it to determine the mechanical equilibrium of the system, as we did above for the air spring.

Our prescription is only indirect, however, because we have no idea what entropy is. Either we have to resort to calorimetry to determine the function $S(U, V, N)$ by integrating measured values of $\Delta Q/T$, or find some way to measure the conjugate forces p/T and μ/T directly. The former is often chosen by chemists, because entropy changes due to rearranging molecules in chemical reactions are substantial, so that chemical balance is dominated by the second term in Eq. (2.29). Physicists prefer the latter, because pressures are easily measured, unlike the entropy changes due to movement of mechanical parts.

2.6.1 *Minimization of Internal Energy*

The principle of virtual work becomes much more practical when one realizes that the same mechanical-balance equation appears in the first law (2.27) under the same restriction of thermal insulation, $\eth Q \equiv 0$, i.e., $dS = d^2S = \ldots \equiv 0$ because the entropy is a fixed number. Then the remainder of dU is also zero,

$$\eth W = -pdV + \mu dN = 0, \tag{2.31}$$

because it is equal to Eq. (2.29), only multiplied by $-T$. Thus, we find that the internal energy is extremal, $dU = 0$, at fixed entropy, because the entropy is maximal at fixed internal energy, and fixing either means the same, thermal insulation.

It takes but a moment to realize that this extremum must be a minimum, because putting dU on the left-hand side instead of dS multiplies all work differentials with $-T$, not just the first. In particular, $d^2S < 0$ at fixed U implies $d^2U > 0$ at fixed S. The internal energy is minimal *because* the entropy is maximal. It is just as good a thermodynamic potential as the entropy when the system is isolated, which fixes its three variables, $U = U(S, V, N)$.

The principal advantage of internal energy as thermodynamic potential is that it is typically a known function of V, the mechanical variables. We know it because we have constructed the machine! The equations of state derived from it are

$$T = \frac{\partial U}{\partial S}, \quad p = -\frac{\partial U}{\partial V}, \quad \mu = \frac{\partial U}{\partial N}. \tag{2.32}$$

These equations branch engineering into material, mechanical, and chemical, respectively. Advances in the second branch, which benefits the most from this reformulation, have historically driven the other two. The later addition of electronics to the third is based on understanding the dependence on N in semiconductors. It has moved feedback loops from the second to the third branch. This shift is epochal because feedbacks are a key part of machine design.

2.6.2 *Enthalpy*

The analysis of the air spring in Fig. 2.2, at the level of forces, can be repeated at the level of potentials now. The hard wall that fixed the volume is replaced by a moving wall that fixes the pressure. Changing the weight moves the wall, so we must include the potential of the weight in the total potential which determines the stable position:

$$H = U + mgh = U + pAh = U + pV, \tag{2.33}$$

where we have inserted the equilibrium condition $mg = pA$ in the second step. This thermodynamic potential is called *enthalpy*. It is extremal when the variables (S, p, N) are fixed. Indeed,

$$dH = dU + d(pV) = TdS + Vdp + \mu dN, \tag{2.34}$$

which shows explicitly that these three are the independent variables of the enthalpy as a function, $H = H(S, p, N)$. Focusing on the changed variable only,

$$H(p) = U(V^\star) + pV^\star, \quad V^\star = \frac{\partial H}{\partial p}. \tag{2.35}$$

The (usually omitted) star means that V on the right-hand side is not a variable, it is a function $V^\star(p)$, the equilibrium volume at the fixed pressure. It is either measured or given by the equation of state above. If we know the function $U(S, V, N)$ explicitly,[3] inserting the second expression into the first gives a differential equation for $H(p)$, which we can solve to obtain an explicit expression for the enthalpy as well.

The enthalpy is sometimes called the heat function, because an isobaric heat transfer is equal to a change in enthalpy:

$$\Delta H|_{p=const.} = \Delta Q, \tag{2.36}$$

assuming that the number of particles does not change. This arrangement reflects measurements in a calorimeter.

[3] Meaning that T, p, or μ do not appear in the formula for U, cf. Eqs. (2.32).

2.6.3 Legendre Transforms. The Gibbs-Duhem Relation

The gravitational potential mgh in the enthalpy is the interaction energy of the system with its surroundings: The mass is fixed externally, the height is the response of the system. Whatever mechanism fixes the pressure, its contribution to the potential, or interaction energy, must be equal to pV, because it cannot matter how it is designed (Fig. 2.6). That understood, we can forget the mechanism and proceed formally: A term like Xdx in the differential of the thermodynamic potential means that a term Xx must be subtracted from the potential itself to change the variable:

$$Xdx - d(Xx) = -xdX. \tag{2.37}$$

The mathematics of the variable change imposes that physical interaction contributions must always be of the form Xx, where X and x are conjugate state variables. Otherwise, one could not describe the system without referring to the surroundings, thus contradicting the separation between them, characteristic of thermodynamic equilibrium. This operation to change the variable is called a Legendre transform. It illustrates the power of thermodynamic phenomenology: We can write practical expressions for interactions without understanding the underlying physics.

It remains to write down the equilibrium potentials for any particular situation, using Legendre transforms. For a system in a thermostat, the temperature is fixed, replacing the entropy. We thus obtain the *Helmholtz free energy*:

$$F = U - TS, \quad dF = -SdT - pdV + \mu dN, \quad S = -\left(\frac{\partial F}{\partial T}\right)_{V,N}, \tag{2.38}$$

where we have introduced the convention of writing the complete variable set in the derivative. When that is omitted, the "natural" variables are assumed, i.e., the independent ones in the differential. The name free energy refers to its change being equal to the work along an isotherm,

$$\Delta F|_{T=const.} = \Delta W, \tag{2.39}$$

so it is the part of internal energy "free to do work."

For fixed temperature and pressure, we have the *Gibbs potential*

$$G = F + pV, \quad dG = -SdT + Vdp + \mu dN, \quad V = \left(\frac{\partial G}{\partial p}\right)_{T,N}. \tag{2.40}$$

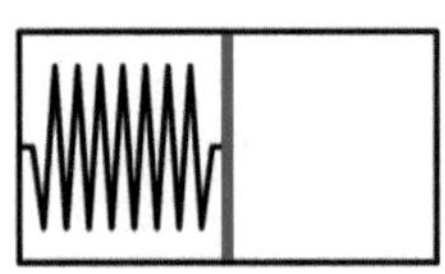

Fig. 2.6 A different air spring

Notice the difference with enthalpy, where the volume played the role of restoring force when the independent conjugate variable (pressure) was changed at fixed entropy (adiabatically), while it is changed at fixed temperature (isothermally) here. One can also write

$$G = H - TS, \tag{2.41}$$

a relation useful to chemists, because both enthalpy and entropy can be measured calorimetrically, while the Gibbs potential is extremal at laboratory temperature and pressure, i.e., for reactions in a test-tube. In this way, one can separate deterministic and random contributions to reaction barriers.

The obvious next step is to replace the dependence on N with a dependence on μ, to obtain a potential $\mathcal{P}(T, p, \mu) := G - \mu N$ as a function of intensive variables only. However, such a potential vanishes identically, because it must double when the system doubles in size, but also cannot change because its variables do not change: $2\mathcal{P} = \mathcal{P} \implies \mathcal{P} \equiv 0$. This cumulative change of all extensive variables reads

$$U - TS + pV - \mu N \equiv 0. \tag{2.42}$$

It is known as the Gibbs-Duhem relation. Taking the differential, we find

$$-SdT + Vdp - Nd\mu \equiv 0, \tag{2.43}$$

the differential form of the Gibbs-Duhem relation. Differentials are vectors, so it means that these three lie in the same plane. Physically: Forces are not independent at equilibrium because they cancel there. Geometrically: Solutions of the equations of state are a two-dimensional subspace of the three-variable space. The relation (2.43) defines the two-dimensional plane tangent to that subspace.

The Gibbs-Duhem relation implies that the Gibbs potential may be written

$$G(T, p, N) = \mu(T, p) \cdot N, \tag{2.44}$$

which can also be argued directly: Because it has only one extensive variable, G must be proportional to it, otherwise it would not be extensive itself. By the same token, we can regroup the terms in Eq. (2.42) to obtain

$$\Omega(T, V, \mu) := -p(T, \mu) \cdot V, \tag{2.45}$$

the so-called *grand potential*. The related change of variable is explicitly

$$\Omega = F - \mu N, \quad d\Omega = -SdT - pdV - Nd\mu, \quad N = -\left(\frac{\partial \Omega}{\partial \mu}\right)_{T,V}. \tag{2.46}$$

One may wonder how to fix the chemical potential. For charged particles, electricians call it "ground" or "earth." When we connect the chassis of an appliance to a conductive rod driven into the ground, the number of electrons on it is no longer fixed, but fluctuates invisibly to keep it at the fixed earth potential. These are equilibrium fluctuations, just like heat fluctuations at fixed temperature, or volume fluctuations at fixed pressure due to molecules impinging on the movable wall in Fig. 2.6.

For a gas of neutral particles, imagine a wall coating on which they are adsorbed. When molecules stick to the coating at the same rate as they are released by it, the chemical potential is fixed, again at the price of fluctuations in the number of molecules in the gas. Clothes and shoes made of artificial fibers are usually packaged with sachets of silica beads, which fix the chemical potential for water such that the vapor concentration is kept very low, to prevent mold. Their microscopically porous structure offers a large effective surface for adsorption, which may be reversed by raising the temperature, without changing either the silica or the water. Thus, adsorption is a physical rather than chemical phenomenon.

2.6.4 Massieu Potentials

All the thermodynamic potentials above were obtained by adapting the internal energy to different laboratory arrangements. Alternatively, we could have started with the entropy. It is useful to rewrite Eq. (2.28) as

$$d\sigma = \beta dU + \beta p dV - \beta\mu dN, \tag{2.47}$$

where $\sigma := S/k$ and $\beta := 1/kT$. To repeat, entropy is maximal for an isolated system, $\sigma = \sigma(U, V, N)$. The corresponding potential for a system at fixed temperature is the *canonical Massieu potential*:

$$\Psi_C = \sigma - \beta U, \quad d\Psi_C = -U d\beta + \beta p dV - \beta\mu dN, \quad U = -\left(\frac{\partial \Psi_C}{\partial \beta}\right)_{V,N}. \tag{2.48}$$

For fixed temperature and chemical potential, we have the *grand Massieu potential*:

$$\Psi_G = \Psi_C + \beta\mu N, \quad d\Psi_G = -U d\beta + \beta p dV + \beta N d\mu, \quad N = \frac{1}{\beta}\left(\frac{\partial \Psi_G}{\partial \mu}\right)_{V,N}. \tag{2.49}$$

They are related algebraically to the free energy and grand potential:

$$\Psi_C = -\beta F, \quad \Psi_G = -\beta\Omega. \tag{2.50}$$

Clearly, what can be done with one pair can be done with the other. It turns out that the Massieu potentials are better adapted to the underlying statistical interpretation, so we shall often prefer them once it is developed. Interestingly enough, the canonical Massieu potential appeared historically before the Helmholtz free energy, just as the second law of thermodynamics preceded the first. The conceptual preference of energy over entropy spread gradually through the physics community in the second half of the nineteenth century.

Minor confusion may arise over our use of the products βp and $\beta\mu$ in Eq. (2.47). It would have been more general to introduce some $\alpha := \beta\mu$, say, and write $\Psi_G = \Psi_C + \alpha N$, $N = \partial\Psi_G/\partial\alpha$, instead of Eq. (2.49). However, there is a difference between varying α and varying μ only if the temperature and chemical potential both vary when α is varied. It is more common to vary μ independently of β, so we assume that by default, similarly for the pressure.

2.6.5 *Practical Calculations. Susceptibilities*

If a thermodynamic potential is known explicitly as a function of its natural variables, the formalism above suffices to obtain the equations of state simply by taking derivatives. Some information may be required in practice that is not available directly from the equations of state.

A typical situation is when we need some potential that is not extremal under the circumstances. Suppose that it is the internal energy of a system in a thermostat. Then we can invert the Legendre transform (2.38) to write[4]

$$U(S^\star) = F(T) + TS^\star, \quad S^\star = -\left(\frac{\partial F}{\partial T}\right)_{V,N}. \tag{2.51}$$

Clearly, the equation of state is used as additional information here, not contained in $F(T)$ explicitly. It is contained implicitly, because the equilibrium value of the entropy at fixed temperature $S^\star(T)$ is obtained from the functional dependence $F(T)$. It is still true that first derivatives of F contain all the information needed.

What if we consider an isothermal change in volume? We have

$$\left(\frac{\partial U}{\partial V}\right)_{T,N} = \left(\frac{\partial F}{\partial V}\right)_{T,N} + \left(\frac{\partial}{\partial V}TS^\star\right)_{T,N} = -p - T\left(\frac{\partial^2 F}{\partial V \partial T}\right)_N, \tag{2.52}$$

[4] $U(S^\star(T)) =: \mathfrak{U}(T)$ is another function, $\mathfrak{U} = U \circ S^\star$, but no one distinguishes that in notation, writing $U(T)$ and listing all independent variables explicitly in partial derivatives instead.

because $\partial T/\partial V = 0$ at fixed T, i.e., when both quantities are independent. We can take the mixed derivative in any order, so we get

$$\left(\frac{\partial U}{\partial V}\right)_{T,N} = -p + T\left(\frac{\partial p}{\partial T}\right)_{V,N}. \tag{2.53}$$

For an adiabatic change (fixed S), the result would have been just $-p$, cf. Eq. (2.27). Because pressure rises with temperature, the second term is positive, so isothermal expansion does not lower the internal energy as much as adiabatic does. The physical reason is that the thermostat adds heat to maintain the temperature. For that reason, the quantity

$$l_V := T\left(\frac{\partial p}{\partial T}\right)_{V,N} = T\left(\frac{\partial S}{\partial V}\right)_{T,N} \tag{2.54}$$

is called *latent heat*. Even if it has the dimension of pressure, the name heat is not misplaced: $l_V \Delta V$ is the heat contribution to the internal energy with an isothermal change in volume, while $-p\Delta V$ is the mechanical contribution.

We could have obtained the same result purely formally, as a change of variables. Dropping N for compactness, we write the chain rule for two variables:

$$\left(\frac{\partial U}{\partial V}\right)_T = \left(\frac{\partial U}{\partial x}\right)_y \left(\frac{\partial x}{\partial V}\right)_T + \left(\frac{\partial U}{\partial y}\right)_x \left(\frac{\partial y}{\partial V}\right)_T. \tag{2.55}$$

Setting $(x, y) = (V, S)$ gives the same as above, with the help of

$$\left(\frac{\partial S}{\partial V}\right)_T = -\frac{\partial^2 F}{\partial V \partial T} = \left(\frac{\partial p}{\partial T}\right)_V \tag{2.56}$$

again. This relation is known as Euler's reciprocal relation. It is valid when equations of state define a smooth surface, so the second derivatives are well defined.

The mixed derivative (2.56) shows why equations of state are not sufficient for all situations. They are first derivatives of the free energy, while we require the second as well to describe the system completely. Changing volume affects pressure directly, changing temperature affects it indirectly. The reciprocal of the direct second derivative is called the *isothermal compressibility* κ_T:

$$\kappa_T^{-1} := -V\left(\frac{\partial p}{\partial V}\right)_T. \tag{2.57}$$

The derivative itself is always negative for a stable system, which resists compression by a rise in pressure. Generally, the derivative of an extensive quantity by its conjugate intensive variable is called a *susceptibility*. Because it is extensive, it is often divided by the same extensive variable to obtain a "specific" susceptibility, which can be looked up in tables of material properties. Analogously, the *heat capacity* is the second derivative of F with respect to temperature:

$$c_V := \left(\frac{\partial U}{\partial T}\right)_{V,N} = \left(\frac{\partial F}{\partial T}\right)_{V,N} + \left(\frac{\partial}{\partial T} T S^\star\right)_{V,N} = -S^\star + S^\star - T\left(\frac{\partial^2 F}{\partial T^2}\right)_{V,N}, \tag{2.58}$$

which is better written

$$c_V = T\left(\frac{\partial S}{\partial T}\right)_{V,N}, \tag{2.59}$$

a more useful formula than the definition in terms of U above. Instead of dividing by S, specific heat capacities are expressed per mass, volume, or mole, whatever is practical. The expression (2.59) shows explicitly that $c_V = 0$ at $T = 0$, as we concluded in general from Carnot's analysis of equilibrium fluctuations (2.20).

2.6.6 Stability. Le Chatelier's Principle

Analytically, the second derivatives of thermodynamic potentials are coefficients of the quadratic form describing the potential paraboloid around its extremum. Geometrically, they carry information about the curvature of that paraboloid. Physically, they are the constants of elasticity of the equivalent harmonic oscillator. Second derivatives in the same variable are called diagonal susceptibilities, mixed derivatives are off-diagonal susceptibilities.

Evidently, the extremum has to be such that the system is stable, maximum for the entropy, minimum for the internal energy. It is easy to see physically that a susceptibility of the wrong sign indicates an instability. For example, if the pressure decreases when the system contracts, it contracts further, so it collapses. Here, we shall derive a necessary requirement on susceptibilities for the system to be stable.

Start with the differential Gibbs-Duhem relation (2.43), $-SdT + Vdp - Nd\mu \equiv 0$, which describes the balance of forces at equilibrium. These forces are themselves derivatives of the potential U. Writing $T = \partial U/\partial S = U_S$ etc., we have

$$SdU_S + VdU_V + NdU_N \equiv 0. \tag{2.60}$$

This is a vector equation in the three differentials, which we can write in components by turning differentials into derivatives with respect to S, V, and N. The component equations are written together as one matrix equation:

$$\begin{pmatrix} U_{SS} & U_{SV} & U_{SN} \\ U_{VS} & U_{VV} & U_{VN} \\ U_{NS} & U_{NV} & U_{NN} \end{pmatrix} \begin{pmatrix} S \\ V \\ N \end{pmatrix} \equiv 0. \tag{2.61}$$

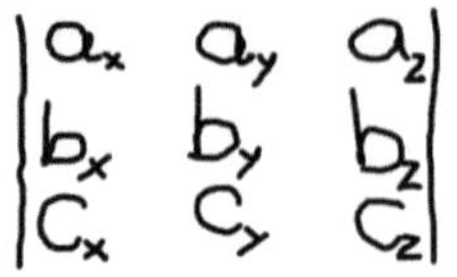

Fig. 2.7 Signed volume

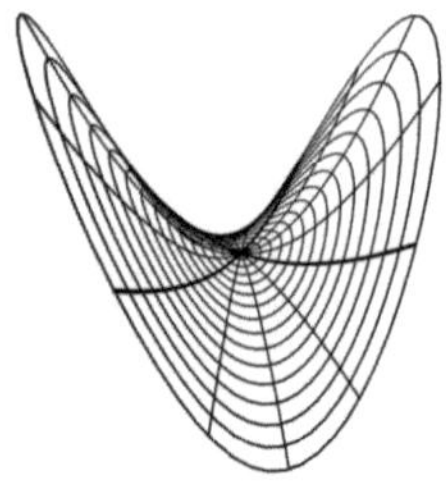

Fig. 2.8 Not good enough

The determinant of this matrix is the signed volume spanned by three co-planar vectors, so it vanishes (Fig. 2.7). That is lucky for thermodynamics, or it would be forced to claim that non-existence, $S = V = N = 0$, is the only stable form of existence.

Previously, we found the internal energy to be minimal at fixed entropy. Generally, whichever variable we fix, stability means that the internal energy is minimal in the others. Traditionally one fixes N, referring to a hermetically sealed system. Then minimality (positivity of the quadratic form) in the other two is mathematically equivalent to all *nested minors* being positive, starting with one diagonal element:

$$U_{SS} > 0, \qquad \begin{vmatrix} U_{SS} & U_{SV} \\ U_{VS} & U_{VV} \end{vmatrix} > 0. \tag{2.62}$$

Geometrically, the conditions (2.62) mean that there is no way downward from the apex of the paraboloid $U(S, V, N = N_0)$. They imply $U_{VV} > 0$. Had we required the latter, they would have implied $U_{SS} > 0$. Requiring $U_{SS}, U_{VV} > 0$ is not good enough, because the determinant could still be negative (Fig. 2.8).

It is easy to see what the first of these conditions means physically:

$$0 < U_{SS} = \frac{\partial T}{\partial S} = \frac{T}{c_V}, \tag{2.63}$$

so the heat capacity must be positive, because the absolute temperature is positive.

The second is analyzed in three steps. First we write the susceptibilities as derivatives of the forces, like we just did for $U_{SS} = T_S$:

$$0 < \begin{vmatrix} T_S & T_V \\ -p_S & -p_V \end{vmatrix} = \frac{\partial(T, -p)}{\partial(S, V)} = \frac{\partial(T, -p)}{\partial(T, V)} \cdot \frac{\partial(T, V)}{\partial(S, V)} = \frac{1}{V\kappa_T} \cdot \frac{T}{c_V}. \tag{2.64}$$

Note the Euler relation in the mixed derivatives. In the second step, we interpret this determinant as a Jacobian to change variables from $(T, -p)$ to (S, V). In the third, we express this Jacobian as a succession of two single-variable changes, each involving a known susceptibility. Combined with Eq. (2.63), condition (2.62) becomes

$$c_V > 0, \quad \kappa_T > 0, \tag{2.65}$$

called *Le Chatelier's principle*: If a system with fixed particle number is stable, heating increases its temperature, and isothermal compression increases its pressure.

Analytically, the Jacobian is a signed volume element appearing in a change of variables. The defectiveness of the susceptibility matrix (2.61) means that the equilibrium is a singular point of the change from (T, p, μ) to (S, V, N). It is like the point $\rho = 0$ in a change of Cartesian coordinates $(x = \rho \cos \phi, y = \rho \sin \phi)$ to polar coordinates (ρ, ϕ). Just as the angle is not determined at radius zero, the size of a system at equilibrium is not determined by the forces alone. The interplay between the restoring forces near equilibrium and the susceptibilities at equilibrium is subtle and important, so we shall devote much of Chap. 11 to it.

2.7 Useful Engines. Feedbacks

As we have seen above, reversible engines model microscopic fluctuations that keep a system at equilibrium, in which all thermodynamic forces are balanced, so that no work is needed for the system to persist in time. The usual notion of a machine is quite different, however. A machine works to make things that we can carry home. Once that happens, it is obviously impossible for it to revert to its initial state. Thus, any useful process must be irreversible.

Production is repeatable by assumption, so the question is how to maintain a cyclic process despite irreversibility. The general answer is a source of fuel that overcomes dissipation by replacing free oscillations with forced ones. Assuming that we have one, the cyclic process can be modelled by a macroscopic Carnot engine, with the irreversibility relegated to the fuel-burning that keeps the two reservoirs at their fixed temperatures. This framework is sufficient to derive an important general principle of useful machine operation.

The closed loop in Clausius' theorem (2.26) can be parametrized in terms of periodic functions, where the angle plays the role of time:

$$0 \geq \oint \frac{\eth Q}{T} = \int_0^{2\pi} \frac{\eth Q(\sin \phi, \cos \phi)}{T(\sin \phi, \cos \phi)} d\phi. \tag{2.66}$$

In order for this integral to be less than zero, the numerator and denominator should be *synchronized* in such a way that the ratio subtends a larger surface below the axis than above it (Fig. 2.9). To be specific, let the temperature oscillate around some

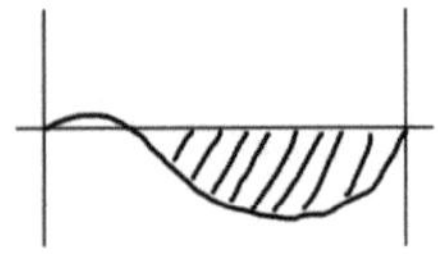

Fig. 2.9 Irreversible cycle

average value, $T = T_0 + \Delta T(\sin\phi, \cos\phi)$. Expanding $\eth Q$ and $1/T$ in a series and multiplying them, the lowest terms have the form

$$\int_0^{2\pi} \sin\phi \sin(\phi + \alpha) d\phi. \tag{2.67}$$

Clearly, the value of α determines the sign of the integral. It is a fixed delay between the moment of heat exchange and some point in the temperature cycle. Temperature is related to pressure by the equation of state. In a combustion engine, the fuel must ignite at the moment of maximum compression in a cycle. To fix α is to synchronize the engine so that this happens. Synchronization is achieved by the timer belt, which is a *noiseless feedback* between the mechanical (power) and electrical (signal) components of the engine.

Feedbacks are essential to usefulness, defined as the ability to persist in a state out of thermodynamic equilibrium. They provide a classification scheme for various forms of natural and social organization.

The essence of feedback is noiseless synchronization. Noiseless means that the feedback can be strictly deterministic without losing its function, such as a timer belt without friction. Synchronization refers to the way feedbacks keep the system out of thermodynamic equilibrium. For example, if the timer belt breaks, the motor stops in less than a second.

A bear must fatten *before* winter. The mechanism triggering this behavior is noiseless. But if fattening is not synchronized with autumn, the bear dies, just as a motor stops if a spark plug fires out of turn.

Generally, feedbacks allow *adiabatic decomposition*, meaning that phenomena at different time scales may be considered separately even if they are coupled, such as the motions of the Earth and the Sun. They carry this notion even further in the sense that they themselves enhance the decoupling. For example, feedbacks which keep cells alive are separate from those which keep the organism alive. This example points to an important difference with adiabatic decomposition at equilibrium. In the latter case, the heavy (slow) subsystem guarantees the stability of the light (fast) one, absent phase transitions. In a hierarchy of feedbacks, one cannot count on the slow ones always providing restoring forces to damp fluctuations in the fast ones. Thus, an organism can be destabilized by disruptions of the cellular metabolism. There may be time to react, but no mechanism in place to do so.

Such a multitude of examples beg for organizing principles. Here we consider two qualitative classifications. One is the presence, or absence, of an invisible control in feedback operation. The other is a distinction between useful and purposeful.

2.7.1 Invisible Control

The water cycle is a classic example of a circular process in nature. It is controlled by a feedback between rainfall and evaporation. This control is visible in the sense that it is driven by the macroscopic circumstances at each point in the cycle. An explanation of a particular cycle is in terms of locally measurable thermodynamic forces.

Ecological cycles, natural evolution, and living beings are different than climate. They cannot be understood without invoking feedbacks with the genetic code, which is an invisible control. Otherwise, the simplest observations, such as shedding leaves in autumn, must be taken as given. If one persists that way, the number of givens increases without bound because of the flexibility and universality of genetic control.

Similarly, human societies are governed by laws and customs. Although laws are overt, it cannot be inferred from them alone which behavior is forbidden, and what would be the consequence if someone did something original. There is an invisible control layer that is instilled by upbringing.

Combustion-engine feedbacks have evolved from visible (climate-like) to invisible (genome-like). The distributor firing the spark plug was mechanical until the late twentieth century, so its inertia had to be included with all other moving parts. In this respect, feedback was no different than the steam-engine regulator studied by Maxwell: At each point in the cycle, macroscopic forces determined the behavior of the engine completely. With microprocessors, feedbacks became without inertia and arbitrarily programmable.[5] In the twenty-first century, a car can surprise its owner no less than a dog.

Cows are more closely related to whales than to pigs. That statement is true in the *parameter space* of the genetic code, however improbable it appears in the *laboratory space* of our observations. It is also true that cows look more like pigs, so calling one true and the other "mere" appearance means that we prefer the invisible to the visible. Is that preference objective? After all, to sequence the cow genome, we take tissue from a visible cow. Who is to say which determines the other?

Genetics is reducible to chemistry, so one could say, here is one more triumph of the molecular hypothesis. However, one cannot invoke a principle to resolve a doubt about that principle. The question is whether the molecular hypothesis uncovers the truth in this case, or is just a habit of thought. An animal uses biochemistry both to reproduce and to feed, so why does one explain its existence and the other not? The question is important because understanding the relation between a particular microscopic parameter space and always-the-same macroscopic laboratory space is required for statistical explanations of macroscopic phenomena.

Philosophical essentialists would say that the genome contains the essence of an animal, which comes into existence as a body, like expanding a compressed

[5] To quote a recent textbook: Engineers have been liberated from the terror of impedance.

file. Existentialists would say that the genome is just one part of the existing body. Thermodynamics breaks the tie by invoking the hierarchy of time scales, or adiabatic decomposition. As the Earth follows the Sun around the galaxy, so animals' bodies follow their gene pool through evolution. As the Earth goes around the Sun many times while the Sun moves a little, so many generations of animals elapse while their genomes change a little. Feeding is in the opposite limit. A feeding cycle is shorter than an animal's life, like a day is less than a year. A similar hierarchy appears in the evolution of machines, ecosystems, and societies. When the slow feedbacks are programmable and without impedance, i.e., invisible, a great variety of forms can be stabilized by faster visible feedbacks, all explained in terms of the slower ones.

One might object that the only difference between climate and genetics is that in the former case, macroscopic thermodynamics is simple enough to avoid mentioning the microscopic parameter space of quantum mechanics, replacing it with the theory of functions of several variables. In this sense, our distinction between visible and invisible feedbacks appears illusory: All complete explanations are in terms of the invisible. However, that is a false dilemma. Microscopic parameters and macroscopic variables coexist harmoniously. The connection between them are the susceptibilities, which can be either measured, or calculated on the basis of a microscopic understanding. Either way, once they are in hand, macroscopic thermodynamics suffices to describe all observations. True, there is no macroscopic description of evolution at present in which evolutionary pressure has a precise mathematical meaning like physical pressure, but it is not necessarily impossible. The essential difference between feedbacks remains: Visible ones are prone to friction and inertia at the scales of the process they control, invisible ones are not.

2.7.2 *Spontaneous Symmetry Breaking. Purposeful Design*

Usefulness has been defined objectively above: It is the persistence of irreversible cycles, keeping a system away from thermodynamic equilibrium. Staying away from equilibrium is profit in the broadest sense. Some of the fuel sustaining the process is spent overcoming friction, so there is a net negative balance to justify. So why does it make sense to stay out of equilibrium? Does it have to make sense?

The answer to the latter question is no. An example is the Belousov-Zhabotinsky reaction, in which the color of a solution of several chemicals oscillates as long as it is exposed to light. Shining light on the solution with any fixed ratio of constituents destabilizes that ratio. The oscillation happens just because it can. Some proposals about the origin of life are similar. Under certain circumstances, a uniform soup of chemicals precipitated into droplets, which went on to become the first living cells. It happened because it could, and has been self-sustaining ever since, making use of ever more sunlight, either direct or stored in the form of fossil fuels, and of heat from teratons of radioactive elements in Earth's mantle. The latter also enables plate tectonics, another useful process, happening just because it can.

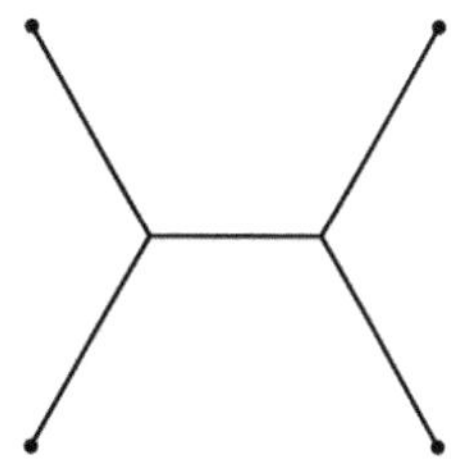

Fig. 2.10 Spontaneous symmetry breaking

This "just because it can" effect is called spontaneous symmetry breaking. Imagine four towns on four corners of a square (Fig. 2.10). Suppose we want to connect them all by the shortest possible road. The shortest symmetric solution is X-shaped, with total length $2\sqrt{2} \approx 2.8$ side-lengths. However, if one puts a horizontal strut in place of the crossing point, one can optimize it to obtain the shortest length $1 + \sqrt{3} \approx 2.7$, which is the solution of the problem. Evidently, a vertical strut is just as good, which is the essence of the phenomenon: The solution has a lower symmetry than the problem. Ouzo turns cloudy when a little water is added: Fragrant oils dissolved in alcohol break translation symmetry when they precipitate into droplets.

Spontaneous symmetry breaking appears generically in optimization. The trigger was the requirement that the length of the road be minimal, or, in the case of ouzo, the free energy.[6] Reducing square to rectangular symmetry forces one to choose between two orientations. Imagine four town mayors discussing that. It motivates a thermodynamic definition of politics: Social moderation of spontaneous symmetry breaking. The definition covers both human and animal societies, so is there any objective difference? Beavers also build houses and dams, and resolve resource conflicts along the way, so how are humans different, if they are?

A natural process can choose between orientations by any random fluctuation, because the symmetric solution is untenable. So if the mayors fight it out, however politely, that is no different in principle than how animals would handle the dilemma.[7] But suppose that they agree on an objectively suboptimal decision based on some hope for the future. Such behavior is hypothetical, or hopeful. Physical forces are local in time and space, while hypothetical behavior is based on non-local considerations. This non-locality defines purpose objectively. Hypothetical behavior is a functional distinction between humans and animals. It is a uniquely human way to engage physical reality by including non-existent forces in the force-balance equation, acting as if an unknown future were the present, or an invisibly distant thing near.

Paleontological evidence suggests that hypothetical behavior is very old in the human lineage. There is a two-million-year-old find of stone tools carried over 13 km to hunting grounds for the purpose of butchering animals. Carrying them

[6] Extremes are found by differentiation, which lowers the symmetry of a function, e.g., x^2 is invariant under the replacement $x \to -x$, but $2x$ is not.

[7] Which town has the dominant mayor is a random fluctuation.

was a considerable struggle, while the catch was not certain. As of this writing, the only way to claim that animals show such behavior is to claim that only *Homo sapiens* is human. Conversely, observing hopeful behavior in giant squid, say, would force us to rethink humanity even more radically than including all the genus *Homo*. It is unlikely that any functional definition of humanity corresponds precisely to any genetic definition.

One should distinguish hypothetical behavior from strategizing based on experience. The latter is how genes optimize adaptation. A dog may be put down because it attacked a human, so it goes badly for that dog, but its behavior is not irrational, given that it cannot imagine courts and laws. The dog's genome programs it to maximize reproductive success. Its existence is physical proof that the program makes good bets in the long run. An animal is a rational robot. Calling it irrational is misunderstanding the probabilistic nature of genetic optimization, which models the future on the past. Humans escape this optimization by hypothetical behavior, which is not rational in any physical sense: Intrinsic motivation is a force from an unknown future, or distance. Behaviorally, it is an adaptation to the unknown, a contradiction in terms. When they take into account only past experience and real forces, humans revert to animals or robots. For example, international politics resembles animal politics more than internal politics does, the latter being more human so far as it is more hopeful. Similarly, corporations are robots if they optimize only computable object functions, such as profit.

We can answer the first question above now. Reduction of entropy is the physical profit in staying out of equilibrium. We mean the entropy of the interesting part of the system, even if total entropy cannot decrease. Spontaneous symmetry breaking offers a choice between solutions, each of which is more ordered than the symmetric one, where order is defined objectively as lower symmetry. Nature also profits from such choices, e.g., making crystals out of liquids. To call it profit in the human sense, we have to give the crystals purpose, such as weapons or jewelry. Distinct from physical profit, which is just a difference in thermodynamic potentials, human profit is order with purpose.

2.8 Purposeful Engines

Human originality is primarily a situational and only secondarily a mental category. It is putting oneself in a situation where the unexpected may happen. Writers, farmers, laboratory heads, and parents are all in an original situation: They do as they decide and face what they have done. But if someone else tells them what to do, or bears the consequences, they are in a normal situation.

Action and observation are one-way links between the invisible world of thoughts and the visible world of deeds. They are a practical realization of the relation between philosophy and revelation, compared to walking on two legs in the Introduction. We can replace that comparison with an abstract engine now. The critical feature of the engine is synchronization. Even if the process is cyclic,

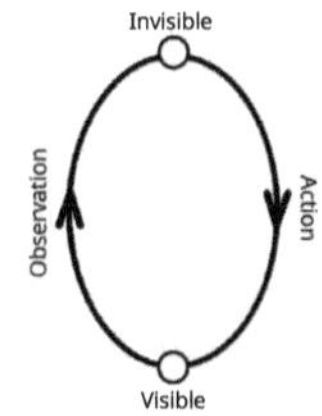

Fig. 2.11 Universal purposeful engine

it matters which part of the cycle occurs before the other. Writers who write a book *before* considering the reaction of the public behave hypothetically, as if a non-existent force were acting on them. They can bring human profit—order with purpose—from the invisible into the visible world. Those who think about the public first are doing the same kind of probabilistic calculation as an animal genome. They maximize sales—physical profit—as the genome maximizes progeny. Their books are determined by the past and by real forces. They cannot change the contemporary equilibrium because they are self-consistently part of it: A desynchronized engine stands still. The universal purposeful engine in Fig. 2.11 has created all purposeful engines. They all began as ideas in the invisible world and were realized in the visible world after many cycles of action and observation. Even a weak social engine, bringing little change per cycle, can go very far if allowed to turn for a long time, as witnessed by the purposeful refinement of tools and crafts within traditional guilds. But if the visible determines the invisible by action—one arrow in the figure is reversed—the engine is no longer synchronized, so purposeful change reverts to the same evolutionary equilibrium as environmental forces randomly sculpting the gene pool.

Economic engines follow the same pattern. Money is created when a bank extends credit to a customer.[8] The question of purpose is whether certain types of credit should be extended *before* others. For example, credit from new money could be allowed only for investments, with consumer credit available only from money repaid by the former. As of this writing, all credit is equal, raising the question what, if anything, synchronizes overall economic activity. Should new money be purposeful, or is it enough to be useful?

Natural evolution provides an interesting perspective on that question. Crocodiles were once warm-blooded animals that hunted in packs on land, like wolves. As in all warm-blooded animals, their blood circulated so that it took in oxygen *before* delivering it to tissues. When they changed tactics to ambush prey from water, warm-bloodedness became a liability. Crocodiles with a heart defect—an aortal shunt between lung and body circulation that desynchronized oxygenation—could stay in the water longer. Today, that shunt is an evolved valve, allowing a crocodile to optimize the mix reflexively. Such an adaptation is useful, but not purposeful, because it evolved under pressure of real forces that were determined by the past.

[8] In common idiom: Every bank prints money.

The crocodiles did not discuss relative merits of warm- and cold-bloodedness over slide presentations, nor could they stay warm-blooded "just in case."

Bankers organize such discussions, but the salient question is whether their outcome is determined by the real forces of the moment, or they are able to bring purpose—a hope for the future—to bear on policy. The relation between financial and real sectors is like the relation between lung and body circulation. To repeat, should that relation be purposeful, or is it enough to be useful?

Life on Earth would be much the same if the Sun were a wandering star, instead of moving in orderly fashion in a galaxy. It is possible to have purpose on one time scale and no purpose on another. Which time scales are optimal to control, and how to control them, are among foundational questions of human history. While the above discussion may give a physical perspective on them, we are still at the level of literary comparisons. To make them literal, we have to show that thermodynamics is really happening in human society. The missing ingredients are extensivity, to be discussed in Chap. 4, coherence, in Chap. 5, and dissipation, in Chap. 11.

Leibniz described the eternal present of the evolutionary equilibrium philosophically by his "best of all possible worlds" concept. It was an early attempt to explain stability in terms of self-consistency. Voltaire famously mocked this idea, without appreciation of what Leibniz was trying to do. This polemic pointed to the substantive question of the material origin of forces stabilizing that equilibrium. To answer that, we must return to elementary physical considerations first.

Chapter 3
Statistical Foundations

The Second Law can never be proved mathematically by means of the equations of dynamics alone.

Ludwig Boltzmann

As an autonomous discipline, thermodynamics has interesting limitations. First, it cannot understand its own basic observations: Why can the temperature be defined independently of any system? Why is there an absolute zero? What is entropy? What is thermal contact?

Second, thermodynamics can connect various properties of the same material, but cannot explain why various materials have different properties: Mercury expands upon heating, rubber contracts; some materials are good thermal and electric insulators, others are not.

The resolution of these difficulties was found in the molecular hypothesis. One could show that properties of interest to thermodynamics belong in some sense to the aggregate itself, not to its constituent parts. Put another way, the aggregate suppresses many properties that would be manifest if the system consisted of only a few particles. When it is large enough, its thermodynamic properties may be described by functions depending only on the mechanical invariants of the whole multitude: Number of particles, total energy, linear and angular momentum. The larger the system, the less probable are departures from the thermodynamic description.

The key to the connection between microscopic and macroscopic properties is the entropy. Once it is shown that a certain function of the microscopic states of the system is equivalent to the thermodynamic entropy, the program to explain thermodynamics physically is essentially complete. It begins by an attempt to describe the deterministic evolution of a very large number of particles statistically, rather than in detail, from one particle to another. In order to provide an overview of the whole program, we construct a simple example.

D. Sunko, *Conceptual Thermodynamics*,
https://doi.org/10.1007/978-3-032-20877-4_3

3.1 Random Equalization of Concentrations

If we connect two closed bottles of gas by a tube, the concentrations will equalize. One may ask how the molecules know where to go to make that happen. We shall show that it will happen even if the molecules do not know anything.

Assume that there is no dynamics among the N molecules. Instead, they are randomly distributed among two boxes by tossing a coin anew N times at every interval τ. They are all distributed at $t = 0$, then again at $t = \tau$, again at $t = 2\tau$, every time as if nothing had happened before. The question is what someone will see who looks at the state of the system at any given moment.

If M molecules end in the first box, there will be $N - M$ in the other. Because the order in which they arrived does not matter, the probability of such an outcome is a normalized binomial coefficient:

$$P_N(M) = \frac{1}{2^N}\binom{N}{M}, \tag{3.1}$$

given that the total number of tossing sequences is 2^N. If it does not matter which box is which, there are only two different partitions for $N = 3$, with probabilities $P_3(0) + P_3(3) = 1/4$ and $P_3(1) + P_3(2) = 3/4$. Evidently, the more symmetric partition is more probable. Table 3.1 shows the behavior of these probabilities for slightly larger numbers of molecules. According to it, at any moment, the observer will probably see one of 2/3 of all partitions for $N = 11$, and one of only 1/5 of all partitions for $N = 101$: The fraction of probable partitions in the total number of partitions has *decreased* when N increased. In other words, as a function of the ratio M/N, the distribution (3.1) has *sharpened*. We can see that clearly if we normalize the distribution so that $P_N(N/2) = 1$. A graph of the binomial distribution, normalized in this way, is given in Fig. 3.1 as a function of M/N, for a few values of N.

The figure shows the prediction of the random model: At any given moment, the observer has a very low chance of seeing anything except equal concentrations in both containers. The larger the number of molecules, the smaller the fraction of partitions which have a noticeable probability of appearing. For the vast majority of partitions, this probability is negligible.

A certain number of partitions, whose probabilities are in the spike of the binomial distribution in Fig. 3.1, always appears with a non-negligible probability.

Table 3.1 Partition probabilities for balls in boxes. For a total of $N = 2k + 1$ balls, probabilities are listed that exactly $M = k$, $M = k - 1, \ldots,$ balls will be found in one of two boxes

$N =$	$2P_N(M)$ for $M =$						
$2k + 1$	k	$k - 1$	$k - 2$	$k - 3$	$k - 4$	...	39
11	0.45	0.32	0.16	0.05	0.01		
101	0.16	0.15	0.14	0.12	0.11	...	0.01

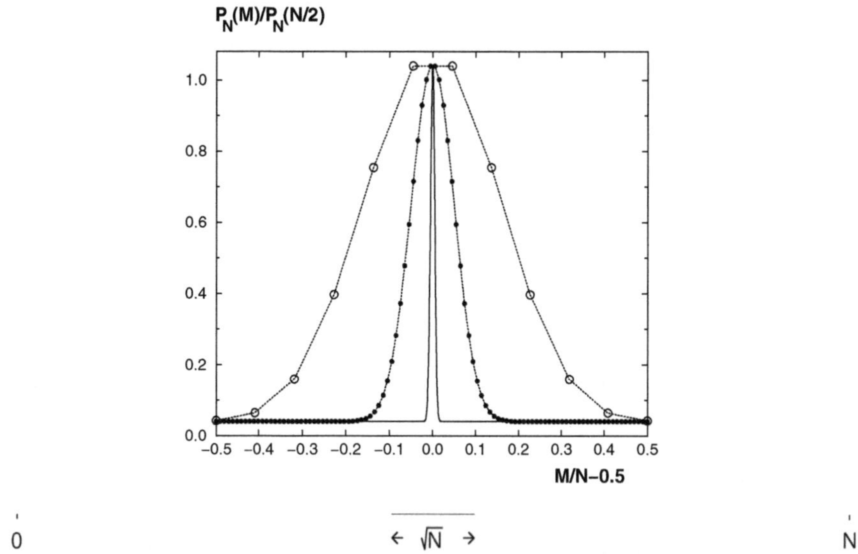

Fig. 3.1 Top: Binomial distribution in relative scale, for $N = 11$, 101, and 10,000, in that order. Bottom: The black rectangle is the distribution in absolute scale for $N = 100$

We can estimate the probability of the most probable partition, i.e., the very peak of the spike, by Stirling's formula $N! \simeq \sqrt{2\pi N}(N/e)^N$:

$$P_N(N/2) \xrightarrow{N\to\infty} \sqrt{\frac{2}{N\pi}} \approx \frac{1}{\sqrt{N}}, \tag{3.2}$$

which gives $0.08 \approx 0.1$ for $N = 100$, in excellent agreement with Table 3.1. Thus, for a very large N, the probability of any *given* partition, even the most probable one, becomes very small. The question is how many partitions are there in the spike, which exhaust almost all that can be encountered. The rough answer is simply $\sqrt{N}$. Namely, if all partitions in the apex of the distribution have an equal probability, and they have the probability 1 all together, then their number is equal to the reciprocal of that probability. This result is shown in the bottom panel of Fig. 3.1. If N were just a little larger, the black rectangle would be wholly invisible. It would vanish horizontally because $\sqrt{N} \ll N$, and vertically because $1/\sqrt{N} \ll 1$.

The basic result of the random model is that equal volumes contain equal numbers of particles, which is precisely Avogadro's law. The real problem for one who would derive such thermodynamic results from the laws of motion is not how the molecules know where to go, but quite the opposite, how do they behave as if they knew nothing, not even their own laws of motion! We shall deal with that problem in the next section.

To repeat the reason why the spike has appeared: Coin-toss sequences with roughly equal numbers of heads and tails, $M \approx N/2$, vastly outnumber all others.

(Conversely, there is only one sequence with $M = N$.) Any one of these can be taken as a representative of the *equilibrium state*: It is the state encountered by the observer at an arbitrary moment. However, because they are equally probable, the observed state will change all the time, so we say that it *fluctuates*. The observed value of M can thus be written

$$M = \frac{N \pm \sqrt{N}}{2}, \tag{3.3}$$

according to the lower panel of Fig. 3.1. These fluctuations become relatively negligible when N is large. For example, if $N = 10^{24}$, the observer has a roughly equal probability 10^{-12} to observe any one of 10^{12} values of M around $M = N/2$. The probability to encounter any one of the remaining $10^{24} - 10^{12} \approx 10^{24}$ values of M outside that range is practically zero, even if every individual sequence of coin tosses has an equal probability to occur.

Although they are negligible macroscopically, fluctuations are responsible for adapting the system to changes of circumstances. Thus, if we were to increase the size of one box, more particles would be found in it, i.e., the spike of the distribution would shift, say to the right. The random model will discover that by partitions *randomly* encountered on the right side of the old spike becoming more probable than those on the left side.

The random model becomes physical when we manage to connect the microscopic deterministic dynamics of a many-particle system with the appearance of a time τ after which the state of motion is chosen at random anew, i.e., the system forgets its initial conditions. The time τ is called the thermalization time. This connection is derived in three steps. The first is to show that the statistical assumption is compatible with the laws of motion. The second is to point out a physical process by which the system forgets its initial conditions. The third is to be convinced that this process is efficient enough in specific situations in which thermalization is observed.

3.2 Thermalization in Classical Mechanics

Isaac Newton speculated that gravitational forces and forces directing chemical reactions had the same origin. He may have been inspired by alchemy, where chemical elements have symbolic representatives in the heavens: Silver–Moon, mercury–Mercury, copper–Venus, gold–Sun, iron–Mars, pewter–Jupiter, lead–Saturn. He was perspicacious enough to suggest that chemical forces had to include "repelling Powers" as well. Roger Boscovich claimed that matter consisted of point-like atoms with mass, understanding, however, that a universal gravitational attraction between them necessarily led to a collapse of matter. He postulated an elaborate short-range repulsive force, with several intermediate positions where it changed sign. Materials differ according to which of these equilibrium positions are occupied by atoms, which corresponds in principle to the modern understanding of atomic orbitals and chemical bonds. His reasoning is realized directly in the birth of stars.

The molecular hypothesis sharpens the program of integrating physics and chemistry, because it allows one and the same particles to be understood in two ways. One is microscopically mechanical, where the motion of the multitude of particles is reversible in time and determined once and for all by the initial conditions. The other is macroscopically thermodynamical, where motion is irreversible and always leads to the same equilibrium states, irrespective of the initial conditions. It seems that we must choose between two paradoxes: Either deterministic evolution somehow ceases to be deterministic with time, or equilibrium states have some strange attraction, such that deterministic motion always leads to them. We shall show that both these statements are correct when properly understood, so that both paradoxes disappear.

3.2.1 Compatibility of Classical Mechanics and Statistics

The main idea of the random model is that the equilibrium macroscopic state (observed value of M) will be the most probable one, when one assigns equal a priori probabilities to all microscopic states (individual sequences of coin-tosses). To connect it with physics, we must first say what is one microscopic state of a system of N particles. Classical mechanics has a proposal ready: It is a point in phase space, a $6N$-dimensional vector of all three-dimensional momenta $\boldsymbol{p}(t)$ and positions $\boldsymbol{q}(t)$ at one instant in time:

$$(\boldsymbol{p}_1(t), \ldots, \boldsymbol{p}_N(t); \boldsymbol{q}_1(t), \ldots, \boldsymbol{q}_N(t)) =: (\boldsymbol{p}(t); \boldsymbol{q}(t)), \tag{3.4}$$

where the second expression is an abbreviation of the first. To describe dynamics statistically is to introduce some distribution function in phase space,

$$P(\boldsymbol{p}, \boldsymbol{q}, t)d\Gamma, \quad d\Gamma := d^3 p_1 \cdots d^3 q_N, \tag{3.5}$$

such that statements about individual trajectories $(\boldsymbol{p}(t); \boldsymbol{q}(t))$ are replaced by statements about their distribution $P(\boldsymbol{p}, \boldsymbol{q}, t)$. Nothing is lost by such a reformulation, nor gained, as long as we allow the distribution P to depend arbitrarily on time. Indeed, the choice

$$P(\boldsymbol{p}, \boldsymbol{q}, t) = \delta(\boldsymbol{p}(t) - \tilde{\boldsymbol{p}}(t))\delta(\boldsymbol{q}(t) - \tilde{\boldsymbol{q}}(t)), \tag{3.6}$$

where δ is the Dirac delta-function, reverts all statistical expressions to deterministic ones, where the prescribed trajectory $(\tilde{\boldsymbol{p}}(t); \tilde{\boldsymbol{q}}(t))$ appears with probability 1, and all others have probability 0.

The random model claims, on the contrary, that physical reality can be described by a stationary distribution, which has no explicit dependence on time:

$$\frac{\partial P}{\partial t} = 0. \tag{3.7}$$

The distribution (3.6) is not like that, because the functions $\tilde{\boldsymbol{p}}(t)$ and $\tilde{\boldsymbol{q}}(t)$ depend on time, but not on $\boldsymbol{p}$ or $\boldsymbol{q}$. Therefore, the claim (3.7) is substantial, and it makes sense to check whether the laws of motion conserve it. If they do, that proves the compatibility of the statistical description of the system, in which time does not appear, with the physical reality of particles that move in time. Specifically, if we begin with a uniform distribution, the question is, does it stay uniform as the system evolves by the laws of motion.

Because probability is conserved, the distribution function satisfies the continuity equation:

$$\frac{\partial P}{\partial t} + \nabla \cdot \boldsymbol{j} = \frac{\partial P}{\partial t} + \boldsymbol{v} \cdot \nabla P + P \nabla \cdot \boldsymbol{v} = 0, \tag{3.8}$$

where $\boldsymbol{j} = P\boldsymbol{v}$ is the probability current, and $\boldsymbol{v}$ is the velocity of a point moving in phase space, $\boldsymbol{v} = (\dot{p}_{1x}, \ldots)$, similarly $\nabla = (\partial/\partial p_{1x}, \ldots)$. We expand $\nabla \cdot \boldsymbol{v}$ into components:

$$\frac{\partial P}{\partial t} + \boldsymbol{v} \cdot \nabla P + P \sum_{i=1}^{3N} \left[\frac{\partial}{\partial p_i} \dot{p}_i + \frac{\partial}{\partial q_i} \dot{q}_i \right] = 0. \tag{3.9}$$

The velocity divergence vanishes because of equations of motion according to a Hamiltonian H:

$$\dot{p}_i = -\frac{\partial H}{\partial q_i}, \quad \dot{q}_i = \frac{\partial H}{\partial p_i} \implies \frac{\partial}{\partial p_i} \dot{p}_i + \frac{\partial}{\partial q_i} \dot{q}_i = 0. \tag{3.10}$$

Thus, the remaining expression, the so-called hydrodynamic derivative, vanishes by itself:

$$\frac{\partial P}{\partial t} + \boldsymbol{v} \cdot \nabla P = 0, \tag{3.11}$$

which is known as Liouville's theorem. It means that the evolution of a system by conservative equations of motion cannot change the volume of phase space accessible to it: This volume behaves like an incompressible liquid. If we insert the stationarity condition (3.7) into Liouville's theorem, we obtain

$$\boldsymbol{v} \cdot \nabla P = 0. \tag{3.12}$$

This is the result we sought. It says that a system with a stationary distribution P can evolve only along trajectories in phase space which are orthogonal to its gradient. By definition of the gradient, these are precisely the surfaces of constant P. Thus, if the distribution was uniform at the beginning, it will stay uniform: The laws of motion do not prefer some states over others.

The derivation above has several subtle points. First, we wrote a local continuity equation for a globally conserved quantity. In real (laboratory) space, this is justified by the theory of relativity. A conserved quantity cannot disappear at one point and simultaneously reappear at another, because what is simultaneous to some observers is not to others, moving relatively to them. The argument extends to phase space, because Hamiltonian dynamics is invariant to canonical transformations, which can mix momentum and position variables.

Second, an autonomous system evolves in phase space with conserved energy, so we have proven the conservation of stationarity only for distributions over systems whose total energy is fixed. The derivation also assumes that the number of particles stays the same. Both of these assumptions correspond to an isolated system, however, an autonomous system is not isolated in the thermodynamic sense. It is a universe unto itself, without surroundings, which is essentially different than real isolated systems, which are separated from their surroundings by walls.

Third, the uniform distribution is only one possible stationary distribution, and it is not clear from the laws of motion why it should be privileged. Just as they do not prefer special states, they do not prefer special distributions. The significance of the result (3.12) is that if we start with a given stationary distribution, the laws of motion will not change it.

Finally, classical mechanics and statistics are compatible only in phase space, of coordinates and momenta, where Hamilton's equations of motion hold. In the space of coordinates and velocities, where Lagrange's equations of motion hold, stationary distributions do not remain stationary. This opens the question of a privileged position for Hamilton's approach. It is emphasized by quantum mechanics, which extends naturally from Hamilton's, but not from other descriptions of classical mechanics. We shall discuss the compatibility of quantum mechanics with statistics later.

3.2.2 Classical Mechanics of a System in a Container

Imagine a thermodynamically isolated system as a gas in a bottle, separated from the surroundings by impermeable, hard, thermally insulated walls. The gas consists of molecules which bounce off the walls, which fixes the volume from the outside. As long as the collisions with the walls are ideal, the gas is an autonomous mechanical system with deterministic evolution.

However, the collisions with the walls are not ideal. Whenever a gas molecule hits a wall, it moves an atom on the surface, which moves its neighbors, so that a sound wave is excited in the wall, whose energy is subtracted from the energy of the rebounding molecule. These waves spread inside the wall, so that a molecule hitting a wall will occasionally encounter an already moving atom, so it can also recoil with a greater energy than it had incoming.

Thermodynamically, these processes are reversible, because no work is needed to keep the gas in the bottle: It is kept by static friction. Indeed, every collision with the wall conforms to the microscopic laws of motion, which are reversible in time. The exchange of momenta with the walls gives rise to fluctuations in the energy of individual molecules. Modelling these fluctuations by Carnot engines gives an estimate of their size, $\pm kT$ per collision.[1]

The question for classical mechanics is how such fluctuations affect the evolution of the system. Given that we cannot control individual collisions with the walls, we must accept that the trajectories of individual molecules change randomly. If we want to keep the deterministic description, we have to show that it is stable against such fluctuations. This is the same question by which Newton's mechanics triumphed: Small perturbations in planetary positions, with respect to deterministic calculations, indicated the existence of unknown planets, so they were not random.

Generally, things are different. The successes of celestial mechanics depend critically on the Sun having a much greater mass than all the planets together. Thus, the motion of each planet in the Sun's gravitational field is practically reducible to the two-body problem, which is mechanically stable: Small errors in measuring the initial conditions give rise to small errors in predicting trajectories, which do not increase with time.

The three-body problem is already not like that. Figure 3.2 shows a three-body system with gravitational interactions in which the timing of orbital transits can simulate a random-number generator. The law of gravity could hardly be discovered in planetary systems around binary stars.

Operatively, the issue of forgetting the initial conditions can be posed as an *inverse scattering problem*. Let $(\boldsymbol{p}(t); \boldsymbol{q}(t)) =: R(t)$ be the vector of all momenta and coordinates at time t. The evolution of the system is presented as an operator shifting it in time:

$$R(t) = O_t R(0). \tag{3.13}$$

If we can solve the inverse problem $R(0) = O_t^{-1} R(t)$, then the system did not forget its initial conditions, i.e., it evolved deterministically between times 0 and t.

Formally, the inverse scattering problem is always solvable, because the mechanical equations of motion are time-reversible, $O_t^{-1} = O_{-t}$. However, for more than two bodies, it is *ill conditioned*: Small errors in measuring the final state lead to large errors in reproducing the initial state, which grow exponentially with time. This sensitivity can be pictured intuitively by imagining a finite ball around the point $R(0)$, and shifting all points in it by O_t. The volume of the ball does not change (says

[1] Who does not believe that temperature is real must still accept that Carnot's estimate (2.20) is realistic, because thermodynamics is an observational science, like astronomy.

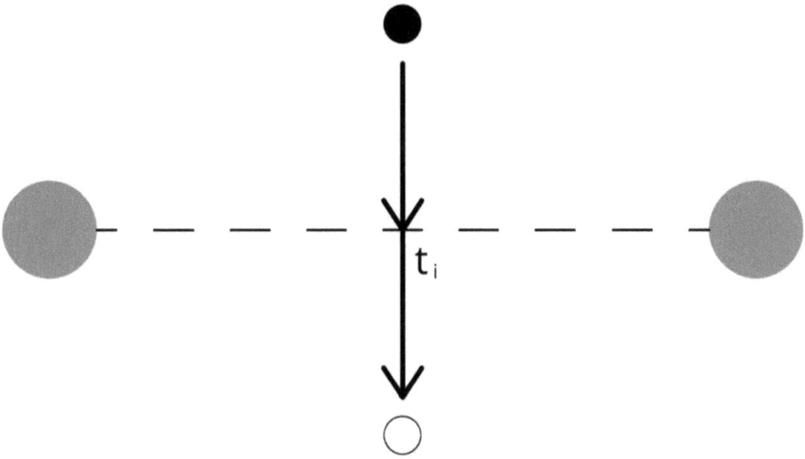

Fig. 3.2 Random gravity. For an arbitrary sequence of numbers, there exists an initial condition such that the intervals $t_{i+1} - t_i$ of the transit times of the small sphere between the two large ones, orbiting each other in the horizontal plane, are that sequence

Liouville's theorem), but it is significantly deformed, so that it becomes very thin in some directions, like a sheaf of leaves, or bundle of threads. Then even a small mistake in determining $R(t)$ in one of these directions will send the point $O_{-t}R(t)$ outside the initial ball. Because of that, the inverse problem is practically unsolvable already for a small number of particles, let alone Avogadro's number.

This unsolvability remains only practical as long as the collisions with the walls are perfect. However, if they are not, as described above, the practically impossible becomes really impossible, because information on the initial conditions is physically lost from the system. Formally, we can still say that it is not really lost, but is transferred into the container as sound waves, which reach the outer side of the walls eventually, to be scattered into the rest of the universe at last. But thermodynamics regards the gas in the bottle as separated from the surroundings, which, as such, finally and irrevocably forgets its initial conditions.

In brief, even if the container fixes the number of particles and the energy, it cannot fix the initial conditions, but exchanges them reversibly with the surroundings. This permeability is the physical mechanism of thermalization, which destroys periodic and chaotic trajectories equally well. The remaining question is how efficient it is.

3.2.3 The Thermalization Time. Choice of Statistical Distribution

The time scale of thermalization can be measured directly. In Chap. 11, we shall see that the spectrum of thermal noise of a system with thermalization time τ is given by

$$G(\nu) \propto \frac{1}{1 + (2\pi\nu\tau)^2} \propto \begin{cases} 1 & \text{for } \nu \ll 1/\tau, \\ 1/\nu^2 & \text{for } \nu \gg 1/\tau. \end{cases} \tag{3.14}$$

For low frequencies, which average the evolution of the system over long intervals $\Delta T \gg \tau$, the noise is white, i.e., it does not depend on frequency. For high frequencies, the noise is reduced in so far as deterministic motion is maintained over short intervals. The transition from one regime to the other begins at $\nu \approx 1$ THz for electrons in a resistor at room temperature, which amounts to $\tau \sim 10^{-13}$ s. In other words, the electrons forget their initial conditions about 10^{13} times per second. Deterministic evolution really ceases being deterministic very quickly when the system is isolated from the surroundings, by which control over the initial conditions is lost. Thus, we have explained the first paradox at the beginning of this section.

The next open question is which statistical distribution can describe 10^{13} random choices of the state of motion of the whole system per second. With such a fast exchange of states, we expect that measuring any quantity f over longer times averages all accessible states of given energy and particle number:

$$\overline{f(t)} = \frac{1}{\Delta T} \int_{t-\Delta T}^{t} f(t')dt' \xrightarrow{\Delta T \gg \tau} \langle f \rangle , \tag{3.15}$$

where the pointed brackets denote the average over all states with the chosen distribution. The time t has disappeared because in the limit $\Delta T \gg \tau$ all measurements repeat the same average many times. If we do not know anything about the system except its energy and particle number, there is no reason to use any distribution other than the uniform one, because any other choice would imply that we know something we have not said, e.g., that the gas has condensed into a liquid.

This argument is known as the "principle of insufficient reason." Its basis in physics is neither axiomatic nor philosophical. This physical foundation was most clearly stated by Heisenberg in the context of quantum mechanics, call it *Heisenberg's razor*: I am not obliged to answer any question that cannot be posed as a measurement. Who thinks that molecules collide perfectly with walls should show how to prevent them from exciting sound waves. Who thinks that the uniform distribution is not adequate should show how to avoid some states of motion.

The choice of a uniform distribution explains the second paradox, the strange attraction of the equilibrium state. It is simply the most probable macroscopic state when the microscopic states are chosen at random. We recall that it is not one microscopic state, but a number of states with similar properties in the peak of the distribution in Fig. 3.1. One can thus understand the stability of the system in time, its physical existence as if time had stopped. Time did not stop, but that is not visible from the outside.

We have chosen the uniform distribution after including all our knowledge about the system, e.g., its volume. Obviously, it makes no sense to assign equal probabilities to states occupying some other volume. That seems so obvious that it does not bear emphasizing, however, historically there was some confusion about how much the conscience of the observer affects measurements, and how one's knowledge can change the entropy. Such confusion has only increased with the advent of quantum mechanics. We can all admire the fact that the statement above is obvious to us, which testifies nicely to the compatibility of our mental states with the physical world. However, we did not fix the volume telepathically, but by real walls that we put where we wanted, engaging a swarm of Carnot engines, in the form of sound fluctuations in the walls, to work ceaselessly and reversibly to keep the gas inside. If we look into the bottle then and see a liquid where we expected a gas, we have discovered the phenomenon of condensation, thus, we must modify our assumptions so that miracles stop happening to us, and the assumption of a uniform distribution becomes valid again for the new description of the system.

The essence of a wall is that it must be regarded stochastically, otherwise the system would encompass it as well. It follows that any part of the system is a wall for the rest. The oxygen in air is a wall for the nitrogen in air. The freedom to regard arbitrary subsystems as systems for themselves reflects the destruction of deterministic trajectories: Once destroyed, they cannot be destroyed all over again. This arbitrariness is of course limited by the interactions between subsystems being weak and random, as with a wall. For example, if a solvent is inert, it is a wall for dissolved substances, so that chemical reactions in solution may be described by ideal-gas equations of state.

On the other hand, general statistical arguments do not exclude special states which behave differently, like the planetary system. Much research is motivated by the vision of noiseless electronics, in which information would be propagated deterministically, without emission of heat that limits the capabilities of processors. The efficiency of thermalization is an obstacle to finding microscopic states which persist reliably at the gigahertz scale (a thousand times longer than thermalization!), but that does not mean that they are impossible.

Finally, note that we have not shown quantitatively that the measured thermal noise is a consequence of the random perturbations by walls, which would require a theoretical analysis of these effects. The important qualitative insight is that attempts to derive thermalization for autonomous systems do not refer to systems physically separated from their surroundings, because these are not autonomous. The only physically autonomous system that we know is the universe as a whole. The issue of its thermalization is not resolved yet, but there is no observation indicating that it is in a box which someone is watching from the outside.

3.3 Statistical Independence and Additive Invariants

If we interpret two systems, such as oxygen and nitrogen in air, as a single larger one, we can write the distribution of the composite system as

$$P_{12}d\Gamma_{12} = P_{12}d\Gamma_1 d\Gamma_2, \tag{3.16}$$

because every configuration of the composite system is obtained by combining the configurations of the subsystems. This operation is called a direct product of vector spaces: Coordinates of air particles are written by listing all coordinates of nitrogens, then of oxygens, and similarly for the momenta. The physical question is whether all configurations of both gases are equally probable after the mixing as before. If they are, we can write $P_{12} = P_1 P_2$, which is the statistically precise expression of the claim that the interactions of oxygen and nitrogen are weak and random: The combined distribution behaves like a distribution of independent events, i.e., the correlation between oxygen and nitrogen vanishes:

$$P_{12} - P_1 P_2 = 0. \tag{3.17}$$

This independence is the statistical model of thermal contact: If nothing happens when two bodies touch, they are in mutual thermodynamic equilibrium. Equilibrium implies the additivity of the logarithm of the distribution over subsystems:

$$\ln P_{12} = \ln P_1 + \ln P_2. \tag{3.18}$$

Conversely, imagine that there is hydrogen instead of nitrogen, and that there has been a chemical reactions in which water appeared. In water, every oxygen atom is followed by two hydrogen atoms at a fixed distance. A random mixture of hydrogen and oxygen has many more configurations than water. Taking into account the stoichiometry of the reaction $2H_2 + O_2 \rightleftharpoons 2H_2O$, we have

$$\left(\int P_{H_2O} d\Gamma_{H_2O}\right)^2 \ll \left(\int P_{H_2} d\Gamma_{H_2}\right)^2 \int P_{O_2} d\Gamma_{O_2}, \tag{3.19}$$

reflecting the fact that the oxygen and hydrogen atoms in water are correlated. The number of states of motion available to water vapor is far smaller than the number available to the random mixture. One can say that interactions order the system. Taking them into account as well, the logarithm of the distribution is *subadditive* over subsystems:

$$\ln P_{12} \leq \ln P_1 + \ln P_2, \tag{3.20}$$

which gives an objective value to our feeling that motion in which two hydrogens always follow one oxygen is more ordered than one where all move randomly, even if the water molecules themselves do move randomly. The more ordered system is one that has fewer states of random motion at its disposal. A perfectly ordered system has only one, so that motion is deterministic.

We have shown by Liouville's theorem that the distribution function is constant during the evolution of a stationary system, so it can depend only on conserved quantities, the mechanical invariants of motion. Additivity over subsystems then means that its logarithm can be at most a *linear* function of *additive* invariants of motion:

$$\ln P = \alpha N - \beta E + \boldsymbol{\gamma} \cdot \mathbf{P} + \boldsymbol{\delta} \cdot \mathbf{L}, \tag{3.21}$$

respectively the total particle number, energy, linear and angular momentum. Because we do not usually consider systems which move as a whole, we write

$$\ln P = \alpha N - \beta E, \tag{3.22}$$

assuming that the total linear and angular momentum are equal to zero. This result is compatible with the physical conclusion of the previous section, that the distribution P is constant for a given number of particles $N = N_0$ and energy $E = E_0$. However, one should refrain from feeling that we have proven that in some mathematical sense now, as if nature were obliged to act like that. We have only checked that our model of thermal contact is consistent with the assumption of a uniform distribution. It is of course good that it does not allow other distributions for a given energy and particle number, or we would have had to correct it to eliminate cases that we have already understood to be irrelevant. It is even better that it allows us to compare distributions with varying E and N by such a simple formula as (3.22).

3.4 Statistical Entropy

There are two ways to model entropy statistically. One is derived directly from the considerations above. It is a measure of the size of the phase space of states among which the system chooses at random during each thermalization interval. That model is called the microscopic entropy, which cannot change in time. It is the largest possible for a given system, so it corresponds to thermodynamic equilibrium. The other is appropriate for systems close to equilibrium, whose entropy increases with time. For example, speech disturbs the thermodynamic equilibrium of air in a room, and when it dies out, the air has warmed a little by Joule heating. This model is called the coarse-grained or relevant entropy.

3.4.1 Microscopic Entropy

If the distributions in Eq. (3.19) are constant, the integrals will simply give the volume of phase space. A larger volume means that the system has more microscopic states of motion available, among which it chooses at random during each thermalization interval. Thus, a microscopic model of entropy is motivated:

$$\sigma(N, E) = \ln \frac{\Gamma(N, E)}{h^r} =: \ln W(N, E), \tag{3.23}$$

where we introduced a unit of measure of phase space h whose dimension is equal to the dimension of action, or the product of momentum and coordinate $[pq]$, or energy and time $[Et]$. The number r is called the number of degrees of freedom, which is the number of products $dpdq$ in the differential $d\Gamma$, $r = 3N$ for a monoatomic gas. The arbitrary choice of h means that our model defines entropy up to an additive constant.

This function corresponds well to an intuitive concept of entropy. It is extensive in thermal equilibrium, because it is additive over systems in thermal contact:

$$\begin{aligned} \sigma(N_1{+}N_2, E_1 + E_2) &= \ln \frac{\Gamma(N_1 + N_2, E_1 + E_2)}{h^r} \\ &= \ln \frac{\Gamma(N_1, E_1)\Gamma(N_2, E_2)}{h^{r_1+r_2}} = \sigma(N_1, E_1) + \sigma(N_2, E_2). \end{aligned} \tag{3.24}$$

Furthermore, it increases when some obstacle is removed, because the volume of available phase space increases. Similarly, interactions reduce the entropy because they act as a constraint on motion, as noted in Eq. (3.19). For the example of balls in boxes in Sect. 3.1, we have a total of $W(N) = 2^N$ coin-toss sequences, so the entropy is

$$\sigma(N) = N \ln 2. \tag{3.25}$$

It is extensive in the number of tosses, or measurements on one coin.

3.4.2 Relevant Entropy

Imagine a system out of equilibrium now, e.g., a sound wave in air. Divide the air into many small cells, each one of which is a large thermodynamic system by itself. Specifically, there are about 10^{25} molecules in a cubic meter of air, so if our cells have a side of 0.1 mm, each will contain 10^{13} molecules. The main idea of this model of a non-equilibrium system, called the *hydrodynamic limit*, is that each cell is in thermodynamic equilibrium by itself, even if the whole system is not, as long as the thermodynamic parameters vary from cell to cell. Because the cells are in thermal contact, they will equilibrate at some moment, but we have to wait for that, because the equilibration of the whole is slower than of a small part.

The statistical distribution $\rho(A)$ of a given macroscopic parameter A is generically shaped like the spike in the upper panel of Fig. 3.1, so it can be normalized like the rectangle in the lower panel:

$$\rho(A^\star)W(A^\star) = 1, \tag{3.26}$$

where $A^\star$ is the equilibrium value of the parameter A, and $W(A^\star)$ the phase-space volume occupied by microscopic states whose macroscopic $A = A^\star$. Therefore, we can write the entropy of the macroscopic state of one cell as

$$\sigma(A^\star) = \ln W(A^\star) = -\ln \rho(A^\star), \tag{3.27}$$

according to the microscopic expression (3.23). Its expectation value with the same distribution ρ is interpreted as the entropy of the whole system of $\mathcal{N}$ cells:

$$\sigma(A^\star) = -\sum_{i=1}^{\mathcal{N}} \rho(A_i) \ln \rho(A_i), \quad A^\star = \sum_{i=1}^{\mathcal{N}} \rho(A_i) A_i, \tag{3.28}$$

where A_i is the value of $A^\star$ for the i-th cell. This entropy is called *relevant*.

The relevant entropy (3.28) is maximal when all the A_i are equal, as expected for the thermodynamic entropy when the out-of-equilibrium system evolves toward equilibrium, producing Joule heat $T\Delta S$ along the way. Indeed, the extremum with the additional condition of normalization is given by

$$\frac{\partial}{\partial \rho_j}\left[-\sum_i \rho_i \ln \rho_i - \lambda\left(\sum_i \rho_i - 1\right)\right] = 0 \implies -\ln \rho_j = 1 + \lambda = const., \tag{3.29}$$

and the second derivative of the expression in square brackets is $-1/\rho_j < 0$. The normalization then gives $\rho_i = 1/\mathcal{N}$.

For our example of balls in boxes, $\mathcal{N} = 2$, so this maximum is

$$\sigma(A^\star) = -\sum_{i=1}^{2} \frac{1}{2} \ln \frac{1}{2} = \ln 2, \quad A^\star = \frac{1}{2} \cdot (-1) + \frac{1}{2} \cdot 1 = 0, \tag{3.30}$$

where $A^\star$ is the equilibrium value of the heads-tails variable. This result is the same as the previous (3.25), only normalized to one coin. It confirms that the maximal value of the relevant entropy is precisely the microscopic entropy. Generally, the relevant entropy becomes equal to the microscopic one when the averaging in the expression (3.28) is refined to the point that the index i refers to microscopic states of motion, so that A_i is the mechanical value of the variable A in the i-th state. In that case, the formula (3.28) is called Gibbs' entropy.

This result can be puzzling because it seems that the entropy of a part of the system (microscopic states which reproduce the equilibrium state) is equal to that of the whole system (all microscopic states). This has happened because our introducing an expectation value per toss tacitly assumed that the number of tosses was infinite, or arbitrarily large. In the limit of large N, it indeed holds asymptotically that

$$\ln \binom{N}{N/2} \simeq N \ln 2, \quad N \to \infty. \tag{3.31}$$

Even if the difference between the binomial coefficient and 2^N diverges when $N \to \infty$, it is easy to show that the difference of their logarithms is negligible as $\ln N \ll N$. Poincaré called this practical equality of the part and the whole "insensitivity of thermodynamic functions." It can be observed in Fig. 3.1 as the vanishing of all probabilities outside the spike.

3.5 Connection with Thermodynamics

The general formulae in the previous section can be connected with the molecular hypothesis in two ways. One takes the microscopic entropy (3.23) and calculates directly the volume of phase space available to a specific system with given mechanical invariants and boundary conditions. This approach is called Gibbs' method[2] or ensemble theory, which Gibbs himself called "rational models of thermodynamics." The other takes the density ρ in the coarse-grained expression (3.28) as the density of physical particles which scatter according to the laws of motion. It is called kinetic theory, founded by Maxwell and Boltzmann. Boltzmann was also the first to propose the expression (3.28) as a mechanical formula for the entropy in such an interpretation. He showed that its value can only increase over time, even if the microscopic scattering is reversible, if one introduces a certain statistical assumption. Namely, it is impossible to obtain an irreversible function from a reversible one solely by operations which preserve reversibility. The kinetic approach became obsolete already in the early twentieth century, so that Gibbs' is used almost exclusively today. It is nevertheless worthwhile to understand how Boltzmann introduced irreversibility in his calculation, although he assumed that collisions with walls were perfect, so he modelled the gas as an autonomous mechanical system.

[2] Gibbs' entropy is equal to the microscopic entropy up to insensitivity of thermodynamic functions.

If the density ρ were a phase-space density, it would satisfy Liouville's theorem (3.11). However, Boltzmann regards it as a density in real space, which evolves by two-particle scattering, without taking into account that part of the information about initial conditions is propagated by three- and many-body correlations, even if all collisions are two-particle. He obtains *Boltzmann's transport equation*, an expression similar to Liouville's theorem, but there appears a contribution from collisions on the right-hand side, instead of zero, which he calls the collision integral. He knows that the density is not stationary in such a case, so he asks under which conditions the collision integral is zero. The answer is, precisely when there are no correlations, which also makes the relevant entropy extremal. This extreme can be either a maximum or a minimum in time, so we are either in equilibrium, or moving towards it from a minimum, which means that the mechanical entropy is increasing. In brief, he formulates the problem of macroscopic equilibration by neglecting all correlations higher than two-body, and solves it by neglecting them as well. This result is called Boltzmann's H-theorem.

The assumption that there are no correlations in the gas is called *molecular chaos*, and it introduces irreversibility. When molecules enter individual cells without memory of what happened in the cell they left, the imagined walls between the cells have the same effect in the calculation as physical walls that introduce randomness. Generally, as soon as a system is divided into cells without mutual correlations, that division itself introduces irreversibility. That is the reason why the expression (3.28) appears in a similar role in very different contexts. Mathematically, the logarithm is the only continuous function that satisfies the functional equation $f(xy) = f(x) + f(y)$, so it linearizes expressions for independent probabilities, while its average is hardly more specific than that. Just as a cough is a certain sign of illness, but one cannot tell which, the expression (3.28) is a certain sign that something is divided into uncorrelated cells, but one cannot tell what.

It is notable how physics can hide behind apparently benign technical assumptions in a mathematical approach. What can be more innocuous than to say, let two particles collide with initial velocities $\boldsymbol{v}_1$ and $\boldsymbol{v}_2$? Let them, but is it the first collision for them both, or the millionth? The assumption that the velocities are arbitrary at an arbitrary moment is precisely the assumption of molecular chaos. Mathematical generality, which seeks all consequences of a model assumption, is opposite to generality in physics, which summarizes observations into model-independent insights.

Although Boltzmann's analysis contains circular reasoning logically, it is well founded physically. Namely, correlation functions of higher order are more sensitive to random perturbations than those of lower order, so they disappear faster. This hierarchy of time scales is the physical basis of the hydrodynamic limit. Because of it, departure from equilibrium for times shorter than thermalization, but not much shorter, can really be described by two-body correlations alone. The perennial usefulness of Boltzmann's transport equation is due to its crudeness reflecting the crudeness of nature itself. It is the physical counterpart of Liouville's theorem in laboratory space, once thermalization destroys the symplectic (Poisson-bracket) structure of $6N$-dimensional phase space, which constrained Hamiltonian dynamics to particular $3N$-dimensional hypersurfaces by preserving $3N$ so-called Poincaré invariants, thus opening phase space to be uniformly probed by thermal fluctuations. The two-body scattering which reverts the system to equilibrium is macroscopically averaged into susceptibilities, giving rise to forces which resist changes of equilibrium. All these insights of kinetic theory are implemented more easily in ensemble theory, in which approach they can also be transferred to quantum mechanics.

Three distinct physical processes connect microscopic dynamics, thermodynamics, and statistics, each with its own time scale. *Thermalization* is the establishment of a temperature by static friction according to Carnot's formula (2.20). It reflects uncontrolled reversible processes, with time scale measured by white noise according to Eq. (3.14). *Dissipation* is the increase in relevant entropy, accompanied by Joule heating. It is manifested by dynamic friction and irreversible evolution from less probable to more probable states. Its time scale is called equilibration time, which may be very long, like in fossils. *Decoherence* is the loss of information about initial conditions by decay of many-particle correlations, with its own decoherence time. In real systems, these processes are usually concurrent. When they are widely separated, interesting phenomena may appear. If the equilibration time of a supercooled liquid is much longer than the thermalization time, we get a glass instead of a crystal. If the decoherence time is much longer, we get frictionless flow at finite temperatures. We shall see later how quantum mechanics resolves the mystery that a macroscopic system can be both coherent and thermalized.

3.6 The Ensemble: A Universal Random Model

The basic idea of ensemble theory is that the random model represents a real system in thermodynamic equilibrium, because in equilibrium there is a thermalization time τ, such that the distribution of particles at a given moment is not correlated with their distribution at moments separated from it by more than τ. That means that the measured average values of any quantity behave in accord with Eq. (3.15), as if they were obtained by averaging over randomly chosen states of motion.

With this idea, Gibbs freed the rational foundations of thermodynamics from the simplification of a rarefied gas, needed by kinetic theory. He considers the sample on a laboratory table as randomly chosen among a great many identical samples. He calls this statistical space of samples an ensemble. He shows by Liouville's theorem that the ensemble can be constructed as the volume of phase space of all microscopic states with given energy and particle number, from which a sample is taken as a point with uniform probability. The entropy is the logarithm of that volume. The connection between ensemble theory and thermodynamics is established by showing that this definition of entropy gives rise to thermodynamic equations of state and plays the same role as the thermodynamic entropy in the first law of thermodynamics.

3.6.1 *Thermodynamic Equations of State*

Consider an isolated system made of two subsystems. The entropy of the equilibrium state of the composite system depends on the number of particles, energy, and mechanical variables:

$$\sigma = \sigma(N_i, U, x_\nu), \tag{3.32}$$

where N_i are the numbers of various particles, $U = \langle E \rangle$ is the expected (most probable) value of the energy, and x_ν are mechanical variables. By additivity, we have

$$\sigma = \sigma_1 + \sigma_2. \tag{3.33}$$

Because the total entropy does not change in equilibrium, that of the components can vary only if one compensates the other:

$$\delta\sigma = \delta\sigma_1 + \delta\sigma_2 = 0. \tag{3.34}$$

Now, we specialize this expression to cases where the entropy variations are related to variations of some particular state variable.

3.6.1.1 Thermal Equilibrium

If the partition between the subsystems does not insulate heat, but is otherwise hard and impermeable, they can exchange energy only. Then

$$\delta\sigma = \frac{\partial\sigma_1}{\partial U}\delta U_1 + \frac{\partial\sigma_2}{\partial U}\delta U_2 = 0. \tag{3.35}$$

Because the whole system is insulated, we must have

$$\delta U = \delta U_1 + \delta U_2 = 0, \tag{3.36}$$

from which it follows that

$$\delta\sigma = \left(\frac{\partial\sigma_1}{\partial U} - \frac{\partial\sigma_2}{\partial U}\right)\delta U_1 = 0, \tag{3.37}$$

or

$$\frac{\partial\sigma_1}{\partial U} = \frac{\partial\sigma_2}{\partial U}, \tag{3.38}$$

because the energy variation was arbitrary. Thus, if we denote the derivative with respect to energy

$$\frac{1}{\tau} = \left(\frac{\partial \sigma}{\partial U} \right)_{N, x_\nu}, \tag{3.39}$$

then the condition of thermal equilibrium can be written

$$\tau_1 = \tau_2. \tag{3.40}$$

What if we bring two systems into contact when $\tau_1 \neq \tau_2$? By removing thermal insulation, a mechanical obstacle to microscopic motion, the entropy increases, so we have

$$\delta\sigma = \left(\frac{1}{\tau_1} - \frac{1}{\tau_2} \right) \delta U_1 > 0, \tag{3.41}$$

whence

$$\tau_1 < \tau_2 \implies \delta U_1 > 0, \tag{3.42}$$

meaning that thermal contact will cause the energy of the body with smaller τ to increase. Evidently, τ behaves like the temperature.

3.6.1.2 Mechanical Balance

Let the partition be movable now, in addition to allowing energy exchange. Then the only interesting mechanical variable is the volume, $x_\nu \to V$, so

$$\delta\sigma = \frac{\partial \sigma_1}{\partial U} \delta U_1 + \frac{\partial \sigma_2}{\partial U} \delta U_2 + \frac{\partial \sigma_1}{\partial V} \delta V_1 + \frac{\partial \sigma_2}{\partial V} \delta V_2 = 0. \tag{3.43}$$

In thermal equilibrium, the first two terms add to zero like before, while the total volume is unchanged:

$$\delta V = \delta V_1 + \delta V_2 = 0, \tag{3.44}$$

so that

$$\delta\sigma = \left(\frac{\partial \sigma_1}{\partial V} - \frac{\partial \sigma_2}{\partial V} \right) \delta V = 0, \tag{3.45}$$

or

$$\frac{\partial \sigma_1}{\partial V} = \frac{\partial \sigma_2}{\partial V}, \tag{3.46}$$

because the volume variation was arbitrary. We can define

$$\frac{\Pi}{\tau} = \left(\frac{\partial \sigma}{\partial V}\right)_{U,N} \tag{3.47}$$

so the condition of mechanical balance in thermal equilibrium may be written

$$\Pi_1 = \Pi_2. \tag{3.48}$$

Evidently, Π behaves like the pressure. Generally, for any mechanical variable (generalized coordinate) x_ν, we can define the corresponding generalized force X_ν by

$$\frac{X_\nu}{\tau} = \frac{\partial \sigma}{\partial x_\nu}, \tag{3.49}$$

whence the condition of mechanical balance at equal temperatures is that the generalized forces be equal.

Like with thermal equilibrium, we can ask how the system evolves towards mechanical equilibrium. Let $\Pi_1 \neq \Pi_2$ initially, with equal temperature. When we remove the mechanical obstruction which prevented the partition from moving, the entropy will increase,

$$\delta\sigma = \frac{1}{\tau}\left(\Pi_1 - \Pi_2\right)\delta V_1 > 0, \tag{3.50}$$

whence we see that

$$\Pi_1 > \Pi_2 \implies \delta V_1 > 0, \tag{3.51}$$

meaning that the subsystem with larger Π expands at the expense of the other.

3.6.1.3 Chemical Balance

Let the partition be a porous membrane now, through which particles can diffuse. In thermal equilibrium we have

$$\delta\sigma = \frac{\partial \sigma_1}{\partial N}\delta N_1 + \frac{\partial \sigma_2}{\partial N}\delta N_2 = 0. \tag{3.52}$$

Define the *chemical potential*

$$-\frac{\mu}{\tau} = \left(\frac{\partial \sigma}{\partial N}\right)_{U,V}. \tag{3.53}$$

The equilibrium condition follows from the condition that the total number of particles does not change:

$$\delta N_1 + \delta N_2 = 0 \implies \mu_1 = \mu_2. \tag{3.54}$$

If $\mu_1 \neq \mu_2$ initially, and we make the partition porous, particle-number equilibrium will be established by

$$\delta\sigma = -\frac{1}{\tau}(\mu_1 - \mu_2)\,\delta N_1 > 0, \tag{3.55}$$

or

$$\mu_1 > \mu_2 \implies \delta N_1 < 0, \tag{3.56}$$

so that the component with *higher* chemical potential *loses* particles.

The essential difference between a porous membrane and a hole in the wall is that a molecule can pass through the hole by a deterministic trajectory. Conversely, in a porous membrane, each molecule spends a time longer than the thermalization time, so that it forgets from which side it came. Both subsystems in equilibrium perceive such a membrane as an adsorbate on the wall with equal chemical potential. The same reasoning applies when the two subsystems are two sides of a chemical reaction. The reaction is in equilibrium when molecules produced in one direction are balanced with those consumed in the other.

3.6.2 *Statistical Derivation of the First Law of Thermodynamics*

According to the definition (3.23), the entropy is a function of the energy, particle number, and external (mechanical) variables that delimit the available phase space. If we change them a little, we obtain a new equilibrium state, whose entropy differs from the former by

$$d\sigma = \left(\frac{\partial\sigma}{\partial U}\right) dU + \sum_\nu \left(\frac{\partial\sigma}{\partial x_\nu}\right) dx_\nu + \sum_i \left(\frac{\partial\sigma}{\partial N_i}\right) dN_i. \tag{3.57}$$

When the definitions of the various derivatives from the previous section are introduced, this becomes

$$d\sigma = \frac{1}{\tau}dU + \frac{1}{\tau}\sum_\nu X_\nu dx_\nu - \frac{1}{\tau}\sum_i \mu_i dN_i. \tag{3.58}$$

Specializing this expression to only one mechanical variable, volume, and only one kind of particle, and moving the energy differential to one side, we get

$$dU = \tau d\sigma - \Pi dV + \mu dN, \tag{3.59}$$

where we recognize the first law of thermodynamics, $dU = TdS - pdV + \mu dN$. The identification of thermodynamic variables rests on $\tau d\sigma$ appearing in the expression (3.59) as that part of the change in internal energy which cannot be expressed by macroscopic variables. That is called heat in thermodynamics. Thus, we have the correspondence

$$\tau d\sigma = TdS. \tag{3.60}$$

Because the statistical entropy σ is additive over subsystems, like the thermodynamic entropy S, i.e., both are extensive, while the variables T and τ are equal for systems in equilibrium (intensive), this correspondence implies such a proportionality for each pair separately,

$$\frac{\tau}{T} = const. = \frac{S}{\sigma}, \tag{3.61}$$

that this ratio must be a universal constant, so that its numerical value can only depend on the choice of units. If we take the Kelvin scale for T, then

$$\tau = kT, \tag{3.62}$$

where k is Boltzmann's constant, so that

$$S = k\sigma, \quad \Pi = p. \tag{3.63}$$

Therefore, ensemble theory reproduces all thermodynamic formulae. However, when Gibbs published his capital monograph on thermodynamics in 1902, a short time before his death, no one knew how to measure Boltzmann's constant directly. Thus, the molecular hypothesis itself was theoretically still under question. In that context, Gibbs expressed himself hypothetically, and kept insisting on the distinction between his model temperature τ and the measured laboratory temperature T. Such caution became obsolete already in 1905, when Einstein showed in his article on Brownian motion that thermalization is a physical process, liberating statistical physics from the need to apologize for the assumption that random events are real. Unfortunately, many later textbooks inherited Gibbs' reticence as an authoritative style, thus puzzling beginners.

3.7 The Probability of a Fluctuation

The expression (3.23) allows a quantitative expression of the amount of information about a system. For example, if we do not know where a gram of methane is in a room whose volume $V = 30\ \mathrm{m}^3$, we have to assume that it is equally distributed; but if someone tells us then that it is actually all in a bottle of volume V' of one liter, the entropy function of the new ensemble will differ from the old by

$$\Delta\sigma = \ln\frac{V'^N}{V^N} = N[\ln(0.001\mathrm{m}^3) - \ln(30\mathrm{m}^3)] \approx -10.3N, \tag{3.64}$$

where N is the number of methane atoms in one gram. The decrease of entropy corresponds to the gain in information. Note, again, that our consciousness did not change reality, it only conformed to it better. The thermodynamic entropy of the methane, either in the bottle or in the room, was the same all the time, only we did not know what it was. The calculation (3.64) is a general estimation of the value of information, how much we should pay someone who resolved our doubt. A greater disparity with expectations means more valuable information. A scientific discovery is also valued more if the surprise is greater.

The same calculation can be interpreted in another way. We can ask what is the probability for a gram of methane to concentrate *spontaneously* in a container with a volume of one liter. It is equal to the number of such configurations, divided by the number of all configurations:

$$\frac{W(0.001\mathrm{m}^3)}{W(30\mathrm{m}^3)} = e^{\Delta\sigma} \approx e^{-10N}, \tag{3.65}$$

which is very small if $N \sim 10^{24}$, of the order of Avogadro's number. By comparison, say some book has 300 pages, 50 rows per page and 80 symbols per row. Then the probability that someone will reproduce such a book without a misprint by punching keys at random is roughly

$$\left(\frac{1}{101}\right)^{1200000} \approx e^{-6\times 10^6}, \tag{3.66}$$

assuming that the keyboard has 101 keys.

The latter interpretation means that we may calculate the entropies of non-equilibrium states, or the probabilities of random fluctuations. What is the whole phase space when the gas is in the bottle are non-equilibrium configurations when the gas is everywhere in the room. The probability p of a fluctuation is simply the ratio of the volume of phase space W' in which this fluctuation appears (desirable states) and the volume W of the whole phase space (all states):

$$\boxed{p = \frac{W'}{W} = e^{\Delta\sigma}.} \tag{3.67}$$

The formula for the probability of a fluctuation is the most far-reaching formula in physics. It is so because experiments measure departures from expectations, while it says which are significant. It gives the response of a system to external perturbations quantitatively precisely, so it enables one to distinguish correlation and response functions. The first equality in (3.67) is the definition of probability, while the second is the definition of entropy, so the formula is mathematically trivial. But when entropy is understood physically, as a thermodynamic potential, it acquires a universal reach, which we shall demonstrate on a few examples now.

3.7.1 Volume Fluctuations

It is clear that the methane in the room will not spontaneously concentrate in a bottle. Small fluctuations of the volume are more interesting, $\Delta V = V - V_0$, where V_0 is the equilibrium volume and $|\Delta V| \ll V_0$. Then

$$e^{\Delta\sigma} = e^{\sigma(V_0+\Delta V)-\sigma(V_0)} = e^{\frac{1}{2}\sigma''(V_0)\Delta V^2+\dots}, \tag{3.68}$$

where the linear term vanishes because the first derivative of the entropy is zero at maximum, $\sigma'(V_0) = 0$. Thus, the formula for the probability of a fluctuation gives the shape of the distribution of equilibrium volume fluctuations: It is a Gaussian distribution with width $1/\sqrt{-\sigma''(V_0)}$, like in Fig. 3.1. The second derivative of the entropy by volume is

$$\left(\frac{\partial^2\sigma}{\partial V^2}\right)_{U,N} = \frac{\partial}{\partial V}\left(\frac{p}{kT}\right)_{U,N}, \tag{3.69}$$

which is some (reciprocal) susceptibility. It has no name, because it is customary to define susceptibilities starting from the internal energy. We can say that it is the entropic version of the adiabatic compressibility. This expression must be negative for a stable system, because the probability of a fluctuation $p < 1$ when the equilibrium state is more probable. [By similar methods as in the proof of Le Chatelier's principle, one can show that

$$\left(\frac{\partial^2 S}{\partial V^2}\right)_{U,N} = -\frac{1}{VT\kappa_T} - \frac{(l_V - p)^2}{c_V T^2} < 0, \tag{3.70}$$

where κ_T, l_V and c_V are respectively the isothermal compressibility, latent heat, and heat capacity.]

3.7.2 Nonlinear Responses

The example above provides a general statistical interpretation of susceptibilities: They represent the width of the distribution of spontaneous fluctuations in the system. A softer system will have larger volume fluctuations, one of small heat capacity larger temperature fluctuations, and so on. Having understood that, we can take the next step: What would higher orders in the expansion of $\Delta\sigma$ mean, say the third? Physically, $\partial\kappa/\partial V \neq 0$ means that the compressibility changes with volume. Such responses are called nonlinear. Statistically, odd terms in the exponent of the formula (3.67) mean that the distribution is asymmetric. Nonlinear responses imply an asymmetric distribution of spontaneous fluctuations, and vice versa.

Intuitively, we feel that it will be more difficult to notice that the distribution is asymmetric if it is narrower. Thus, we expect nonlinear responses sooner when the system is softer. It is easier to reach large amplitudes in that case, where anharmonicities appear. Experience suggests that anharmonicities often presage plastic deformation and rupture. Indeed, the third-order term becomes larger than the second-order one at some elongation, after which the fluctuation probability begins to rise with elongation, meaning that the susceptibility has changed sign, which signals a phase transition in general (Fig. 3.3). It can be reversible (physical change) or not (chemical change). Rupture can be prevented by a fourth-order term, which is usually negligible at equilibrium, but may become important in the vicinity of a phase transition.

3.7.3 Purposeful Governance

The insight above is important for governance. Everyone will say that it is better to have more good outcomes than bad ones, whatever the matter. However, that cannot be achieved by rules, which are in the nature of second-order correlations: If this, then that. Rules can move the centroid of the distribution, called the first moment in statistics, external potential in physics, or the Legendre transform. They can also narrow the distribution, i.e., harden the system, but it stays symmetric. That is a common complaint against bureaucracy: It constrains the good because there are the bad. If we want the good to have it easier than the bad at a given position of the centroid, we should take care that the distribution be asymmetric.

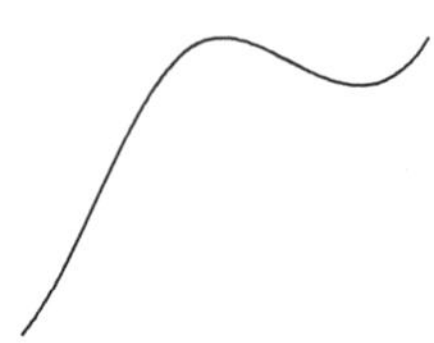

Fig. 3.3 Anharmonicity and instability

Nonlinear responses are sometimes implemented in society by legal discrimination, in which the centroid of the distribution is selectively shifted for some population category, so the total distribution is asymmetric because it really consists of two overlapping distributions. Such a simulation of nonlinear response is open to criticism that both underlying distributions are still symmetric, so shifting one gives an advantage to all in it, which stimulates the others to game the criteria.[3] Statistically, the restoring force in the overlap region is a sum of a positive and a negative number, both of which are much larger than the desired result. The instability of such a calculation requires fine-tuning, which causes additional administration.

A deeper criticism is that this is a category mistake. The third moment of a distribution is not the first! We want the original system to develop a nonlinear response, not to divide it into subsystems. For that, one needs to embed third-order correlations into it. They correspond to three-body scattering in Boltzmann's analysis of relevant entropy. The key physical insight is that higher-order correlations appear dynamically, meaning in real interactions between people. Simply, there need to be three sides in a meeting, two is not enough. An example is the German system of decision-making in industry (*Mitbestimmung*). Certain decisions can be made only if representatives of the government, employers, and unions agree. Tripartition of power is an idea of the same kind. This approach can work if the three share a common purpose. Otherwise, a selfish participant can push the anharmonicity beyond an instability, so that caution with such arrangements is also founded.

Natural evolution can develop useful nonlinear responses by a cumulation of discriminations, i.e., second-order correlations, subject to natural selection. An example is *adaptive radiation*, in which one species is dispersed into many subspecies, each specialized for one environment. Traditional human societies differentiate in a similarly unselfconscious way, transmitted by upbringing. Communities which could not agree on common things, or agreed about the wrong things, fell apart before they became traditional. The appearance of purpose separates the evolution of human societies from natural evolution, which cannot think in advance. Participation of three sides in decision-making is one way to embed purpose into governance.

3.7.4 Mathematical and Theoretical Physics

The denominator of the fluctuation probability (3.67) encompasses everything that may happen, limited by boundary conditions, symmetries, and conservation laws. That means that it is determined kinematically, one could say by bookkeeping. The list of all possible states of motion of a system does not take into account their

[3] As a park ranger put it, when asked why not equip refuse containers with some bolts that grizzlies could not open: The problem is that the smartest bears are much smarter than the dumbest tourists.

dynamics. Classically, the volume of phase space is the same, whether we can solve the equations of motion or not. Quantum-mechanically, the states of motion are vectors in a Hilbert space, so the free-particle basis is formally as good as any other.

Formally, we do not need to know what is happening in the system in order to determine the denominator, it is enough to know what the system is. That might lead one to believe that the calculation of the microscopic entropy is a mathematical problem. Historically, however, such mathematical approaches have not been fruitful physically. The reason is that a physicist wants to understand the outcome of a particular measurement on a particular system, which requires a different methodological approach to the probability of a fluctuation than the mathematical one.

The first step is a reduction of the denominator. A large number of materials consists of a small number of atomic species, which crystallize as various compounds in various crystal lattices. One needs to specify all information about the system in the denominator, which reduces the space of events to those among which we really do not know what will happen. This step is not possible without insight into measurements. Thinking does not replace measurements, but determines which ones are interesting. That is the beginning of the conversation between philosophy and revelation.

The second step is the identification of low-energy excitations, such as sound waves, which determine the nature of fluctuations around equilibrium. It is the criterion for understanding the system, i.e., whether the conversation between philosophy and revelation has reached the truth. If it has, microscopic understanding can be expressed in just a few numbers, which are the measured susceptibilities: Compressibility, heat capacity, conductivity.

The probability of a fluctuation distinguishes mathematical physics (the denominator) from theoretical physics (the numerator). This distinction is real, because no work is needed to maintain the equilibrium state, so the kinematic restrictions in the denominator are satisfied reversibly, without generating Joule heat. For example, electrons in a resistor satisfy the Pauli principle without dynamic friction, because they have no state of motion available which violates it. On the other hand, electric current creates Joule heat, because scattering can degrade momentum in various ways, which are all compatible with the laws of motion. What will really happen when the system is perturbed is determined by the numerator of the fluctuation probability, which is the subject of theoretical physics. We can say that mathematical physics gives a framework for the behavior of a system, while theoretical physics provides understanding what happens within that framework in a specific case.

The expansion of the exponent of the fluctuation formula in a Taylor series indicates this difference at a technical level. Mathematically, Taylor's series is understood as an approximation which is all the better, the smaller the expansion parameter, and if it is larger, one needs to take into account higher orders. Physically, it is the reverse. Too small values, within the width of the distribution, do not trigger response forces, so they are physically equivalent to zero. On the other hand, large values require the inclusion of higher terms only if the system really has a nonlinear response, like in the vicinity of a phase transition, otherwise the harmonic approximation is good for them as well.

3.8 The Denominator of the Fluctuation Probability

The universal reach of the fluctuation probability formula is based on the fact that random events are real, so the thermodynamic entropy is really statistical in nature. The choice of denominator is the most sensitive step in research, because it determines everything that can happen, that is, the framework in which observations are interpreted. It is thus useful to be conscious of it in language, which is the purpose of the following examples.

A liquid does not have a dispersive response to shear, only dissipative. Thus, if we shake the wall of the container transversely, we only heat the liquid inside. But when the temperature is lowered, the liquid solidifies, and we get transverse sound waves, a new phenomenon whose probability was negligibly small above the transition temperature. As microscopic fluctuations, these waves contribute to the stability of the new equilibrium state of the solid. That can be shown by the solid resisting macroscopic shear deformations by macroscopic forces, as distinct from the liquid. Therefore, the denominator which should include all possible sound waves must include the transverse ones for a solid, but not for a liquid. Clearly, the concept "all possible" does not have the same logical and physical content.

There was much discussion about euthanasia in the Netherlands before it was allowed by law. After that, the director of a euthanasia clinic arranged with insurance companies to pay his clinic 3000 € for each of their clients it helps with that, even if the client pulls out at the last moment. From the point of view of statistical physics, it is interesting that no one imagined that happening, in all previous discussions, even though the financial motive is obvious. A measurement was needed to discover transverse sound waves, so to speak. Now, it is of course easy to imagine how they help stabilize the new equilibrium, namely, those who profit from such an arrangement will offer macroscopic resistance to any change of law that forbids it.

Imagine a job advertisement "tall people wanted to work in a warehouse." Five candidates apply, three men and two women. Men are on the average taller than women. Does that justify limiting the choice to men? No, because the advertisement stipulated tall people, and the applicants know how tall they are, while the employer does not, without looking at them. Our interventions in a system—measurement setups—are real only if they change the denominator, and interesting only if we do not know what will happen without looking. Philosophy does not serve to make predictions, but to sharpen the question until revelation is needed.

Yet, what if 500 candidates had applied? The issue of signal-to-noise ratio is precisely the issue of setting up the denominator well. The ideal job advertisement is the one for which only one candidate applies, and that is the one we need. For that, one would have to include the real experience of the job's advantages and disadvantages in the advertisement itself. Ideally, the denominator of the fluctuation probability is specialized on the basis of physical insights until it refers to only one system, and that is the one being measured.

Sometimes, one can similarly embed purpose into governance kinematically. For example, in France the annual receipt for property tax is the only proof of ownership. Thus, it is achieved without dissipation that no one owns property on which tax has not been paid regularly. In human societies, upbringing is the kinematic restriction that reduces the denominator of the fluctuation probability. Just as in France paying taxes delimits ownership, so upbringing delimits a society.

On the other hand, if anyone could do the job, selection by lottery, i.e., white noise, is the statistically appropriate method, however difficult to defend from the point of view of merit. Statistical physics opens the possibility to control the ratio between merit and accident just like the balance between energy and entropy. For example, a job interview may serve to choose candidates deterministically, among which the final selection is made by lot. If there exists an optimal ratio of one approach to the other, that means that the temperature is real, and that it is neither zero nor infinity, as mystics of the first or second kind might imagine.

Setting up the denominator of the fluctuation probability in accord with all available information about a specific system is crucial for a statistical description of macroscopic phenomena. That is the subject of the next chapter.

Chapter 4
Ensembles

> *If I have had any success in mathematical physics, it is, I think, because I have been able to dodge mathematical difficulties.*
>
> *Josiah Willard Gibbs*

The physical idea behind ensemble theory has been discussed in the previous chapter. It is that a real system in thermodynamic equilibrium with fixed mechanical invariants of motion—particle number, energy, linear and angular momentum—randomly chooses its state of motion among all possible states having the same value of those invariants, with equal a priori probabilities.

While often sufficient for qualitative discussion, this idea is too limited to be practical, in two related ways. First, the actual system under consideration usually does not have a fixed energy, but rather a fixed temperature. Similarly, instead of the particle number, we may have a fixed value of the chemical potential. Generally, the ensemble construction should be as flexible as thermodynamics itself, with respect to which of a pair of conjugate variables is fixed and which is fluctuating.

Second, effective calculation with fixed energy and particle number turns out to be prohibitively difficult. While that is formally just a technical objection, it is practically capable of scuttling the whole program. The success of ensemble theory is due to Gibbs' circumventing the combinatorial technicalities by a statistical trick which made calculations of realistic thermalized systems feasible, thus avoiding both limitations at once.

This trick is well known to opinion pollsters, under the name "controlling a variable" or "weighting the distribution." Physicists call it "polarizing the ensemble." Suppose we ask a question for which we are not certain that men and women hold the same opinion on average. Their proportion in the general population is equal, but our polled sample has, say, twice more men than women. How to take that into account so that the poll reflects the average opinion of the population? Easily: Ponder all mens' answers with a factor of $1/2$. Then the total weight of mens' and womens' opinions in the poll will be the same, in accord with the situation on the ground.

D. Sunko, *Conceptual Thermodynamics*,
https://doi.org/10.1007/978-3-032-20877-4_4

The ensemble with equal a priori probabilities can be polarized just as well, by weighting the probabilities so as to fix values of the temperature, chemical potential, pressure, or whichever other variable is controlled experimentally. To polarize the ensemble in accord with all fixed values of thermodynamic variables is to take into account all available information about the system.

Gibbs introduced a terminology for three standard situations. He called the ensemble at fixed energy and particle number *microcanonical*, at fixed temperature and particle number *canonical*, and at fixed temperature and chemical potential *grand canonical*. We shall consider each one in turn now, using the classical ideal gas as a running example. As we shall see, our statistical ideas are still not quite free of mechanistic illusions.

4.1 Classical Ideal Gas in the Microcanonical Ensemble

We have studied the microcanonical ensemble as the generic representative of ensemble theory in the previous chapter. It remains to demonstrate a practical calculation, specifically of the equations of state of the classical ideal gas.

The volume of phase space available to the gas is

$$\Gamma(N, U, V) = \int_{U,V} d^3q_1 \dots d^3q_N d^3p_1 \dots d^3p_N = V^N \int_U d^3p_1 \dots d^3p_N, \tag{4.1}$$

where the momentum integral is constrained to a surface of given energy U,

$$\sum_{i=1}^{N} \frac{|\boldsymbol{p}_i|^2}{2M} = U. \tag{4.2}$$

It is practical to replace this constraint with a weaker one,

$$U - \Delta U \leq \sum_{i=1}^{N} \frac{|\boldsymbol{p}_i|^2}{2M} \leq U, \tag{4.3}$$

so that we calculate the volume of a spherical shell of thickness ΔU at radius $\sqrt{2MU}$ in $3N$-dimensional momentum space. Our first encounter with the qualitative effects of N being a very large number is that the thickness ΔU does not matter, in fact we can take $\Delta U = U$ and calculate the volume of the whole sphere instead of an arbitrarily thin shell! Indeed, the volume of a sphere of radius r in d dimensions is $V(r) = cr^d$, so the volume of a shell of thickness Δr is

$$V = c[r^d - (r - \Delta r)^d] = cr^d \left[1 - \left(1 - \frac{\Delta r}{r}\right)^d\right]. \tag{4.4}$$

The expression in parenthesis is less than 1, so when $d \gg 1$, the volume of an arbitrarily thin surface shell dominates the volume of the sphere. (The readers can try to draw this for $d = 1, 2, 3$ and $\Delta r/r = 1/2$.)

Next, one must find the constant c. Writing the volume as an integral,

$$V(R) = \int d\Omega_d \int_0^R r^{d-1} dr = \Omega_d \frac{R^d}{d}, \tag{4.5}$$

we see that $c = \Omega_d/d$, where Ω_d is the solid angle in d dimensions. We find it by a trick, calculating a d-dimensional Gaussian integral in two ways. In Cartesian coordinates, it is

$$I = \int_{-\infty}^{\infty} e^{-(x_1^2+\ldots+x_d^2)} dx_1 \ldots dx_d = \left(\int_{-\infty}^{\infty} e^{-x^2} dx \right)^d = \pi^{d/2}, \tag{4.6}$$

while the same in spherical coordinates is

$$I = \Omega_d \int_0^{\infty} e^{-r^2} r^{d-1} dr = \frac{1}{2} \Omega_d \Gamma(d/2) \implies \Omega_d = \frac{2\pi^{d/2}}{\Gamma(d/2)}, \tag{4.7}$$

where Γ is Euler's gamma function. Noting that $n\Gamma(n) = n!$, we find

$$V(R) = \frac{\pi^{d/2}}{(d/2)!} R^d. \tag{4.8}$$

Inserting the radius $R = \sqrt{2MU}$ and dimension $d = 3N$, with Stirling's formula $\ln N! \simeq N \ln N - N$, the microscopic entropy (3.23) is

$$\sigma = \ln \frac{\Gamma(N, U, V)}{h^{3N}} = N \ln[h^{-3} V (4\pi M/3)^{3/2} (U/N)^{3/2}] + 3N/2. \tag{4.9}$$

Unfortunately, this result is wrong, because the volume appears under the logarithm, so it is not additive over subsystems. An extensive expression can be found by dividing the phase-space volume by $N!$:

$$\sigma = \ln \frac{\Gamma(N, U, V)}{N! h^{3N}}. \tag{4.10}$$

Subtracting $N \ln N - N$ from the previous expression, we get the so-called Sackur-Tetrode formula:

$$\sigma(N, U, V) = N \ln[h^{-3} (V/N) (4\pi M/3)^{3/2} (U/N)^{3/2}] + 5N/2. \tag{4.11}$$

In other words, the phase-space volume visible to thermodynamic phenomena is a factor of $N!$ *smaller* than assumed by a mechanical microscopic interpretation of the entropy. We cannot trust the statistical explanation of thermodynamics until we understand this factor, which is the subject of the next section.

In the meanwhile, we derive the equations of state from the Sackur-Tetrode formula following the program of Sect. 3.6.1. Thermal equilibrium gives

$$\frac{1}{\tau} = \left(\frac{\partial \sigma}{\partial U}\right)_{N,V} = \frac{3N}{2U} \Longrightarrow U = \frac{3}{2}NkT, \tag{4.12}$$

and mechanical equilibrium,

$$\frac{p}{\tau} = \left(\frac{\partial \sigma}{\partial V}\right)_{U,N} = \frac{N}{V} \Longrightarrow pV = NkT. \tag{4.13}$$

For particle-number equilibrium, the N inside the logarithm appears as $N^{-5/2}$, so its derivative cancels with the derivative of $5N/2$ outside. The factor h^{-3} balances the units, so V/N in the numerator means that everything else in the brackets has the dimension of volume in the denominator. Thus,

$$-\frac{\mu}{\tau} = \left(\frac{\partial \sigma}{\partial N}\right)_{U,V} = \ln \frac{V/N}{\lambda_T^3} \Longrightarrow e^{\mu/kT} = \left(\frac{\lambda_T}{d}\right)^3. \tag{4.14}$$

We have replaced U/N with $3kT/2$ and introduced

$$d = \sqrt[3]{\frac{V}{N}}, \quad \lambda_T = \frac{h}{\sqrt{2\pi MkT}}, \tag{4.15}$$

as the average distance between particles and so-called thermal wavelength, respectively. Classically, h is an arbitrary measure for the volume of phase space, so the origin of μ depends on the choice of units, and the ratio λ_T/d has no physical meaning. Then the most efficient way to write this equation of state is

$$\mu = kT \ln \frac{N}{V} + f(T) = kT \ln p + g(T), \tag{4.16}$$

where f or g are shifted by constants depending on units, the salient point being that they are functions of temperature only. This equation shows, for example, that changing the pressure is not an efficient way to change the chemical potential. Increasing the pressure by a factor of 3 will increase the chemical potential by barely more than kT. Conversely, for doubly charged ions in an electrolyte, changing the cell voltage by 0.1 V changes the chemical potential by 0.2 eV $\approx 8kT$ at room temperature, increasing their partial pressure on the electrode by a factor of 3000.

We shall see in Chap. 5 that h must be Planck's constant if the ideal-gas equations of state are to be consistent with quantum mechanics. With this insight, the ratio λ_T/d acquires a profound physical meaning. Because the thermal wavelength

Fig. 4.1 Classical limit

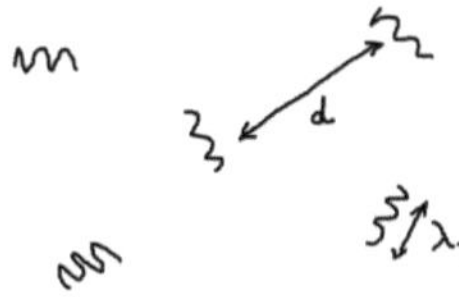

has dimension of length, the denominator in (4.15) has dimension of momentum, motivating us to compare it with the well-known Heisenberg uncertainty relation,

$$\Delta x \Delta p_Q \geq \frac{h}{4\pi}, \tag{4.17}$$

where Δp_Q is the *quantum* momentum fluctuation of a particle localized within a box of linear size Δx. On the other hand, $kT = \Delta p_T^2/2M$ defines a *thermal* momentum fluctuation associated with random kicks of energy kT experienced by a particle of mass M in a thermostat. Because the localization scale $\Delta x \approx d$ in an ideal gas, the physical meaning of λ_T/d is the ratio of quantum to thermal fluctuations:

$$\frac{\lambda_T}{d} \sim \frac{\Delta p_Q}{\Delta p_T}. \tag{4.18}$$

The limit $\lambda_T \ll d$ is the limit of point particles, where the classical model is adequate. In brief, Planck's and Boltzmann's constant compete in the thermal wavelength (4.15), with Boltzmann's winning at high temperature (Fig. 4.1).

Because the ratio λ_T/d is physically small in the high-temperature limit, it follows from Eq. (4.14) that the chemical potential of a classical ideal gas is negative. This result is easiest to understand starting from the internal energy:

$$\mu = \left(\frac{\partial U}{\partial N}\right)_{S,V}. \tag{4.19}$$

Adding particles at fixed internal energy increases the entropy, because the number of available states of motion has increased. Then one must *decrease* U to return to the adiabate, in other words, adiabates in the $U(N)$ plane have a negative slope. This slope is the chemical potential (Fig. 4.2). Thus, it is not enough to say that the chemical potential is large at high temperature. It is better to say that it is many kT's less than zero (about 10 for ^{4}He slightly above liquefaction, generally more). Such a statement is physical because an independent scale λ_T competes with the density. Absolute scales are the essential feature of quantum mechanics that distinguishes it from classical: One is so many hydrogen atoms tall.

The heat capacity of a classical ideal gas is independent of temperature: $c_V = \partial U/\partial T = 3Nk/2$, or $3R/2$ per mole. This property is in contradiction with fundamental thermodynamics, which stipulates that heat capacities of all substances vanish at absolute zero, cf. Eq. (2.20). The discussion above points to the resolution

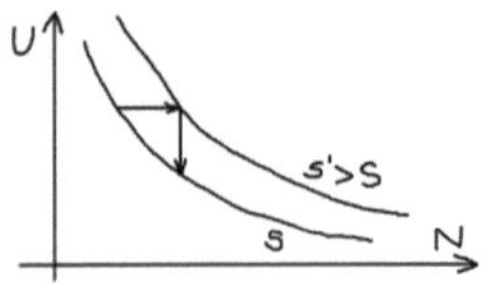

Fig. 4.2 Negative chemical potential

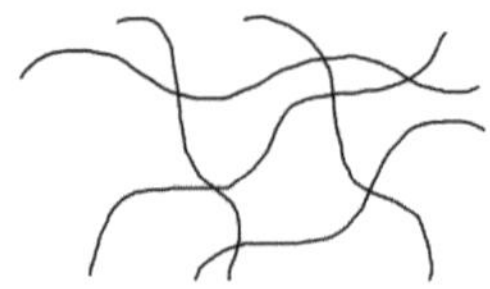

Fig. 4.3 Quantum limit

of this puzzle: $\lambda_T \gg d$ at sufficiently low temperature, so the model is no longer in the classical regime. When h is Planck's constant, the classical model of an ideal gas knows its own limits, like all good physical models. The quantum regime may be imagined after de Broglie as an overlapping of wave packets of size λ_T, which makes the gas behave as a coherent whole, so that the number of its degrees of freedom becomes smaller than the number of particles (Fig. 4.3).[1]

4.2 Indistinguishability of Identical Particles

The factor of $N!$, introduced in Eq. (4.10) to make the microscopic entropy extensive, has a simple explanation. The labels $1, 2, \ldots N$ of particle coordinates and momenta in the calculation do not correspond to physical reality. The reason is that molecules have a discrete energy spectrum, so labeling them physically—by ionization, or attaching another atom—causes a large disturbance at the scale of an individual molecule. Such labeling is quite different than painting a billiard ball, where the influence of the paint on the ball is negligible. Because the only way to make a microscopic particle distinguishable from another of the same kind is to change its kind, particles of the same kind are indistinguishable *in principle*.

Such indistinguishability is a subtler phenomenon than can be handled by an ad hoc reduction of the classical phase space by a factor of $N!$. However, this factor is the only part of the quantum correction which survives in the classical limit $\lambda_T \ll d$, as we shall see in Chap. 5. The classical model of an ideal gas with the reduction by $N!$ is reproduced by an exact quantum-mechanical calculation in the same limit. No more can be required of a model.

In 1875, Gibbs gave an explanation of the factor $N!$ that did not require indistinguishability in principle. Although obsoleted by quantum mechanics for physically indistinguishable particles, it can explain observations of extensivity in biological evolution and human societies, economics in particular. It also clarifies

[1] Although obsolete as a physical description, de Broglie's matter waves are still helpful to the imagination.

the concept of entropy in cases where we ignore a real microscopic distinction, such as between oxygen and nitrogen in air. We consider these various aspects in turn.

Notably, extensivity is not required either theoretically or physically. It is simply most often observed. Even when a system is not extensive as a whole, it may be reasonable to assume it locally, for example in describing the outer layers of a star by pressure and temperature fields, connected by an equation of state at each point. Less exotically, non-extensivity is commonly observed as surface effects, such as surface tension in water droplets. Surface effects are the reason why it is difficult to grow large monocrystals, nevertheless they can be grown in many cases. The thermodynamic limit is attainable in practice against apparently cogent theoretical objections, so it is useful to understand how they lose their force, the ideal gas being the generic example in point.

4.2.1 Gibbs' Explanation

The factor of $N!$ provides extensivity, but it seems to introduce an inconsistency, known as Gibbs' paradox. Suppose that there are two gases separated by a partition in two equal compartments of volumes $V_1 = V_2 = V/2$, with equal particle numbers $N_1 = N_2 = N/2$. Removing the partition, they mix, so the entropy increases by the so-called entropy of mixing:

$$\Delta\sigma = \ln V^N - \ln V_1^{N_1} - \ln V_2^{N_2} = N \ln 2. \tag{4.20}$$

On the other hand, imagine that both compartments were filled with $N/2$ molecules of the same gas. In that case, nothing happens: The entropy of the combined system is equal to the sum of the entropies of the subsystems, before and after, $\Delta\sigma = 0$. But labels like "oxygen" and "nitrogen" are not part of the calculation (4.20), we just imagined them, so how can the real physical entropy be different in the two cases?

The question is not idle, because a difference in entropy at fixed temperature is a difference $T\Delta S$ in free energy. Does that mean that useful work can be extracted from the difference of two calculations in one's head? Yes, sometimes, but we need a few steps to see when. Consider what happens when the partition is put back. In the case of two gases, both compartments contain a mixture. For one gas, both *appear* the same as before, but there is no guarantee that the actual molecules in a compartment are the same ones as were in it before the partition was removed. Indeed, for them to be the same is just as improbable as a spontaneous separation of the two gases. In other words, the value of $\Delta\sigma$ is the same $N \ln 2$ in both cases if we can follow the microscopic trajectories, but we cannot. Hence, $\Delta\sigma = 0$ for one gas, and we must stipulate that all permutations of the molecule labels are *macroscopically* the same state of motion, hence divide the phase space by $N!$.

For two gases, we can sharpen the issue further. Imagine that there are two gases, but we do not know that, so we claim that $\Delta\sigma = 0$. When someone tells us otherwise, we duly change the calculation and get $\Delta\sigma = N \ln 2$ on paper. So what, in reality? Nothing, unless there is a *macroscopic* way to make use of the difference. Gibbs' explanation is an early version of Heisenberg's razor: The calculation is not only in one's head if it can be realized as an experiment. If we have membranes that are permeable to either kind of atom, but not the other, we can use them to extract mechanical work precisely equal to $T\Delta S$. We shall show that explicitly below, but the claim should be plausible already.

In his 1902 monograph, Gibbs restated the explanation in abstract terms. To make his point vividly, it is impossible for a piston pressing on a gas to feel a permutation of the particles inside, if all the positions and velocities are exactly the same. Conversely, it is impossible for an experimenter to induce such a net permutation by any manipulation of the piston. Thus, the available phase space is effectively reduced by $N!$ for all macroscopic potentials, entropy in particular.

Gibbs' argument shows that extensivity is only a side effect of introducing the factor $N!$. The overcounting by $N!$ reflects a mechanistic illusion, that permutations of particles in the container can be manipulated by macroscopic means. On the contrary, equivalent states that differ only by permutation are generally far from each other in phase space, so they are inaccessible thermodynamically for any given sample, starting from that sample. They remain so even if the system is not extensive, for example when the entropy is subadditive because of interactions.

A condensed version of Gibbs' explanation is that pushing a partition repeatedly in and out of a container with one gas does not change the statistical entropy only if one divides by $N!$. It leads to the final subtlety: Is our inability to manipulate the microscopic permutations practical or fundamental? It is fundamental, due to thermalization. No amount of static friction with the walls will reproduce the same state with particles permuted, just as it will not cause spontaneous separation of two gases. Gibbs' argument is physically founded on the reality of random events, not particle indistinguishability, or inability to manipulate (long-destroyed) trajectories. Thus, we can expect extensivity whenever particle movement is diffusive, without regard to details of microscopic interactions, or any other theoretical considerations.

4.2.2 Extensivity in Society

Gibbs' argument lends itself readily to observations of extensivity in ecology. For example, a meadow twice as large contains twice as many blades of grass. By Gibbs, this statement is the same as saying that a random permutation of blades leads to the same meadow, even if the blades are distinguishable in principle.

A random permutation of anchovies in a shoal leads to the same shoal of anchovies, even if the anchovies show some individuality on closer examination. Those at the edges constantly strive to push themselves inside, where it is safer, leading to a continuous exchange of individuals. Yet, this observation only reinforces our attitude, shared by other predators, that it does not matter which fish is which.

In human society, Gibbs' argument may be translated to say that it does not matter whether Peter has a car-wash and Paul a diner, or vice versa. The economy would not feel anything if they exchanged places. Conversely, no macroeconomic measure could force them to do so—it is like a piston to individual atoms. Any observation of extensivity, in the form of a uniform density of car-washes or diners, is an indication that merit does not play an important role. (Most people go to the nearest place.) Evolution, both biological and social, is survival of the good enough, not the fittest. Adaptation is successful when everyone is good enough to survive.

Voting provides a detailed example of Gibbs' whole reasoning. When women were given the right to vote, the null prediction, that the electorate would simply double with no change in voting patterns, was by definition extensive. On the other hand, had they voted as a bloc, women, with an arbitrary percentage of men who shared their opinion, would have had more than 50% of the vote on any issue. The actual outcome in the United States was more nuanced. A net effect on fiscal policy can be ascribed to the enfranchisement of women, on the basis of historically documented efforts to organize and influence women voters. In thermodynamic terms, work was expended to produce a concentration of voters with elevated partial pressure on a particular set of issues. The entropy of mixing does not know that we are calculating it, and the tag man/woman means no more in politics than nitrogen/oxygen in physics, unless there is a *macroscopic* way to make use of the distinction.

One can extend this insight by additional tags, as in identity or party politics. If the tagged fraction is f, the entropy of mixing (4.20) becomes

$$\Delta\sigma = N\left[-f \ln f - (1-f)\ln(1-f)\right], \tag{4.21}$$

which retains about half its maximal value at $f = 1/10$ (Fig. 4.4). Thus, the thermodynamic potential of partisan politics is non-negligible even if f is fairly small, which helps understand the fracturing of parliaments with proportional representation. The conversion of the entropy of mixing into useful work (legislative pressure) depends on the efficiency of political engines, so when a tagged group does not vote as a single bloc, it indicates that an effective political temperature is greater than zero.

Finding a way to measure political and economic temperatures would disentangle effects of merit and chance in society objectively. The qualitative discussion above shows how the thermodynamic language itself frames an issue in an operative way. The equation $\Delta F = -T\Delta S$ means that more work can be extracted from

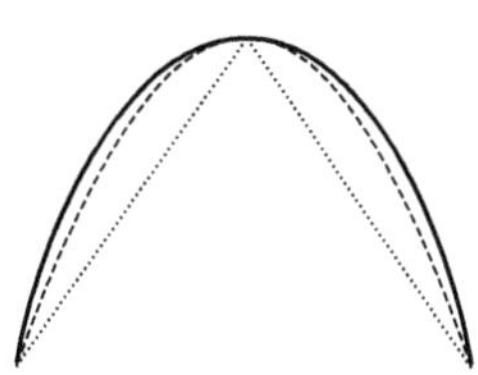

Fig. 4.4 Line, parabola, entropy of mixing

a given entropy of mixing if the temperature is higher. One can think in terms of changing the political temperature even without having an objective way to measure it. Another approach, quite popular when the tag is a political party, is to introduce thresholds for representation, so the entropy of mixing is zero by definition when $f < 0.05$, say. Reducing the entropy reduces noise, but when the entropy is zero (one party), there is no space for change. This conundrum is the same in physics, biology, and society.

4.2.3 Macroscopic Manipulation of Microscopic Distinctions

The entropy is physically a macroscopic quantity. The attenuation of thermal noise at high frequency shows that it has no meaning for too short times. It comes into being only as the effect of the averaging process (3.15), physically realized as static friction. Humans have a psychological difficulty with such emergent quantities. It is against natural intuition that a real event can occur at the level of a distribution without anything special happening at the individual level, so mechanistic (conspiratorial) narratives sometimes emerge.

The name "microscopic entropy" for the expression (3.23) misleadingly suggests that the entropy is an input to thermodynamic optimization. Actually, it describes the *outcome* of the time-averaging that determines the universe of events in the denominator of a fluctuation probability. Every application of the formula (3.23) requires a conjecture, not what will happen, but what already happened to obtain the sample under consideration. The formula cannot tell us what this physical situation is, any more than a grammar can tell us what to say. The question whether to include the distinction between oxygen and nitrogen in air is no different in principle than whether to include transverse sound waves in a solid. A reduced description may be appropriate for some experiments, but not for others. Technically, if there is no plan to manipulate the distinction, the entropy of mixing is an additive constant which may as well be put to zero. There is no guarantee that such an omission will not haunt us later as an apparent miracle—that is how discoveries are made.

If we have a membrane permeable to nitrogen, but not oxygen, we can make use of the distinction between the two. Let the fraction of oxygen be f. The reduction in free energy due to mixing of both gases may be written

$$\Delta F = -T\Delta S = -T\left(\Delta S_f + \Delta S_{1-f}\right), \tag{4.22}$$

where the two terms refer to the two terms in Eq. (4.21). If we replace the partition dividing the volume $V_f = fV$ from V_{1-f} by our membrane, nitrogen will diffuse irreversibly at the price of an entropy increase $T\Delta S_{1-f}$, reducing the free energy by the second term in the expression above. On the other hand, oxygen will expand by moving the membrane as if the nitrogen were not there, until it occupies the whole volume. The work done by this expansion is

$$W_f = \int_{V_f}^{V} p(V)dV = fNkT \int_{V_f}^{V} \frac{dV}{V} = fNkT \ln \frac{V}{V_f} = T\Delta S_f, \tag{4.23}$$

which is equal to the first term, but now the reduction in free energy is due to useful work, not random diffusion. We could similarly convert the other term with a membrane permeable the other way. If we did not know that such membranes existed, or thought that there was only one gas, the experiment would look like a *perpetuum mobile*, extracting work from heat without other effects. Such a miracle would herald the discovery of such a membrane to some, or indicate two gases to others. Gibbs-like paradoxes appear whenever the line between ontology (a thing exists) and gnoseology (what we know about the thing) is not drawn properly.

4.2.4 The Sackur-Tetrode Formula Revisited. Non-extensivity

Instead of the statistical derivation above, the Sackur-Tetrode formula can also be derived thermodynamically, by direct integration of the equations of state. The derivation clarifies several conceptual issues about the ideal gas, extensivity, and absolute value of the entropy.

Thermodynamically, at fixed particle number we have in general

$$S(T, V, N = N_0) = \int \left[\left(\frac{\partial S}{\partial V} \right)_T dV + \left(\frac{\partial S}{\partial T} \right)_V dT \right]. \tag{4.24}$$

Euler's relation and the expression for the heat capacity also hold in general:

$$\left(\frac{\partial S}{\partial V} \right)_T = \left(\frac{\partial p}{\partial T} \right)_V, \quad c_V = T \left(\frac{\partial S}{\partial T} \right)_V. \tag{4.25}$$

The specialization to the ideal gas is that c_V is a constant. Inserting the appropriate values from the equations of state (4.12) and (4.13), we have

$$S(T, V, N) = Nk \ln V + \frac{3}{2} Nk \ln T + kf(N), \tag{4.26}$$

where we have replaced the number N_0 with a functional dependence on N after noticing that the integration constant may depend on it. It appears in the entropy as a function $f(N)$, which can be determined only by experiments in which N changes. Such are chemical reactions in the gas phase, in which Avogadro *observed* the ideal gas to be extensive. A minimal adjustment of $f(N)$ to reflect that observation gives

$$S(T, V, N) = Nk \left(\ln \frac{V}{N} + \frac{3}{2} \ln T + C \right), \tag{4.27}$$

where C is truly a constant now, which can be used to balance the units under the logarithm. Scaling with the molecular mass and Boltzmann's constant leaves an arbitrary constant h with the dimension of action:

$$C = a + \ln \frac{(2\pi Mk)^{3/2}}{h^3}, \tag{4.28}$$

which recovers the Sackur-Tetrode formula (4.11) up to the last term $5N/2$, fixed by setting $a = 5/2$. That choice adjusts the origin of the chemical potential to be zero, which is not required by thermodynamics. A quantum-mechanical calculation in the classical limit shows that it is indeed zero for the ideal gas, and h is Planck's constant. Then the value of the entropy is absolute because $S = 0$ at $T = 0$. The reason for that lies outside thermodynamics: There is only one state of motion at zero temperature, so $S = k \ln 1 = 0$.

The principal insight from the derivation above is that extensivity is an observational fact, not a fundamental requirement. A non-extensive system can be treated in the same way, by adjusting $f(N)$ to fit the observed behavior, for example to account for surface tension. Similarly, Hawking derived a non-extensive equation of state for the entropy of a black hole,

$$S \propto A, \tag{4.29}$$

where A is the surface of the event horizon. Here, the static friction responsible for thermalization is due to fluctuations in which one of a particle-antiparticle pair is absorbed by the black hole, while the other becomes asymptotically free. Thus, the black hole can lose mass as well as gain it, so the thermalization mechanism becomes reversible, as required by Carnot's analysis (2.20). Hawking's identification of pair fluctuations with Carnot engines makes his equation of state part of standard thermodynamics, where the temperature means that past dynamics is forgotten. Nevertheless, there have been alternative interpretations of black-hole entropy as a quantity of information accumulated inside.[2] To resolve this issue, measurements were made on an analogous laboratory system, which detected an analogue of Hawking radiation, supporting the notion that black-hole temperature is ordinary thermodynamic temperature. Strictly speaking, the matter is open until someone measures black-hole thermal noise directly, or finds an error in Hawking's reasoning.

[2] Note the parallel with Gibbs' paradox.

4.3 Fixed Temperature: Canonical Ensemble

We model a system in a thermostat as a small subsystem of a large insulated system with a fixed energy E_t. The small system can only exchange energy with the remainder (environment), so they are at equal temperature. The energy balance at all times is

$$E_t = E + E_0, \quad E \ll E_0, \tag{4.30}$$

where E is the energy of the subsystem, and E_0 the energy of the environment, without the subsystem. (Think of a sample on a holder.) Thermal contact means that the total phase space available to the system may be written $\Gamma_0(E_0)\Gamma(E)$. To calculate the probability distribution $\rho(E)$ that the small subsystem is found in some state of energy E, we count all possible states of the environment compatible with this division of energy. In the differential limit of the small system, the product of phase-space volumes becomes the distribution

$$\rho(E)d\Gamma = c\Gamma_0(E_0)d\Gamma, \tag{4.31}$$

where c is some normalization constant. According to the microscopic formula for the entropy,

$$\frac{\Gamma_0(E_0)}{h^r} = e^{\sigma_0(E_0)} = e^{\sigma_0(E_t - E)}. \tag{4.32}$$

Because $E \ll E_t$, we can expand the entropy of the environment to first order; the derivative is not zero because it is evaluated at E_t, not E_0:

$$\sigma_0(E_t - E) \approx \sigma_0(E_t) - \left(\frac{\partial \sigma_0}{\partial E}\right)\bigg|_{E_t} \cdot E = \sigma_0(E_t) - \frac{E}{kT}. \tag{4.33}$$

Thus, the *canonical (Gibbs) distribution* is

$$\rho(E)d\Gamma = Ae^{-E/kT}d\Gamma = Ae^{-\beta E}d\Gamma, \tag{4.34}$$

where A is the overall norm. It counts all the states in the denominator, so it satisfies the relation $A_{12} = A_1 A_2$ for two systems in thermal contact. It follows that

$$\ln \rho_1(E_1) + \ln \rho_2(E_2) = \ln \rho_{12}(E_1 + E_2), \tag{4.35}$$

so the logarithm of the distribution is linear over subsystems, as expected on general grounds.

Recall that each microscopic state appeared with equal a priori probability in the microcanonical distribution. This probability can be written $1/W(N, E)$, where

$$W(N, E) = \sum_{\{i:N_i=N, E_i=E\}} 1 \tag{4.36}$$

is the total number of microscopic states i with the given energy and particle number. (It is conceptually clearer to write phase-space integrals as sums, so we shall revert to integrals only for calculation.) By the same token, we write $A = 1/Z_C$, where

$$Z_C(N, \beta) = \sum_{\{i:N_i=N\}} e^{-\beta E_i}, \tag{4.37}$$

by definition of the norm. It is called the *canonical sum over states*. The index i runs over all states with given N, but without the restriction $E_i = E$, at the price of weighting each state with the exponential (Gibbs) probability distribution, instead of unity. The latter, sum of ones (4.36), is called the *microcanonical sum over states*. The expression (4.37) is paradigmatic of the ensemble polarization mentioned above. The canonical ensemble is weighted so that all its members have the same temperature, as fixed externally.[3]

The norm can be identified with a thermodynamic potential, just like $\sigma = \ln W$ is the entropy. Collect all terms with equal energy in the sum (4.37):

$$\sum_{\{i:N_i=N\}} e^{-\beta E_i} = \sum_E e^{-\beta E} \left[\sum_{\{i:N_i=N, E_i=E\}} 1 \right] = \sum_E W(N, E) e^{-\beta E}. \tag{4.38}$$

For the classical ideal gas, we have found in Eq. (4.11) that $W \sim E^{3N/2}$, so we expect W to rise very quickly with E when N is large. Because W rises like a power, the exponential term eventually drives the summand to zero as E increases. The product of a power-like increase and exponential decrease has a peak, which is all the higher and narrower, the larger N is. Asymptotically, the sum is dominated by a single term at some energy $E = E^\star(\beta)$:

$$\ln Z_C(N, \beta) \simeq \ln W(N, E^\star) e^{-\beta E^\star} = \sigma(N, E^\star) - \beta E^\star. \tag{4.39}$$

It remains to identify $E^\star$ with the internal energy U to recognize the canonical Massieu potential (2.48):

$$\ln Z_C(N, \beta) = \sigma(N, U) - \beta U = \Psi_C(N, \beta). \tag{4.40}$$

[3] For historical reasons, the sum over states is often called a partition function.

Hence, the normalized canonical distribution can be written physically as

$$\rho(E)d\Gamma = e^{-\Psi_C - \beta E}d\Gamma = e^{\beta(F-E)}d\Gamma, \tag{4.41}$$

where F is the Helmholtz free energy.

[The calculation of the norm explains how the variable is changed mathematically. The expression (4.38) is the generating function of the distribution $W(N, E)$, so β replaces E by summing over it as a dummy variable. The corresponding integral expression is called the Laplace transform. The asymptotic evaluation above may be summarized as

$$\ln Z_C(N, \beta) = \max_E \left[\sigma(N, E) - \beta E\right], \tag{4.42}$$

which shows that the macroscopic Legendre transform arises as a saddle-point evaluation of the microscopic Laplace transform in the thermodynamic limit.]

4.3.1 Classical Ideal Gas at Fixed Temperature

The sum over states of the classical ideal gas is calculated easily, because the kinetic energy $E = p_1^2/2M + \ldots$ gives rise to a $3N$-fold product of Gaussians:

$$Z_C(N, \beta) = \frac{1}{N!h^{3N}}\left(\int d^3q \int e^{-p^2/2MkT}d^3p\right)^N = \frac{1}{N!}\left(\frac{V}{\lambda_T^3}\right)^N. \tag{4.43}$$

The form of this expression is quite instructive. First, the sum over states for one particle is

$$Z_C(1, \beta) = \frac{V}{\lambda_T^3}. \tag{4.44}$$

This result is exact, because there is no difference between classical and quantum thermodynamics for one particle. The size of its phase space is the volume of the room measured by the thermal wavelength. Next, it is raised to the N-th power for N particles in thermal contact. Finally, the division by $N!$ is interpreted interestingly with the help of Stirling's formula $N! \simeq (N/e)^N$:

$$Z_C(N, \beta) = \left(\frac{eV/N}{\lambda_T^3}\right)^N. \tag{4.45}$$

This expression looks like a one-body result raised to the N-th power, but the one-body phase space measures the volume per particle, not the total volume! So somehow each individual particle "knows" how many there are. By quantum mechanics, yes, indistinguishability induces kinematic correlations between particles even without interactions, that is why the ground state is unique even for non-interacting particles. By Gibbs, yes, extensivity is emergent: Particles exchange positions only with the nearest ones, they do not dash all over the place just to satisfy the definition of the factorial. (Neither do fish in a shoal.) These interpretations are materially different because indistinguishability-in-principle induces correlations which a practical division by $N!$ does not describe, only they disappear at high temperature. Thus, the latter interpretation is correct for air at room temperature, because the former appeals to something that is physically irrelevant.

The canonical Massieu potential is

$$\Psi_C(N,\beta) = \ln Z_C(N,\beta) = N\left[1 + \ln\left(\frac{V/N}{\lambda_T^3}\right)\right], \tag{4.46}$$

whence the equations of state follow easily, e.g.,

$$U = -\frac{\partial \Psi_C}{\partial \beta} = -N\frac{\partial}{\partial \beta}\ln \beta^{-3/2} = \frac{3}{2}NkT, \tag{4.47}$$

and similarly for the other two:

$$\frac{p}{kT} = \frac{\partial \Psi_C}{\partial V} = N\frac{\partial}{\partial V}\ln V = \frac{N}{V}, \tag{4.48}$$

$$-\frac{\mu}{kT} = \frac{\partial \Psi_C}{\partial N} = 1 + \ln\left(\frac{V/N}{\lambda_T^3}\right) + N\frac{\partial}{\partial N}\ln N^{-1} = \ln\left(\frac{V/N}{\lambda_T^3}\right). \tag{4.49}$$

4.3.2 Heat Capacity and Latent Heat

Evaluation of the internal energy with Eq. (4.37) for Z_C gives

$$U = -\frac{\partial}{\partial \beta}\ln Z_C = \sum_i p_i E_i =: \langle E\rangle, \quad p_i = \frac{e^{-\beta E_i}}{Z_C}. \tag{4.50}$$

The internal energy at fixed temperature is statistically the average of the mechanical energies of all microscopic states i of the whole system, pondered by the canonical distribution. It is a fluctuating quantity, because the second moment (variance) of this average is not zero. The variance in units of $(kT)^2$ is just the heat capacity:

$$c_V = \frac{\partial U}{\partial T} = k\beta^2\frac{\partial^2 \Psi_C}{\partial \beta^2} = k\frac{\langle E^2\rangle - \langle E\rangle^2}{(kT)^2}. \tag{4.51}$$

A susceptibility is physically the elasticity constant of the equivalent oscillator, geometrically the curvature of the thermodynamic potential, statistically the variance of the distribution of the fluctuating variable.

Similarly, we can identify the latent heat with a covariance:

$$\frac{\partial U}{\partial V} = \left\langle \frac{\partial E}{\partial V} \right\rangle + \beta \left[\langle E \rangle \left\langle \frac{\partial E}{\partial V} \right\rangle - \left\langle E \frac{\partial E}{\partial V} \right\rangle \right], \tag{4.52}$$

which should be compared with Eq. (2.53). Diagonal susceptibilities are variances, off-diagonal ones are covariances.

4.3.3 Entropy and Heat

The entropy at fixed temperature is easily found by inverting the Legendre transform for the Massieu potential:

$$\begin{aligned} \sigma(N, U(\beta)) &= \Psi_C(N, \beta) + \beta \langle E \rangle \\ &= \ln Z_C + \sum_i p_i(-\ln p_i - \ln Z_C) \\ &= -\sum_i p_i \ln p_i, \end{aligned} \tag{4.53}$$

where we have expressed βE_i in terms of p_i and invoked $\sum_i p_i = 1$. This expression has the form of the relevant entropy (3.28), but with i denoting a microscopic state of motion, not an aggregate cell. It is called Gibbs' entropy, to differentiate it from the microscopic entropy (3.23). The two are equal up to the insensitivity of thermodynamic functions:

$$\sigma(N, \langle E \rangle) = \left\langle \ln \frac{1}{p_i} \right\rangle = \left\langle \ln \frac{1}{\rho(E_i)} \right\rangle = \ln \frac{1}{\rho(\langle E \rangle)} = \ln W(N, \langle E \rangle), \tag{4.54}$$

where we have invoked the linearity of $\ln \rho$ in energy. This derivation corresponds physically to equilibrating a system at a given temperature first, and wrapping it in thermal insulation next. Then $W(N, E = \langle E \rangle)$ is the phase-space volume of the equivalent microcanonical ensemble.

The (inexact) heat differential is obtained from the entropy differential:

$$\begin{aligned} d\sigma &= -\sum_i (\ln p_i + 1) dp_i = -\sum_i \ln p_i dp_i = \beta \sum_i E_i dp_i \\ &\Longrightarrow T dS = \sum_i E_i dp_i, \end{aligned} \tag{4.55}$$

because $\sum_i dp_i = 0$. It is one of two terms in

$$dU = \sum_i E_i dp_i + \sum_i p_i dE_i = \eth Q + \eth W. \tag{4.56}$$

The other is the work term. This expression explains why heat transfer is invisible. There is literally nothing to see, only some microscopic states are becoming more or less probable. A change in the distribution is invisible at the individual level. To accept the reality of this explanation is to accept the reality of the physical world.

More broadly, it explains the intransigence of some conflicts. If one is betting on the shape of a distribution, one cannot be persuaded by any individual event. Such entropy-driven conflicts seem to be common historically, so deterministic narratives do not always help understanding. For example, given the migration pressures on Visigoths, one could have bet on the sack of Rome without knowing about Alaric. Alaric, the man who can eat 120 grams of chili peppers in a minute, and the death-camp commander are all improbable, until one becomes probable. Polarization of the ensemble is like a job advertisement. Statistical physics prompts us to ask how the ensemble became polarized, before asking how there can be such people.

4.4 Fixed Chemical Potential: Grand Canonical Ensemble

Suppose now that the small system in a thermostat is separated from it by a permeable membrane, so they exchange particles at fixed chemical potential. The derivation of the canonical distribution can be repeated verbatim, only adding the particle-number balance to the energy balance:

$$E_t = E + E_0, \quad N_t = N + N_0, \quad E \ll E_0, \quad N \ll N_0. \tag{4.57}$$

By the same arguments as above, the distribution in N and E reads

$$\rho(N, E)d\Gamma \propto \Gamma_0(N_0, E_0)d\Gamma \propto e^{\sigma_0(N_t - N, E_t - E)}d\Gamma, \tag{4.58}$$

and the expansion of the entropy is in two variables now:

$$\begin{aligned}\sigma_0(N_t - N, E_t - E) &\approx \sigma_0(N_t, E_t) - \left(\frac{\partial \sigma_0}{\partial N}\right)\bigg|_{N_t, E_t} \cdot N - \left(\frac{\partial \sigma_0}{\partial E}\right)\bigg|_{N_t, E_t} \cdot E \\ &= \sigma_0(N_t, E_t) + \frac{\mu N - E}{kT}.\end{aligned} \tag{4.59}$$

Thus, the *grand-canonical distribution* is

$$\rho(N, E)d\Gamma = Ae^{(\mu N - E)/kT}d\Gamma = Ae^{\beta(\mu N - E)}d\Gamma, \tag{4.60}$$

where A is the overall norm. Writing $A = 1/Z_G$, by definition of the norm the denominator is the *grand-canonical sum over states*

$$Z_G(\mu, \beta) = \sum_i e^{\beta(\mu N_i - E_i)}, \tag{4.61}$$

where the sum is over all microscopic states without restriction. It can be interpreted statistically by collecting all terms with the same particle number:

$$\sum_i e^{\beta(\mu N_i - E_i)} = \sum_N e^{\beta\mu N} \left[\sum_{\{i:N_i=N\}} e^{-\beta E_i} \right] = \sum_N e^{\beta\mu N} Z_C(N, \beta). \tag{4.62}$$

Thus, the grand-canonical sum over states is the generating function of the canonical sum over states. Its logarithm can be identified with the grand Massieu potential by asymptotic evaluation. In the classical limit, the chemical potential is negative, while Z_C increases rapidly with N, so the sum (4.62) is again dominated by a single term at some value $N = N^\star(\mu)$:

$$\ln Z_G(\mu, \beta) \simeq \ln e^{\beta\mu N^\star} Z_C(N^\star, \beta) = \Psi_C(N^\star, \beta) + \beta\mu N^\star = \Psi_G(\mu, \beta). \tag{4.63}$$

This relation is valid universally, but we shall see later that the chemical potential can be positive at low temperature, so a different argument is needed to establish it there. Thus, the grand-canonical distribution is normalized physically as

$$\rho(N, E)d\Gamma = e^{-\Psi_G + \beta(\mu N - E)} d\Gamma = e^{\beta(\Omega + \mu N - E)} d\Gamma, \tag{4.64}$$

where Ω is the grand potential.

4.4.1 Classical Ideal Gas at Fixed Chemical Potential

The grand Massieu potential of the classical ideal gas is obtained from Eq. (4.62) with (4.43) for Z_C:

$$\Psi_G(\mu, \beta) = \ln \sum_N \frac{e^{\beta\mu N}}{N!} \left(\frac{V}{\lambda_T^3} \right)^N = e^{\beta\mu} \frac{V}{\lambda_T^3}. \tag{4.65}$$

It is numerically equal to the average particle number at fixed μ:

$$\langle N \rangle = \sum_i p_i N_i = \frac{1}{\beta} \frac{\partial \Psi_G}{\partial \mu} = e^{\beta\mu} \frac{V}{\lambda_T^3}, \quad p_i = \frac{e^{\beta(\mu N_i - E_i)}}{Z_G}. \tag{4.66}$$

This is the equation of state (4.14). The variance (susceptibility) of this expectation value is equal to it:

$$\frac{1}{\beta}\frac{\partial \langle N \rangle}{\partial \mu} = \left\langle N^2 \right\rangle - \langle N \rangle^2 = \langle N \rangle . \tag{4.67}$$

The corresponding standard deviation $\sqrt{\left\langle N^2 \right\rangle - \langle N \rangle^2} =: \Delta N$ is called the *number fluctuation*. We have

$$\frac{\Delta N}{\langle N \rangle} = \frac{1}{\sqrt{\langle N \rangle}}. \tag{4.68}$$

Hence, if 10^{24} gas molecules are in equilibrium with an adsorbate, their number will fluctuate by about 10^{12} molecules between thermalization intervals.

The equation of state (4.13) is derived just as easily. The internal energy brings a small surprise. Instead of (4.12), we get

$$U = -\frac{\partial \Psi_G}{\partial \beta} = \frac{3}{2} \langle N \rangle kT - \mu \langle N \rangle . \tag{4.69}$$

The internal energy is the sum of kinetic and potential energies of all particles. The second term is their potential energy with respect to the particle reservoir. The chemical potential is physically the binding energy of an adsorbed molecule, or, for charged particles, the ground potential per particle.

The readers will have noticed that the derivations above became progressively easier as the conditions on the sum over states were relaxed. The grand-canonical ensemble is so much easier to use in practice that it is preferable to calculate at fixed μ even when the physical situation is at fixed N. In that case, one must solve the equation

$$\langle N \rangle = N \tag{4.70}$$

for the chemical potential, to know which value of μ to insert, so it is useful to distinguish its two sides in notation. That is the reason why we have distinguished the fluctuating quantity $\langle N \rangle$ from the fixed N in the equations of state throughout, which is usually superfluous.[4] To recover the canonical results, one only needs to remember that the internal energy at fixed N is $U + \mu N$, if U is the internal energy at fixed $\mu = \mu^\star(N)$. This is an example of Gibbs' dodging the mathematical difficulties.

[4] The equations of state in the various ensembles all look the same. When we unplug a grounded appliance, the electrons in the casing switch from fixed μ to fixed N without visible change.

4.5 The Maxwell-Boltzmann Distribution

The Maxwell-Boltzmann distribution is a one-particle distribution, unlike the ensemble distributions considered so far, which refer to properties of the whole sample. It answers a most elementary question: What is the probability that a single molecule in a system will have some given value of linear momentum, at a given temperature? It can be calculated as the probability of the corresponding fluctuation in the canonical ensemble. We fix one particle to have momentum $\boldsymbol{p}_1 = \boldsymbol{p}$ and let the rest in the sample do as they may:

$$\begin{aligned} P(\boldsymbol{p})d^3p &= \frac{e^{-\beta p^2/2M}d^3p \int d^3q_1 \cdots \int d^3p_N d^3q_N e^{-\beta(p_2^2+\ldots+p_N^2)/2M}e^{-\beta\mathcal{V}}}{\int d^3p_1 d^3q_1 \cdots \int d^3p_N d^3q_N e^{-\beta(p_1^2+\ldots+p_N^2)/2M}e^{-\beta\mathcal{V}}} \\ &= e^{-\beta p^2/2M} \cdot \frac{\lambda_T^3 d^3p}{h^3}, \end{aligned} \tag{4.71}$$

where $\mathcal{V} := \mathcal{V}(\boldsymbol{q}_1, \ldots, \boldsymbol{q}_N)$ refers to all possible interactions between the particles. The result is quite general as long as the interactions do not depend on momenta, so the position integrals cancel. Conservative potentials fulfill this condition.

The Maxwell-Boltzmann distribution (4.71) is normalized to unity by construction. For an ideal gas, one can use the equation of state (4.14) to normalize it to concentration instead:

$$\int e^{\beta(\mu-p^2/2M)}\frac{d^3p}{h^3} = \frac{N}{V}. \tag{4.72}$$

The normalization by $e^{\beta\mu}$ is sometimes more natural than by λ_T^3. In the literature, the distribution (4.71) is often written in terms of the velocity $\boldsymbol{v} = \boldsymbol{p}/M$ or speed $c = |\boldsymbol{v}|$. In the latter case, the volume element is $4\pi c^2 dc$, showing that low speeds are improbable because the phase-space volume they occupy is small.

One may well be uneasy with the simple dismissal of interactions in the derivation above. While the claim, that the distribution of momenta in air and water is the same, does not sound impossible, what about water and ice? Intuitively, we feel that molecules in a gas should attain high velocities more regularly than in a solid. E.g., one easily finds that the most probable value of the speed of a molecule of nitrogen at room temperature is $\sim$ 400 m/s. That seems preposterous for a molecule in a solid, which one would rather imagine standing still. How does the Maxwell-Boltzmann distribution miss that?

The short answer is that it does not, only the single-particle velocity distribution is not easily observed. Propagation in liquids is dominated by scattering, so it is not ballistic at long range as in a rarefied gas, but diffusive, where the concept of velocity loses meaning. The distribution (4.71) is still observed between collisions in molecular-dynamics simulations of dense liquids, where one can analyze the motion at microscopic resolution in a computer.

In solids, the atoms do not diffuse as a rule, but are bound to neighboring atoms. Linear momentum in a bound state is not well defined in quantum mechanics, but a classical model of atoms in a solid as balls connected by springs is surprisingly good, at not too low temperatures. The atoms are heavy enough to neglect Heisenberg's uncertainty relations, so they are like balls. The Coulomb forces on the moving ions turn out to be the same as one would calculate classically for the same charges and displacements,[5] so they are like springs. In this model, the Maxwell-Boltzmann distribution for thermal oscillations of atoms is correct, only the amplitudes are small.

The distribution fails when quantum effects become important. Quantum liquids, such as helium at low temperatures and conducting electrons in metals, have a different momentum distribution because of the indistinguishability of identical particles, even in the limit of no interactions (ideal gas). Ice is an example of a quantum solid, in which the positions of the protons are quantum-mechanically uncertain. Their momentum and position integrals are not independent because of the uncertainty relations, so the cancellation of interactions fails, even if indistinguishability is not an issue because the protons do not exchange places physically.

There is an alternative derivation of the Maxwell-Boltzmann distribution, which is both misleading and revealing. If the gas is ideal, without interactions, one can say that a single sample of the gas is a Gibbs ensemble of molecules, instead of it being a member of the ensemble of samples as in the derivation (4.71). This ensemble distribution has the energy of a single molecule, $E = p^2/2M$, in the general formula (4.34), which gives the result (4.71) by inspection.

The derivation is misleading because it misses the generality of the result (4.71) in the presence of interactions, and risks confusion between the one-body (Maxwell-Botzmann) and sample (Gibbs) distributions. It also reveals the reason why quantum one-body distributions are different. Namely, there are no correlations between samples in the ensemble construction, so the derivation assumes that no interaction implies no correlation. That assumption is not true: Indistinguishability induces kinematic correlations, such as the Pauli principle, even without interactions.

4.5.1 *Maximization of Boltzmann's Entropy. Quasiparticles*

Let the distribution ρ in the formula (3.28) for the relevant entropy be the particle number density, namely the fraction of all particles found in the i-th cell at the macroscopic position $\boldsymbol{r}_i$ in a liquid:

$$\rho(A_i) = \frac{N(\boldsymbol{r}_i)}{N} =: n_i \,. \tag{4.73}$$

[5] This result is known as the Hellmann-Feynman theorem.

This historically first interpretation of the formula (3.28) is called Boltzmann's entropy. Each cell has a local internal energy e_i per particle, so the total energy and particle number are

$$U = N \cdot \sum_i n_i e_i, \quad N = N \cdot \sum_i n_i. \tag{4.74}$$

To find the distribution that maximizes the relevant entropy per particle at fixed total U and N, we have to solve the extremal equation

$$\frac{\partial}{\partial n_j}\left[-\sum_i n_i \ln n_i + \beta\mu\left(\sum_i n_i - 1\right) - \beta\left(\sum_i n_i e_i - \frac{U}{N}\right)\right] = 0, \tag{4.75}$$

where we have denoted the Lagrange multipliers tendentiously. The equation for a given cell and its solution are

$$-\ln n_j - 1 + \beta\mu - \beta e_j = 0 \implies n_j = e^{\alpha - \beta e_j}, \tag{4.76}$$

where $\alpha = \beta\mu - 1$. The solution has the form of the Maxwell-Boltzmann distribution, up to the -1 which is easily fixed by putting $\ln(n_i/e)$ instead of $\ln n_i$ in the expression above, cf. Eq. (4.45), a reminder that thermodynamics does not require extensivity, or set the origin of the chemical potential. It should not detract from the generality of the result: The distribution (4.76) maximizes the relevant entropy. It is only slightly more specific than the formula for the relevant entropy itself, the specialization entering through the Lagrange constraints on the maximum. In particular, the energy e_i is the macroscopic internal energy of a liquid cell, including all interactions, per particle, not a microscopic energy of one particle. Explicitly, it is the variation of the total energy with the local density, i.e., a macroscopic (specific) susceptibility:

$$e_i = \frac{1}{N}\frac{\delta U}{\delta n_i}. \tag{4.77}$$

We can specialize further, to the number of particles with a given momentum $\boldsymbol{p}$, which allows us to write the energy explicitly for an ideal gas:

$$e_i(\boldsymbol{p}) = \frac{p^2}{2M_i}. \tag{4.78}$$

This expression looks like the energy of one particle, but it is not. It is the energy of a macroscopic number of particles in the i-th cell, labeled by a common momentum, divided by their number.

Recalling that the microscopic derivation of the Maxwell-Boltzmann distribution (4.71) did refer to one particle, we may well ask, have we rederived it here, or derived something else, which looks like it only by accident? The short answer is, something else, but the similarity is not accidental.

The expression (4.76) was obtained by maximizing a thermodynamic potential, so it is a *macroscopic equation of state*, connecting the density and internal energy of a cell in real space. On the other hand, the distribution (4.71) is the momentum distribution in phase space of a single particle in contact with a thermostat. The particle's energy is a solution of the *microscopic equations of motion*, which have nothing to do with thermodynamic optimization. The equation of state is the outcome of that optimization. We have jumped to the conclusion that the optimal value for the mass in the macroscopic (output) expression (4.78) is the same as in the microscopic (input) expression for the kinetic energy, but that is not given. The mass in (4.78) could be different, and also depend on the temperature and pressure.

Landau introduced a terminology for this distinction, originally in the context of quantum liquids. A particle before the first collision is a particle. After a thermalization interval, each particle in a large group carries a continuously disappearing memory of the last, say, few million mutual collisions. Such a collective is called a *quasiparticle.* The main idea is that static friction with other particles creates a self-consistent collective motion which can still be described *on the average* as the motion of many single particles, possibly with a changed (effective) value of the mass. For liquid helium, the formula for e_i even becomes linear in momentum, so the quasiparticles are sound waves in that case. Particles are distributed in $6N$-dimensional phase space, quasiparticles in real three-dimensional (laboratory) space. Particles are sometimes called "bare," while quasiparticles are "dressed" particles.

Now, we can look at Boltzmann's efforts to describe the approach to equilibrium with new eyes. Collisions drive the system to equilibrium, which is characterized by the Maxwell-Boltzmann distribution. It is stationary, meaning that the collision integral is zero. Does that mean that collisions between particles have stopped? No, but the collisions between quasiparticles have stopped. The quasiparticles whose scattering drives the system toward equilibrium are collective departures from the equation of state (4.76), whose number decreases as equilibrium is approached. The microscopic motion of individual particles gives no sign of belonging to such a collective, any more than it can tell us whether time is going forward or backward. It is only in a very rarefied gas that the quasiparticles can be identified with individual particles, which is the limit Boltzmann considered. While the identification is physically correct in that limit, it obscures the nature of the approach to equilibrium so that it is difficult to extend it to real dense and interacting liquids, which was Landau's achievement with the quasiparticle concept.

The expression (4.78) is constrained to the microscopic form in the dilute limit. The success of the quasiparticle concept is measured by its evolution, with increasing density, as the self-consistent solution of the equation of state (4.76). If it retains the same form, with only the mass evolving as a function of thermodynamic parameters, the concept is evidently successful, which is why the similarity with the microscopic expression is not accidental.[6] If it changes form, say becoming linear in momentum, the outcome perhaps fails our expectations, but the concept survives as long as the total energy can be expressed as a sum of one-body contributions.

All observations are macroscopic. When we say that we see individual atoms with a scanning tunneling microscope, we actually see some colored pattern on a screen. (An art student would see nothing else.) The currents in the instrument and the currents they trigger in the screen are all macroscopic. In scattering experiments, the target is a source of macroscopic dissipation for the macroscopic beam current. We cannot perceive a single particle in phase space, only masses of quasiparticles in real space. Aristotelians (and Thomists) would recognize Landau's distinction between particles and quasiparticles as one between essence and existence.

4.6 Equipartition of Energy

Statistical physics sets out to rationalize the generality of thermodynamic relations. It achieves that by interpreting them as physical manifestations of general statistical relations, in particular between the norm, first, and second moments of distributions, which are interpreted as thermodynamic potentials, state variables, and susceptibilities, respectively. Sometimes, the statistical results are so general that we seem to be getting too much of a good thing. We saw above how the Maxwell-Boltzmann distribution arises in such a way that it is difficult to see how there can be any exception to it, even if it makes everything look awkwardly like an ideal gas.

The equipartition of energy is another result in the same vein. Assuming that the energy of a microscopic degree of freedom depends on some parameter s quadratically ("quadratic degree of freedom"), $\varepsilon(s) = s^2$, one can calculate its contribution to the internal energy quite generally:

$$\langle \varepsilon \rangle = \frac{\int_{-\infty}^{\infty} s^2 e^{-s^2/kT} ds}{\int_{-\infty}^{\infty} e^{-s^2/kT} ds} = \frac{kT}{2}, \tag{4.79}$$

a result known as the equipartition theorem. Applying it to the classical ideal gas, which has $3N$ such independent terms in the kinetic energy, the equation of state $U = 3NkT/2$ follows. If we imagine a monoatomic crystal as balls connected by springs, each direction in space contributes a one-dimensional harmonic oscillator per atom, so $U = 3NkT$ because the oscillator has two quadratic terms, one kinetic and one potential. Therefore, the heat capacity of the crystal should be $3R$ per mole. This prediction is called the law of Dulong and Petit.

Recall that a constant heat capacity contradicts the definition of temperature itself, Eq. (2.20). Nevertheless, the Dulong-Petit law is observed around room

[6] In liquid gadolinium, the measured atomic effective mass is about 1.65 times larger than in a gas.

temperature for many elements, such as iron and cobalt, but not others, such as diamond, whose heat capacity in addition depends on temperature. Experimentally, we have sporadic success of a law based on general statistical reasoning, while the law itself is in contradiction with thermodynamics, which that reasoning was supposed to explain.

If this situation does not seem puzzling enough, there is worse to come. For a diatomic molecule, such as carbon monoxide, it is easy to count the quadratic degrees of freedom. Three are due to translation of the center of mass, the same as in a monoatomic gas. Two are due to mutual rotation, like longitude and latitude. Two are due to vibration along the axis connecting the two atoms, like a one-dimensional harmonic oscillator. The total, seven, implies a heat capacity of $7R/2$ per mole, but only five are observed in diatomic gases at room temperature.

This failure is of the worst possible kind, because it is in the wrong direction. We already counted 7 quadratic degrees of freedom, so that is the *least* number we expect to observe! Any further analysis of the molecule can only increase it, undermining the reasoning even more.

Normally, if a theory works in some cases, its failure in others is not alarming. The specific problem with the equipartition theorem is that its generality allows no exception at all. Perhaps it appears less convincing in crystals, because we have skipped a discussion of the classical harmonic approximation, but the Dulong-Petit law is confirmed quite often there. Conversely, failure of equipartition in diatomic molecules is spectacular, because that is the simplest application beyond the monoatomic gas.

We have already remarked that the vanishing of heat capacities of all substances at absolute zero is a quantum effect. In order to understand it, and to establish the domain of validity of the equipartition theorem, we turn to quantum statistical physics.

Chapter 5
Quantum Statistical Physics

Natural science does not simply describe and explain nature; it is part of the interplay between nature and ourselves; it describes nature as exposed to our method of questioning.

Werner Heisenberg

From its beginnings, thermodynamics gave hints that the laws of motion of microscopic particles were not in accord with Newton's mechanics, even if two centuries passed before these discrepancies became clear in researchers' minds. The earliest such observation was the existence of an absolute zero of temperature. Next, Carnot's analysis of the cyclic process indicated that heat capacities should vanish at absolute zero, in apparent contradiction with the molecular hypothesis. More precise attempts at a statistical description encountered the problems of extensivity and heat capacity of diatomic gases. Even though extensivity can be explained by diffusive motion at room temperature, absolute zero and the behavior of the heat capacity can only be understood on the basis of quantum mechanics.

The fundamental characteristic of quantum mechanics is the appearance of absolute scales that may be expressed by way of Planck's constant. The size of the hydrogen atom can be used to measure length without invoking arbitrary units. Quantum mechanics reduces to classical in the limit of high temperatures, but does not cease to be valid, because Planck's constant is never zero. Because temperature is an emergent quantity, as well as entropy, one cannot say what temperature is high in an absolute sense, but only depending on the system under consideration. Thus, room temperature is high for the molecules in air, but low for electrons in metals.

Quantum laws of motion can be expressed in several equivalent ways. In order to explain thermodynamic phenomena, we shall always use the most direct one. Qualitative considerations will prove sufficient to understand observations. Just as it was previously sufficient to know that classical states of motion are points in phase space, so it will now be sufficient to know that quantum states of motion, so-called wave functions, are vectors in a countably dimensional vector space over complex-number coefficients (scalars).

D. Sunko, *Conceptual Thermodynamics*,
https://doi.org/10.1007/978-3-032-20877-4_5

5.1 Indistinguishability

Microscopic particles cannot be marked arbitrarily, so that particles of the same kind are indistinguishable in principle: To mark a particle is the same as to change its kind. In the formalism, this property is expressed by the rule that the wave function remains proportional to itself when two particles exchange places, i.e., no new states of motion appear:

$$\Psi(r_1, r_2, \ldots, r_i, \ldots, r_j, \ldots) = c\Psi(r_1, r_2, \ldots, r_j, \ldots, r_i, \ldots). \tag{5.1}$$

Because two such exchanges lead to identically the same function, one must have $c^2 = 1$, so $c = \pm 1$. Both of these possibilities exist in nature. Particles for which the sign is positive are called bosons, those for which it is negative, fermions.

If the sign is negative, and the coordinates of two particles are *numerically* identical, $r_i = r_j = R$, one has

$$\Psi(r_1, r_2, \ldots, R, \ldots, R, \ldots) = -\Psi(r_1, r_2, \ldots, R, \ldots, R, \ldots), \tag{5.2}$$

whence evidently $\Psi \equiv 0$, i.e., two fermions cannot be found in the same state. This constraint is general, because 'r' in the derivation did not necessarily refer to a coordinate, but to any complete description of the particle's state of motion. It is called the Pauli exclusion principle, which is a modern interpretation of the impenetrability condition. It is the main reason for the appearance of solid bodies, so that it indirectly explains traditional impenetrability as well, i.e., the mechanical resistance of solid bodies to mutual overlap.

Indistinguishability means that $N!$ points in the phase space of N particles are mutually equivalent, e.g.,

$$(p_1, q_1; p_2, q_2; \ldots) \sim (p_2, q_2; p_1, q_1; \ldots), \tag{5.3}$$

and similarly for all possible permutations. It is very difficult to visualize such a reduced space. Fock invented a formal apparatus which enables manipulation of many particles without bookkeeping errors. It is called "second quantization." It treats the whole set of equivalent points in phase space as a single state, which can be occupied or empty, i.e., either there are particles with the corresponding coordinates, or not. Fock's description is thus very similar to balls in boxes, where the boxes are distinguishable states of motion, and the balls are particles. One keeps account only of the occupation numbers of individual states, so that wrong counting is avoided automatically: One does not track *which* particles are in a given state, only how many there are. Pauli's principle is included by forbidding occupation numbers greater than one for fermions, while there is no such limitation for bosons.

5.1.1 Quantum Correlations

Indistinguishability introduces correlations among particles of the same kind even if there is no interaction. Imagine two bosons in two possible states, e.g., two photons with horizontal or vertical polarization. Obviously, either both will be polarized horizontally, or one horizontally and one vertically, or both vertically. If they are polarized at random, each of these three possibilities will have the same probability of being observed, thus $1/3$ (Fig. 5.1). Were the photons distinguishable, as a black and white ball, these probabilities would respectively be $1/4$, $1/2$ and $1/4$, because there would be two possibilities in the second case, depending on how which photon is polarized. Evidently,

$$\frac{1}{3} > \frac{1}{4}, \tag{5.4}$$

so the probability of observing two photons in the same state is measurably larger than if they were distinguishable particles. A classically thinking observer would get the impression that there is an interaction between the two polarizations that rotates them, like between two magnets, but that is not true. Quantum correlations are kinematic, not dynamic, in origin.

The relation (5.4) is in contradiction with the claim that the photon is somehow still marked, even if we do not know how. Such claims are called "hidden variable theories." Measurements show that these theories necessarily lead to discrepancies with the predictions of quantum mechanics, so there is no doubt that particles of the same kind are indistinguishable in principle.

If the particles are fermions in projection states "up" and "down," e.g., electrons in a single atomic suborbital, the probabilities are respectively 0, 1, and 0, because only one state of the two particles is allowed by the Pauli principle. Again, a naive observer would think that electrons of the same projection repulse each other, but the truth is that states which violate the Pauli principle *do not exist*. Kinematics defines the limits of the possible, i.e., the denominator of the fluctuation formula, while dynamics chooses within the possible in the numerator. Because of that, the Pauli principle is satisfied noiselessly, just like the conservation of momentum.

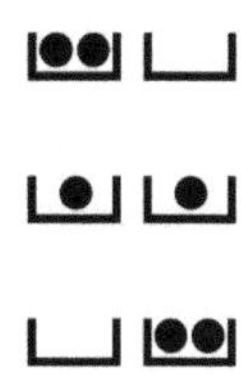

Fig. 5.1 Two bosons in two states

5.1.2 Connection Between Spin and Statistics

The exchange of particles was described combinatorially above, as a permutation of coordinate indices. That operation is allowed even if the system is one-dimensional, when exchange is physically impossible. Thus, the combinatorial model of exchange does not conform to physical reality. It also cannot tell us when to take indistinguishability of identical particles into account, even if we feel that it should not be necessary if some of these particles are in a laboratory sample and others, say, on the Moon. In cases when it is necessary, we say that the particles are *coherent*.

Around 1980, Feynman began disseminating an intuitive model of particle coherence: Two particles remain coherent by circling each other in real (laboratory) space, i.e., by exchanging places physically. In that model, it is obvious that there is no exchange in one dimension, and it is also clear that particles in the sample do not change places with those on the Moon during the observation, meaning that they are distinguishable in practice. A significant success of his model is an explanation of the observation that all fermions in nature have a half-integer spin $s = (2n + 1)/2$, while all bosons have integer spin, $s = n$, where $n = 0, 1, 2 \ldots$ is a natural number. Until then, there was no simple explanation of this so-called connection between spin and statistics.

When particles exchange places, that appears to one of them as if the other has rotated around its axis by a full circle (2π). A particle of spin $1/2$ can be found in two possible quantum states, conventionally called "up" and "down," imagining the spin like an arrow. This means that the vector space of quantum states of spin $1/2$ is two-dimensional, so it can be presented by a Cartesian coordinate system, spanned by two axes labeled "up" and "down."

The essential property of a spin $1/2$ is that the up and down *directions* are opposite in laboratory space, while the *axes* "up" and "down" are at right angles in state space. Because of that, rotation by 2π does not return the "up" state vector to its initial value, multiplying it by -1 instead (Fig. 5.2). Thus, the state vectors before and after rotation add to zero. That destructive interference is precisely the Pauli exclusion principle. It can be shown that all half-integer spins need two full rotations to return to the initial state, whence it follows that all particles with half-integer spins are fermions.

A particle of spin 1 has three possible quantum states, with arrow directions x, y, and z, that can accordingly be spanned by a three-dimensional Cartesian system. In that case, rotations in laboratory space are translated into state space point by point, according to geometric intuition, so a rotation by 2π in the laboratory leads to the same vector in state space. Vectors before and after rotation interfere constructively

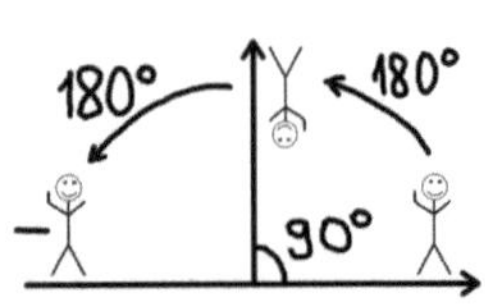

Fig. 5.2 Rotation of a spin $1/2$

now, so the probability of finding two particles in the same state is greater than for classical particles, in agreement with Eq. (5.4). All particles of integer spin behave similarly, so they are bosons.

5.2 Parametric and Laboratory Space

Feynman's interpretation of the spin-statistics connection points to the difference between the space of parameters that describe a physical system and the laboratory (real) space in which the system exists. We have already seen such a difference between the $6N$-dimensional phase space and three-dimensional real space of classical mechanics. It is not immediately clear that Hamilton's description of motion has substantial advantages over Newton's. For a small number of particles, that can appear as a matter of taste. However, ensemble theory indicates that many-particle systems exist as objects in the laboratory by randomly searching their whole available phase space, thus achieving stability in time. From that point of view, phase space is primary, while its reduction to possible solutions of the equations of state appears in laboratory space. By founding thermodynamics on ensemble theory, Gibbs shifts our understanding of physical reality away from modern rationalism, which begins with observations, towards Platonic idealism, which begins with invisible primary forms, without having to deny the physical reality of particles from which objects are composed. The essence of this shift is that the primary forms are not some things, but relations among things, specifically the mathematical structure of the space of solutions of the equations of motion. Because relations are by their nature invisible, truth is at the same time objective, knowable, and invisible.

Quantum mechanics is a step further in the same direction. While we can still try to imagine continuous phase space on the basis of geometric intuitions acquired in real space, we must develop a new intuition for the discrete vector space of quantum states of indistinguishable particles. Thus, the primary space of physical events becomes even more abstract. In the example above, the quantum state of a particle moves in the parameter space of the group of rotations while the particle rotates in real (laboratory) space.

There is direct experimental evidence of this interpretation. Two neutron beams are scattered, one of which has passed through a magnetic field, which rotates the neutrons in real space. As the field becomes stronger, the neutrons' rotation increases, until they rotate through eight full circles in the strongest field. However, at the same time, the intensity of interference between the two beams returns to its original value only four times: The neutrons must rotate through two full circles to appear the same again to those in the other beam!

5.3 Volume of Phase Space

Quantum states are discrete (countable) rather than continuous (uncountable), as suggested by the picture of classical phase space. This means that a finite volume (cell) of phase space corresponds to each quantum state in the classical limit, rather than a point. In other words, phase space has a natural unit of volume. We can find it easily if we accept that the constant h in the expression (3.23) is universal, so we can calculate it once and for all by the simplest example possible.

To wit, take a free particle in a one-dimensional box. The wave functions of such a particle are

$$\Psi_n(x) \sim \sin\left(\frac{n\pi x}{L}\right), \quad n = 1, 2, \ldots, \tag{5.5}$$

where L is the length of the box, and there appear discrete wave numbers

$$k_n = \frac{n\pi}{L}. \tag{5.6}$$

The connection between momentum and wave number is $p = \hbar k$, $\hbar = h/2\pi$. Negative momenta refer to the same state (only the sign of the wave function changes), so the volume element of classical phase space, belonging to the n-th state, is double that which belongs to the n-th positive momentum:

$$\Delta\Gamma = 2\Delta p_n L = 2\hbar(k_n - k_{n-1})L = 2\hbar\frac{\pi}{L}L = h, \tag{5.7}$$

therefore the natural unit h in the ratio (3.23) is precisely Planck's constant.

The absolute measure of phase-space size provides a criterion, when the classical description of a system is good enough. Namely, we expect a priori $W(N, E) = \Gamma(N, E)/h^{3N}$ cells, i.e., quantum states, in a volume $\Gamma(N, E)$ of phase space. If $W \gg N$, it is improbable that the classical principle of a priori equal probabilities will be at odds with the limitation that all points in a cell correspond to the same state: There is no "crowding" in phase space, so the classical limit is probably good enough, up to the indistinguishability of cells with permuted coordinates, which can be compensated dividing by $N!$. On the other hand, if $W \approx N$, we encounter radically different behavior, as a consequence of the indistinguishability of identical particles. For example, if $W = N$, a fermionic system has only one state of motion, the one in which each available quantum state is occupied by one particle.

5.4 The Gauge Degree of Freedom

Quantum states of motion have a degree of freedom which cannot be envisaged classically. It is the phase of the wave function, taken as a complex number. The choice of the origin of phase is called gauging, hence the name. The choice can be understood intuitively as the choice of a dateline on the surface of the Earth. Nothing depends on it physically, but if we have an external reference point, such as the direction of the Sun, we can move in such a way that the Sun stays directly overhead. When we return to the initial position, we will discover that 24 hours have passed without our seeing a single sunset. When Magellan's sailors disembarked on St. James' island in the Cape Verde archipelago after their voyage around the world, they thought it was Wednesday, but it was Thursday. In three years of sailing after the Sun, they saw exactly one sunset less than the locals.

Carnot's understanding of temperature (2.20) depends only on the reversibility of static friction, so it is not undermined by our changed understanding of the microscopic nature of that friction.[1] Thus, Gibbs' ensemble theory is equally valid in quantum as in classical theory, except that sums over states like (4.37) should be interpreted literally as sums now, because quantum states are discrete. The only new question is whether the gauge degree of freedom is thermalized, in which case states with different values of the phase origin should be counted as different. Specifically, if we imagine the phase as a hand on a clock face, the question is whether collisions with container walls push the dial around at random. The answer is that they do not, because of momentum conservation.

In quantum mechanics, the momentum is described by a wave number, which appears in the phase of a wave function of given momentum. The momentum operator is a coordinate derivative, meaning that the momentum of a state of motion depends on how the wave function changes with position:

$$\hat{p}e^{ikx} = -i\hbar\frac{d}{dx}e^{ikx} = \hbar k e^{ikx} = p e^{ikx}. \tag{5.8}$$

Let us mark an arbitrary phase origin (gauge) by adding a number α to the phase, so $i(\alpha + px/\hbar)$ appears in the exponent. The phase of a molecule m at the moment of collision with an atom a at position x_0 on the wall is single-valued, and can be written in two ways, either using quantities before, or those after the collision:

$$\hbar\alpha + p_m x_0 + p_a x_0 = \hbar\alpha' + p'_m x_0 + p'_a x_0. \tag{5.9}$$

If momentum is conserved, α does not depend on x_0, so that $\alpha = \alpha'$. Waves can push a ship around at random, but the date line does not move: The gauge degree of freedom does not thermalize.

For that reason, coherence of a macroscopic number of particles is possible at finite temperature. However, it needs an external reference point, like the direction of the Sun, so that the phases α of their wave functions can change synchronously. Macroscopic coherence then manifests itself as a current of particles without friction, i.e., Joule heating. Superfluidity and superconductivity are such manifestations.

[1] Planck's constant was not zero in Carnot's time either.

5.4.1 Example: The Meissner Effect

The explanation of the Meissner effect, published by Fritz London in 1937, is the first application of gauge symmetry to explain a physical phenomenon. He reduced it to an extremely simple one-electron problem that can be solved practically without calculation.

A piece of lead is shaped as an ellipsoid and cooled in a homogeneous magnetic field. At the transition into the superconducting state, there appears a pair of forces which turn the ellipsoid so that it shows its smallest cross-section to the field. The density of magnetic field lines near the ellipsoid increases, remaining unchanged far from it, so it is clear that the lead has ejected the field from its interior. It achieves that physically by a frictionless current induced on its surface that cancels the field inside the sample precisely, so that the material behaves as a perfect diamagnet.

Electric current is usually understood as the net effect of electron diffusion in an external field. It is limited by incoherent scattering on many microscopic degrees of freedom that do not conduct, such as impurities and sound waves. The transfer of energy to them is manifested as Joule heating. Given that electric conduction is so closely related to dissipation, it is not clear how the Meissner current can persist without contacts, even in the field of a permanent magnet. It is also not clear why it should cancel the external field so precisely. Why not half the field, say?

London asks himself first, how does a *single* atom manage to have a diamagnetic response. What does a single electron do in a single orbital state to shield the magnetic field? Magnetic fields are much too weak to change the electron's orbit, so they can change the shape of the wave function in the absence of a field, Ψ_0, only a little:

$$\Psi \approx \Psi_0 + \boldsymbol{B} \cdot \boldsymbol{B}\Psi_1 + \ldots, \tag{5.10}$$

where it is enough to stop at the first non-vanishing order, which is quadratic because the diamagnetic response does not depend on the field orientation. For a wave function coupled to a vector potential, $\boldsymbol{B} = \nabla \times \boldsymbol{A}$, the quantum current is in general

$$\boldsymbol{j} = \frac{\hbar c}{2iM}\left(\Psi^*\nabla\Psi - \Psi\nabla\Psi^*\right) - \frac{q^2}{Mc}|\Psi|^2\boldsymbol{A}. \tag{5.11}$$

When the wave function in the field is inserted in this expression, the first non-vanishing term is

$$\boldsymbol{j} = -\frac{q^2}{Mc}|\Psi_0|^2\boldsymbol{A} + O(\boldsymbol{A}^3), \tag{5.12}$$

because there is no current without field in a diamagnetic atom, $\Psi_0^* \nabla \Psi_0 - \Psi_0 \nabla \Psi_0^* = 0$. Comparing this result with the general connection between momentum and velocity in a field, the so-called minimal substitution,

$$\boldsymbol{p} = M\boldsymbol{v} + \frac{q}{c}\boldsymbol{A}, \tag{5.13}$$

it becomes clear what the atom is doing. It adjusts the phase of the wave function so that its current becomes "a kind of image of the actual magnetic field" (London's words), so the mechanical current $M\boldsymbol{v}$ cancels the contribution of the field precisely so that the canonical momentum $\boldsymbol{p}$ stays zero as it was in the absence of the field, when the quantum current had only a mechanical component, $\boldsymbol{p} = M\boldsymbol{v} = 0$. The physical essence of this screening is that the phase of the wave function is determined by the canonical momentum $\boldsymbol{p}$, not the mechanical momentum $M\boldsymbol{v}$. The difference of these two momenta compensates the external vector potential by a mechanical current. There is no dissipation because there are no atomic transitions: The current is due to one single quantum state.

Because all electrons see the same vector potential, like the direction of the Sun, London postulates the same outcome for the current, simply multiplying by their number, $|\Psi_0|^2 \to n_s$:

$$\boldsymbol{J} = -\frac{e^2 n_s}{Mc}\boldsymbol{A} =: -\frac{c}{\lambda_L^2}\boldsymbol{A}, \tag{5.14}$$

where $\boldsymbol{J}$ is now the macroscopic current density, n_s is the concentration of superconducting electrons, and the constant λ_L has the dimension of length. It is called *London's length*, or penetration depth. The vector potential is defined up to the value $\boldsymbol{B} = \nabla \times \boldsymbol{A}$, so the same equation is better written invariantly:

$$\nabla \times \boldsymbol{J} = -\frac{c}{\lambda_L^2}\boldsymbol{B}, \tag{5.15}$$

which in combination with Maxwell's equation $c\nabla \times \boldsymbol{B} = \boldsymbol{J}$ gives

$$-\lambda_L^2 \nabla \times \nabla \times \boldsymbol{B} = \lambda_L^2 \nabla^2 \boldsymbol{B} = \boldsymbol{B}, \tag{5.16}$$

whose solution is a function vanishing exponentially inside the sample, $B_x \sim \exp(-x/\lambda_L)$, which is the Meissner effect. Thus, the wave function shifts its phase to establish a noiseless mechanical current which screens the magnetic field completely, without being deformed itself, i.e., the net canonical momentum stays zero as in a free atom. The sample maintains the surface current just as the table supports the sample, without work or heat.

London's explanation of the Meissner effect is macroscopic, even if it appears microscopic. It gives no hint of the wave function which screens the field, nor how it is established, i.e., what is the mechanism of superconductivity, or of what happens physically during the transition itself. It appeals to gauge symmetry as to a conservation law: One knows that it cannot be violated, so one concludes that something must happen to satisfy it in the final state. For electromagnetic fields in the vicinity of a superconductor, London's equation (5.15) appears as a fifth Maxwell equation.

London's length can be measured directly. Thus, it is an important phenomenological parameter, which can be used to determine the concentration of superconducting carriers in specific circumstances.

5.5 Simple Quantum Systems

The microcanonical entropy of a quantum system of N particles with total energy E is

$$\sigma(N, E) = \ln W(N, E), \tag{5.17}$$

where $W(N, E)$ is the number of quantum states of that system with a given energy and particle number. All sums over states which we denoted schematically as such can now be taken literally as sums. In particular, the sum over states of a canonical ensemble is

$$Z_C(N, \beta) = \sum_i e^{-\beta E_i} = \sum_E W(N, E) e^{-\beta E}, \tag{5.18}$$

where the first sums is over all discrete quantum states, while the second is over all distinct values of the energy. Because the gauge degree of freedom is not thermalized, every quantum state enters the sum once, with an arbitrary value of the phase according to the assumption of equal a priori probabilities, barring some special preparation of the system.

Technically, indistinguishability complicates calculations with a given energy or particle number so much that they are viable only in simple cases. These are nevertheless useful to understand the specifics of quantum phenomena, so we begin with them.

5.5.1 *Arbitrary Number of States of the Same Energy*

Consider a fermion system first. If there are L available states of the same energy ε, each of which can take at most one particle, then the number of N-particle states is the number of ways to choose which N of L states are occupied, a total of

$$W(N, N\varepsilon) = \binom{L}{N} \tag{5.19}$$

states.

The analogous calculation for bosons is only slightly harder. A single N-particle configuration can be depicted:

$$|_1 \bullet \bullet |_2 \bullet |_3 \bullet \bullet \bullet \bullet |_4 \bullet \bullet \cdots |_L \bullet \bullet, \tag{5.20}$$

where lines delimit states, and dots are particles. Obviously, all allowed drawings begin with a line, and we can obtain them by deciding which of the remaining $L + N - 1$ lines and dots will be dots, which can be done in

$$W(N, N\varepsilon) = \binom{L+N-1}{N} \tag{5.21}$$

ways. In the limit when N and L are large numbers with a fixed ratio $n = N/L$, the entropy can be written by Stirling's formula:

$$\sigma = \ln W = L\left[-n \ln n \pm (1 \pm n) \ln(1 \pm n)\right], \tag{5.22}$$

where the upper sign is for bosons, and the lower for fermions. It is obvious that this entropy is extensive in L, the number of states available to one particle. In the classical limit, $n \ll 1$, both expressions reduce to the same,

$$\sigma/L \to -n \ln n. \tag{5.23}$$

When there is no crowding in state space, the difference between bosons and fermions disappears. That is understandable, because when the probability of occupation of any given state is small, the probability of double occupation is even smaller, while the difference between bosons and fermions appears only upon multiple occupation of the same state.

5.5.2 Two Particles Without Interaction

Imagine that each particle can have any energy in some rising sequence ε_i. These could be the kinetic energies of free particles in a box. If there is no interaction, the energy of two particles will be $\varepsilon_i + \varepsilon_j$, for given i and j. Because the particles are indistinguishable, every pair of indices (i, j) may be counted only once. If the particles are fermions, one must also have $i \neq j$. Both of these requirements are satisfied by the restriction $i < j$, so the canonical sum over states for two fermions without interaction is

$$Z_C(2,\beta) = \sum_{i<j} e^{-\beta(\varepsilon_i+\varepsilon_j)}. \tag{5.24}$$

For bosons, one should put $i \leq j$ instead of $i < j$, because two bosons can be in the same state. The sums with these restrictions can be rewritten in terms of sums without restrictions. A product is commutative, $f_i f_j = f_j f_i$, so we have

$$\left(\sum_i f_i\right)^2 = \sum_i \sum_j f_i f_j = \left(\sum_{i=j} + 2\sum_{i<j}\right) f_i f_j = \left(2\sum_{i\leq j} - \sum_{i=j}\right) f_i f_j, \tag{5.25}$$

whence

$$Z_C(2,\beta) = \frac{1}{2}\left[\left(\sum_i e^{-\beta\varepsilon_i}\right)^2 \pm \sum_i e^{-2\beta\varepsilon_i}\right], \tag{5.26}$$

where the upper sign is for bosons. The square is subject to two corrections, a rough division of the surface by 2 due to indistinguishability (multiplicative), and a fine adjustment of the diagonal because of the special status of doubly occupied states (additive). The multiplicative correction for N particles is $N!$. We can show that the additive one is negligible at high temperature. Namely, even if these sums are formally infinite for particles in a box, the number of terms in each sum is effectively controlled by the temperature, because of the exponential weight function. E.g., it is not probable that a molecule in the air we breathe will reach the third cosmic velocity. Therefore, the triangle in Fig. 5.3 becomes larger as the temperature rises. The states on the diagonal are the edge of the triangle, so as it becomes larger, the edge becomes negligible with respect to the surface. That is the reason why the difference between bosons and fermions disappears at high temperature.

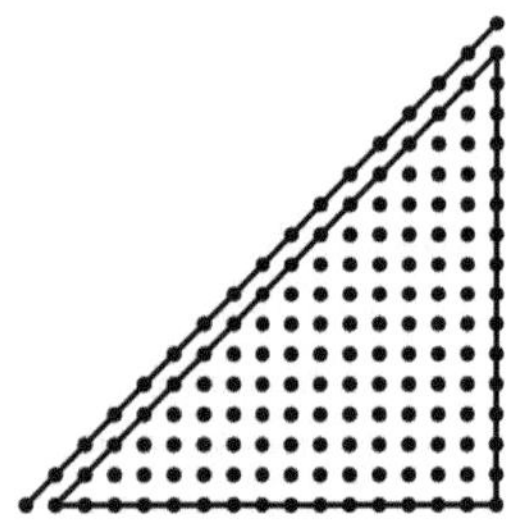

Fig. 5.3 States of two particles

For three particles, iterating Eq. (5.25) gives

$$Z_C(3,\beta) = \frac{1}{6}\left[\left(\sum_i e^{-\beta\varepsilon_i}\right)^3 \pm 3\sum_i e^{-\beta\varepsilon_i}\sum_j e^{-2\beta\varepsilon_j} + \right. \\ \left. + 2\sum_i e^{-3\beta\varepsilon_i}\right]. \tag{5.27}$$

The readers can draw the sixth of a cube that remains after the multiplicative correction, and mark the lines and surfaces corresponding to the additive one. Notably, the three terms in the expression (5.27) correspond to the three partitions of the number 3, respectively $3 = 1+1+1$, $3 = 1+2$, and $3 = 3$. One can show in general that every summand corresponds to one partition of the number of particles N, and there is a closed formula for the coefficients, like 3 and 2 above, called Faà di Bruno's formula.

The number of partitions $p(N)$ grows very quickly with N, so that $p(200) \sim 10^{10}$. Taking logarithms of such expressions is physically not transparent, because of their combinatorial structure. Thus, calculations with a given number of indistinguishable particles are not practical, so that only the grand-canonical formalism is used for them. Nevertheless, we can draw a general conclusion from the example above, that additive corrections due to indistinguishability are always negligible at high temperature, because they relate to the classical term as edges to a volume. The argument for that is both obvious and highly abstract. The dots in Fig. 5.3 are neither in phase space, nor in the space of quantum states as in Fig. 5.2, but in the space of quantum numbers, state labels that are in principle arbitrary, i.e., they can always be mapped onto sequences of natural numbers. However, even in that space, we found a geometric explanation of a physical phenomenon, the vanishing of the difference between bosons and fermions at high temperature. The key to that success was the ordering of quantum numbers on each axis according to increasing energy.

5.5.3 Two States of Different Energies

At most two fermions can be put in two quantum states with energies $\varepsilon_{1,2}$. If there is no interaction, the grand-canonical sum over states is

$$\begin{aligned} Z_G(\mu,\beta) &= \sum_{N=0}^{2} e^{\beta\mu N} Z_C(N,\beta) \\ &= 1 + e^{\beta\mu}\left(e^{-\beta\varepsilon_1} + e^{-\beta\varepsilon_2}\right) + e^{2\beta\mu} e^{-\beta(\varepsilon_1+\varepsilon_2)} \\ &= \left(1 + e^{\beta(\mu-\varepsilon_1)}\right) \times \left(1 + e^{\beta(\mu-\varepsilon_2)}\right) \quad \text{(fermions)}. \end{aligned} \tag{5.28}$$

The factorization of the sum succeeded because the two states are occupied independently of each other. We can read the sign + as a logical or, and × as and, saying: In the first state, there is no fermion or there is, and in the second, there is not or there is. When there is a particle, it contributes 1 to the particle number and ε_i to the energy. Setting $\varepsilon_1 = \varepsilon_2 = 0$ reduces this insight to a combinatorial interpretation of the coefficients 1, 2, 1 in the square of a binomial.

Having understood that, it is easy to write the sum over states for bosons. In each state, there can be an arbitrary number of bosons, such that m bosons contribute m to the particle number and $m\varepsilon_i$ to the energy. Thus, each factor is a geometric series:

$$1 + e^{\beta(\mu-\varepsilon_i)} + e^{2\beta(\mu-\varepsilon_i)} + \dots \tag{5.29}$$

One must have $\mu < \varepsilon_{1,2}$, otherwise some geometric series would diverge. Physically, that corresponds to a thermodynamic instability in which the system draws particles from the reservoir until it exhausts it. Assuming stability, we have

$$Z_G(\mu, \beta) = \frac{1}{1 - e^{\beta(\mu-\varepsilon_1)}} \times \frac{1}{1 - e^{\beta(\mu-\varepsilon_2)}}, \quad \mu < \varepsilon_{1,2} \quad \text{(bosons)}. \tag{5.30}$$

5.6 Sums over States for Free Particles

The example of free fermions in two states is easily generalized to an arbitrary number of states. The grand-canonical sum over states is the product of factors which correspond respectively to all states:

$$Z_G(\mu, \beta) = \prod_i \left(1 + e^{\beta(\mu-\varepsilon_i)}\right) \quad \text{(fermions)}. \tag{5.31}$$

For bosons, the same reasoning gives

$$Z_G(\mu, \beta) = \prod_i \left(1 - e^{\beta(\mu-\varepsilon_i)}\right)^{-1}, \quad \mu < \varepsilon_i \ \forall i \quad \text{(bosons)}. \tag{5.32}$$

It is easy to take logarithms of products, so we finally obtain an operative expression for the grand Massieu potential of indistinguishable free particles:

$$\Psi_G(\mu, \beta) = \mp \sum_i \ln\left(1 \mp e^{\beta(\mu-\varepsilon_i)}\right), \tag{5.33}$$

where the upper sign is again for bosons. The readers can obtain canonical sums over states for small given numbers of particles by taking derivatives of the expression $\exp(\Psi_G)$ with respect to the variable $\xi := e^{\beta\mu}$ (the *fugacity*), in accord with Taylor's theorem for the generating function (4.62).

5.6.1 Particle Number and Occupation Probability

When the number of particles is given, but formulas refer to the chemical potential, one must solve the equation of state which connects the expected number of particles with the chemical potential:

$$\langle N \rangle = \frac{1}{\beta} \left(\frac{\partial \Psi_G}{\partial \mu} \right)_{V,T} = \sum_i \frac{e^{\beta(\mu - \varepsilon_i)}}{1 + e^{\beta(\mu - \varepsilon_i)}} \quad \text{(fermions)}. \tag{5.34}$$

It can be interpreted as $\langle N \rangle = \sum_i \langle n_i \rangle$, where

$$\langle n_i \rangle = \frac{1}{e^{\beta(\varepsilon_i - \mu)} + 1} =: f_\mu(\varepsilon_i) \tag{5.35}$$

is the expected number of fermions in the i-th state. This expression is called the *Fermi-Dirac distribution.* It is directly analogous to the Maxwell one-particle distribution for the classical ideal gas. Because there can be only one or zero fermions in a state, this number is numerically equal to the probability p_i that the i-th state will be occupied: $\langle n_i \rangle = p_i \cdot 1 + (1 - p_i) \cdot 0 = p_i$. As a check, we can calculate this probability directly, as the probability of the corresponding fluctuation. The grand-canonical sum over states with the constraint that the i-th state is occupied is

$$e^{\beta(\mu - \varepsilon_i)} \prod_{j \neq i} \left(1 + e^{\beta(\mu - \varepsilon_j)} \right), \tag{5.36}$$

and when it is normalized by the unconstrained sum (5.31), one obtains (5.35).

For bosons, the equation of state is

$$\langle N \rangle = \frac{1}{\beta} \left(\frac{\partial \Psi_G}{\partial \mu} \right)_{V,T} = \sum_i \frac{e^{\beta(\mu - \varepsilon_i)}}{1 - e^{\beta(\mu - \varepsilon_i)}} \quad \text{(bosons)}. \tag{5.37}$$

It can be interpreted as $\langle N \rangle = \sum_i \langle n_i \rangle$, where

$$\langle n_i \rangle = \frac{1}{e^{\beta(\varepsilon_i - \mu)} - 1} =: n_\mu(\varepsilon_i) \tag{5.38}$$

is the expected number of bosons in the i-th state. (Here, $\mu > \varepsilon_i$ would imply a negative particle number.) The expression (5.38) is called the *Bose-Einstein distribution.* The number of bosons $\langle n_i \rangle$ can be arbitrarily large, so the occupation probability differs from the expected number of particles. It can be calculated again as a fluctuation probability:

$$p_i = \frac{\left(e^{\beta(\mu-\varepsilon_i)} + e^{2\beta(\mu-\varepsilon_i)} + \ldots\right) \prod_{j \neq i} \left(1 + e^{\beta(\mu-\varepsilon_j)} + e^{2\beta(\mu-\varepsilon_j)} + \ldots\right)}{Z_G}$$

$$= e^{\beta(\mu-\varepsilon_i)} \frac{Z_G}{Z_G} = e^{\beta(\mu-\varepsilon_i)}, \quad \mu < \varepsilon_i \quad \text{(bosons)}. \tag{5.39}$$

5.6.2 Classical Ideal Gas

The grand Massieu potential can be written for both fermions and bosons as

$$\Psi_G(\mu, \beta) = -\sum_i \ln(1 - p_i), \tag{5.40}$$

where p_i is the quantum occupation probability,

$$p_i = \begin{cases} f_\mu(\varepsilon_i) & \text{(fermions)}, \\ e^{\beta(\mu-\varepsilon_i)} & \text{(bosons)}. \end{cases} \tag{5.41}$$

This expression is convenient for passing to the classical limit. There, the occupation probability of all one-particle states is small, $p_i \ll 1$, so the logarithm can be expanded to first order in $e^{\beta\mu}$:

$$\Psi_G(\mu, \beta) \to \sum_i p_i \to e^{\beta\mu} \sum_i e^{-\beta\varepsilon_i} = e^{\beta\mu} Z_C(1, \beta), \tag{5.42}$$

and the one-particle sum over states is V/λ_T^3, which gives precisely (4.65). The small occupation probability is consistent with a negative chemical potential. In that limit, the fermion occupation probability is also $f_\mu(\varepsilon_i) \to e^{\beta(\mu-\varepsilon_i)}$, so the difference between bosons and fermions disappears again. Recall that $e^{\beta(\mu-\varepsilon_i)}$ is the Maxwell-Boltzmann distribution (4.71), normalized to the concentration, which is therefore the classical limit of both occupation probabilities (5.41), as can be observed in Fig. 5.4.

This derivation of the equations of state of a classical ideal gas is the only theoretically exact derivation. That is the case because the multiplicative correction $N!$ due to indistinguishability survives in the classical limit, so it is not necessary to introduce it a posteriori as an effect of diffusion (which hinders the exchange of particles within a thermalization interval). However, the question remains open, is air at room temperature extensive because of diffusion, or because of the intrinsic indistinguishability of identical particles with loss of quantum coherence, as suggested by the derivation above. This question can be posed as a measurement. Avogadro's

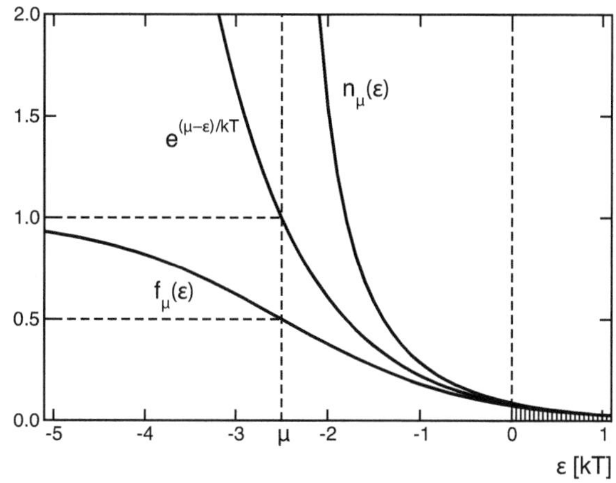

Fig. 5.4 Fermi-Dirac distribution (5.35), Maxwell-Boltzmann distribution (4.71), and Bose-Einstein distribution (5.38), with chemical potential in the classical regime $\mu = -2.5kT$. The shaded surface for physical values $\varepsilon > 0$ is the concentration, according to (4.72)

observation of extensivity in chemical reactions of *different* particles means that atoms bind into molecules with atoms nearby, so it indicates diffusion. It remains as a possible consequence of indistinguishability that a gas stays extensive in the limit of extremely low density, the so-called ballistic regime. One could imagine that in such a limit, quantum coherence due to kinematic exchanges of position would dominate thermal decoherence due to dynamic scattering. The competition of these two processes is a physical path by which indistinguishability crosses over continuously to distinguishability, even if they are logically incompatible concepts. However, we also know that rarefied gases in interstellar space are not extensive, but are subject to the gravitational instability discussed by Boscovich, which gives rise to star formation. We conclude that a physical regime in which a gas would be both classical and extensive due to indistinguishability, while not unstable because of long-range interactions, is narrower than might appear at first sight. Because of that, attempts to derive extensivity in the classical limit by deterministic arguments are more of specialist than of general interest.

We have noted already that extensivity is a side issue. The value of this derivation is that the conformity of the exact quantum expression with the empirical one at high temperature means that the theoretical calibration of the thermometer by the classical ideal gas remains valid when quantum effects are taken into account. This thought was first enunciated by Planck, when he commented on his description of blackbody radiation: Now, light can serve as a thermometer. The physical nature of temperature does not change with temperature, which was in essence understood already by Carnot.

5.7 Time in Quantum Mechanics

Time does not exist in quantum mechanics. The equation of motion of an autonomous system is the Schrödinger equation

$$\hat{H}\psi(\boldsymbol{r}_1, \ldots) = E\psi(\boldsymbol{r}_1, \ldots), \tag{5.43}$$

in which the wave function ψ depends only on the positions $\boldsymbol{r}_i = (x_i, y_i, z_i)$, $\hat{H}$ is the Hamiltonian in which the classical momentum variables are replaced by derivatives with respect to position, and E is the energy of the state described by the wave function, which does not change in time.

The wave function of a quantum system in contact with a macroscopic environment, described by classical equations of motion, satisfies the so-called time-dependent Schrödinger equation

$$\hat{H}\Psi(\boldsymbol{r}_1, \ldots, t) = i\hbar\frac{\partial\Psi(\boldsymbol{r}_1, \ldots, t)}{\partial t}, \tag{5.44}$$

in which time appears because it appears in Hamilton's equations of motion for the environment. The time derivative describes a one-way coupling of the quantum system to the classical environment, such that the latter influences the (small) quantum system, but the reverse influence is negligible. Put picturesquely, the system sees the clock on the wall, but the clock does not see it. In the special case when the time dependence is only in the phase of the wave function,

$$\Psi(\boldsymbol{r}_1, \ldots, t) = \psi(\boldsymbol{r}_1, \ldots)e^{-iEt/\hbar}, \tag{5.45}$$

the time-dependent equation (5.44) reduces to the fundamental one (5.43). That does not mean that the wave function of a real stationary state oscillates in time, but that the phase of a wave function is not a measurable quantity. The absolute square of the wave function $|\Psi|^2$, which represents the probability that the particle will be observed, is measurable, but the phase disappears in it. Thus, the function (5.45) represents a time-independent state of motion, even if it formally depends on time.

One can derive Eq. (5.44) from Eq. (5.43) by separating the quantum system into two parts, one of which is declared to be large and slow, so its coordinates and momenta conform to Hamilton's equations of motion, the only ones in which time appears. Two steps are crucial in this derivation. The first is to turn off quantum coherence between the large and small systems, and the second to claim that the same time variable t refers to all parts of the large system. We say that the small and large system are *separated adiabatically* by any physical procedure that is described by this derivation.

It follows from the above that time is by nature a different variable than position, so that it is an emergent variable in quantum mechanics, like temperature in thermodynamics. However, no one has succeeded so far to rationalize time at the same level as Carnot rationalized temperature by reversible fluctuations of heat. In the derivation above, time appears as an already formed variable in the classical equations of motion, so the step in which one could see how it comes into being is missing. Until such a step is found, the possibility remains open that time is just as fundamental a coordinate as position. Similarly, before Carnot one could have imagined that the temperature is also a fundamental field, like the electromagnetic one.

The theory of relativity throws an interesting light on this possibility. In it, the four variables (x, y, z, t) appear on an equal footing, apparently in contradiction with quantum mechanics. However, relativistic transformations conserve the quantity

$$c^2t^2 - x^2 - y^2 - z^2 = s^2, \tag{5.46}$$

the so-called invariant interval, where c is the speed of light. Just as the equation of a circle $x^2 + y^2 = r^2$ describes a family of one-dimensional objects in the two-dimensional plane, preserving the quantity r^2, so the existence of invariant intervals means that the four variables (x, y, z, t) have only three degrees of freedom together. Therefore, even if there is a formal disagreement between quantum mechanics and the theory of relativity, they agree physically that there are only three independent variables, except that time does not exist in quantum mechanics, while it is not independent in relativity. From the point of view of quantum mechanics, time could have appeared in the early universe when the energy density became low enough to enable adiabatic separation, with parts of the universe becoming clocks for other parts, with which they were not coherent. Hence, the possibility also remains open that an explanation of the emergence of time will also explain the constraint (5.46).

5.8 Coherence and Dissipation

Fixing a moment in time is thermodynamically irreversible, and connected with dissipation, which destroys quantum coherence. Conversely, coherence effects are timeless: The interference pattern of a slit builds up even with a very low-intensity beam, but if we detect the moments when a particle passes through the slit, we will destroy the interference. This observation opens the interesting question, how small can a clock be for the time it measures to have a macroscopic physical meaning.

The smallest quantum system in which this question can be analysed is a system of two states, among which transitions are possible. Specifically, the ammonia molecule NH_3 has the form of a pyramid, with the nitrogen atom at the apex. The three hydrogens are in the same plane, so there are two nitrogen states, above and below the plane. However, there is also a possibility for the molecule to invert itself, like an umbrella, because of which the states "above" and "below" are not stationary. Stationary states can be obtained by solving the time-independent Schrödinger equation, and their wave functions are symmetric (ψ_I) and antisymmetric (ψ_{II})

combinations of the states "above" (ψ_1) and "below" (ψ_2), with a small difference in energy $\Delta E = E_{II} - E_I > 0$:

$$\psi_{I,II}(\boldsymbol{r}) = \frac{1}{\sqrt{2}} \left[\psi_1(\boldsymbol{r}) \pm \psi_2(\boldsymbol{r}) \right]. \tag{5.47}$$

If we solve the same (stationary) problem with the time-dependent Schrödinger equation, the solution will be

$$\Psi_{I,II}(\boldsymbol{r}, t) = \psi_{I,II}(\boldsymbol{r}) e^{-itE_{I,II}/\hbar}, \tag{5.48}$$

whence it is clear that the frequency of these oscillations is fictitious, because we could have chosen the energy origin at will. Thus, our clock does not work.

Things become more interesting if we prepare the system in a non-stationary state, say ψ_1, at the moment $t = 0$ according to the clock on the wall, thus providing the time-dependent Schrödinger equation with an external reference point for the phase. Because the energy of the initial state ψ_1 is different than E_I or E_{II}, and the system is isolated by assumption, it cannot reach a stationary state, instead it is really physically oscillating between two non-stationary states now:

$$\Psi(\boldsymbol{r}, t) \propto \psi_1(\boldsymbol{r}) \cos \omega t - i\psi_2(\boldsymbol{r}) \sin \omega t, \tag{5.49}$$

where the proportionality factor contains the "extra phase" that can be removed by choosing the origin of energy. On the other hand, $\omega = \Delta E/2\hbar$ is the real physical frequency of this oscillation, which can be detected by preparing the system repeatedly and measuring its state at different times. The probability of finding it in the state ψ_1 oscillates according to

$$P_1(t) = \cos^2 \omega t, \tag{5.50}$$

which never ceases, if the system stays isolated. Thus, the clock works.

Unfortunately, it does not work well, because it is difficult to isolate the system. After some time, it will lose all the energy it can into the environment, and come to rest in the stationary state of lowest energy E_I, the so-called ground state. Now, the time dependence of such a damped process is known, it is an exponential reduction of probability

$$P(t) = e^{-\alpha t}, \tag{5.51}$$

with some characteristic constant $\alpha > 0$, which is essentially the coefficient of dynamic friction. Notably, only the expression (5.51) measures the macroscopic time of irreversible events since the system was created in the initial state. The expression (5.50) measures the reversible classical Hamiltonian (microscopic) time since the moment of maximal coherence according to the clock on the wall, which is periodically repeated because of the oscillation induced by the preparation of the system. Finally, the time in the stationary expression (5.48) is fictitious, in accord with the wave function not being physically observable.

Now comes the most interesting question: How does reversible time become irreversible? How can the dependence $P_1(t)$ cross over to $P(t)$, when one of them is quadratic for short times, while the other is linear? That question was studied by Khalfin in the 1950s, continuing upon the work of Krylov and Fock from the 1940s. The answer is, it cannot immediately. When a coherent quantum system coupled to a dissipative thermostat is modelled carefully, the classical exponential solution is valid only for times which are neither too short nor too long. For short times, there is always a finite discrepancy between the parabola and the line. For very long times, the exponential $P(t)$ crosses over into a slower decay, which means physically that the system is intermittently refreshed by rare events in which it absorbs energy from the environment. Nevertheless, classical dissipation (5.51) dominates most of the time, so it is compatible with quantum mechanics at the level of coarse observations. The fundamental incompatibility between the parabola and the straight line for short intervals did not attract the attention of the scientific community at the time.

5.8.1 *Bell's Theorem*

In 1964, Bell rediscovered the incompatibility between the parabola and the line in the much more interesting context of the so-called Einstein-Podolsky-Rosen (EPR) paradox. While Khalfin's result depended on technical assumptions that limited its scope, Bell's reasoning is obviously general, and there is no doubt that it reflects a fundamental property of quantum mechanics.

The simplest (Bohm-Aharonov) version of the EPR paradox considers the decay of a particle of spin 0 into two particles of spin 1/2. Uncorrelated spin states of two particles, with that of particle 1 tilted at an angle α with respect to an arbitrary axis, and that of particle 2 at an angle β, are described by the wave function

$$\Psi_{\alpha\beta} = \begin{bmatrix} \cos \alpha/2 \\ \sin \alpha/2 \end{bmatrix}_1 \begin{bmatrix} \cos \beta/2 \\ \sin \beta/2 \end{bmatrix}_2 . \tag{5.52}$$

Here, we have presented spin states as vectors in a two-dimensional Cartesian plane spanned by the axes "up" and "down," separately for each particle.[2] In the two-particle state created by the decay, the particles are kinematically correlated by angular momentum conservation, such that both their total spin, and its projection on an arbitrary axis, must be zero. Such a two-particle state is described by an antisymmetric spin function

$$\frac{1}{\sqrt{2}}(\Psi_{0\pi} - \Psi_{\pi 0}) = \frac{1}{\sqrt{2}}\left(\begin{bmatrix} 1 \\ 0 \end{bmatrix}_1 \begin{bmatrix} 0 \\ 1 \end{bmatrix}_2 - \begin{bmatrix} 0 \\ 1 \end{bmatrix}_1 \begin{bmatrix} 1 \\ 0 \end{bmatrix}_2 \right) . \tag{5.53}$$

[2] Half-angles appear for the reason visible in Fig. 5.2.

Many-particle states whose wave function cannot be factorized across the individual particles are called *entangled*. The probability that a measurement on this state will find particle 1 in a state tilted by an angle α, and particle 2 by β, is given by the square of the corresponding probability amplitude:

$$P_{\alpha\beta} = \frac{1}{2} \left| \Psi_{\alpha\beta} \cdot (\Psi_{0\pi} - \Psi_{\pi 0}) \right|^2 = \frac{1}{2} \sin^2 \frac{\alpha - \beta}{2}, \tag{5.54}$$

where the scalar product $\cdot$ acts separately in the space of each particle. The arbitrary axis has disappeared in the difference $\alpha - \beta$, so we have obtained the quantum prediction of observing these two states at the detectors. The probability of observing them in the same orientation $\alpha = \beta$ is zero, because the projections along *any* axis must be opposed when the total projection is zero. Conversely, the probability is maximal for opposed states $\alpha = \beta \pm \pi$, for the same reason.

EPR pointed out that this apparently innocuous result is incompatible with the idea that the physical events at the two detectors are independent, no matter how large the distance between them. The probability depends only on the relative orientation of the two detectors, which either of them can influence at will. Thus, the question is how the detectors stay correlated. If the preparation of the experiment somehow informs them of the absolute orientation, which is absent from the calculation, then quantum mechanics is not a complete theory. If quantum mechanics is complete, then such a prediction of correlations over great distances means that it is not a local theory.

Bell's Theorem
No local theory can reproduce the prediction of quantum mechanics.

Proof In a local theory, the probability of events at distant detectors a and b is a probability of independent events, $P_a(\lambda, \alpha) P_b(\lambda, \beta)$, where we have allowed a dependence on hidden variables λ that would be introduced by a more complete theory. If P_a and P_b are different functions, the quantum prediction

$$P_a(\lambda, \alpha) P_b(\lambda, \alpha) = 0 \quad \forall \alpha \tag{5.55}$$

is a zero of some function $F(\lambda) := P_a P_b$, so inserting the same solution $\lambda = \lambda_0$ into both functions P_a and P_b is a collusion of two distant detectors, which is at odds with locality. Thus, P_a and P_b are the same function, which means physically that the detectors are constructed to give the same signal independently when they measure the same thing. That is of course reasonable, but then we have

$$P_a(\lambda, \alpha)^2 = 0 \quad \forall \alpha, \tag{5.56}$$

which is a contradiction, because the probability must be different from zero somewhere. □

Note how the requirement "$\forall\alpha$" undermines locality. The quantum state of two entangled particles is correlated with respect to any axis, which the factorized expression cannot describe, because a product is not a difference. It is easy with a difference: $\alpha - \beta = 0$ for an arbitrary $\alpha = \beta$.

If the other detector is slightly tilted away from the position of maximal correlation for a given $\alpha = \alpha_0$, $\beta = \alpha_0 + \epsilon$, the quantum probability changes quadratically in ϵ:

$$\sin^2 \frac{\epsilon}{2} - \sin^2 \frac{0}{2} \propto \epsilon^2, \tag{5.57}$$

while the locally factorized probability generally changes linearly in ϵ:

$$P_a(\lambda, \alpha_0) P_a(\lambda, \alpha_0 + \epsilon) - P_a(\lambda, \alpha_0) P_a(\lambda, \alpha_0) \propto \epsilon. \tag{5.58}$$

Thus, if we attempt to outwit Bell's theorem by adjusting $\lambda = \lambda_0$ in the local expression to reproduce the quantum prediction for a given position of the detectors, as soon as one is slightly tilted, there will appear a qualitative difference which cannot be arbitrarily small. Bell had originally proven his theorem in this way: He showed that the quantum correlation function was stationary around its maximum, while the local one was linear. This incompatibility between a parabola and a line is essentially the same as was studied by Khalfin.

5.8.2 *Bell's Inequalities*

Bell's theorem proves the incompatibility of local theories with quantum mechanics, but allows that one of them may be correct and quantum mechanics wrong. In that case, the expected outcome of measurement on two detectors would simply be the product

$$P(\kappa, \lambda, M) = \sum_{A,B} AB P_A(\kappa, M) P_B(\lambda, M) = \bar{A}(\kappa, M) \bar{B}(\lambda, M), \tag{5.59}$$

where A and B are the measured values. Assuming that their range is finite, we can normalize measurement outcomes without further loss of generality:

$$|\bar{A}| \leq 1, \quad |\bar{B}| \leq 1. \tag{5.60}$$

The variables κ and λ are known to only one detector, while M are events in the common past. For example, that is how books are correlated when they are printed in two towns κ and λ on the basis of the same template M. Locality is not violated if it is sufficient to agree in advance.

Bell's inequalities are measurable consequences of the expression (5.59), which therefore must be respected by any local theory, but are not respected by quantum mechanics. Locality thus becomes subject to experimental confirmation. While Bell's theorem undermines locality (5.59) by a mathematical contradiction with quantum mechanics, Bell's inequalities undermine it by a physical contradiction with measurements.

Indeed, Eq. (5.59) implies

$$\begin{aligned} P(\kappa, \lambda, M) \pm P(\kappa, \lambda', M) &= \bar{A}(\kappa, M)\left[\bar{B}(\lambda, M) \pm \bar{B}(\lambda', M)\right], \\ P(\kappa', \lambda, M) \pm P(\kappa', \lambda', M) &= \bar{A}(\kappa', M)\left[\bar{B}(\lambda, M) \pm \bar{B}(\lambda', M)\right], \end{aligned} \tag{5.61}$$

where we have introduced the towns κ' and λ'. Because of (5.60), we have

$$\begin{aligned} |P(\kappa, \lambda, M) \pm P(\kappa, \lambda', M)| &\le \left|\bar{B}(\lambda, M) \pm \bar{B}(\lambda', M)\right|, \\ |P(\kappa', \lambda, M) \pm P(\kappa', \lambda', M)| &\le \left|\bar{B}(\lambda, M) \pm \bar{B}(\lambda', M)\right|, \end{aligned} \tag{5.62}$$

Because the right-hand sides are the same, when we flip signs in the second expression, we get

$$|P(\kappa, \lambda, M) \pm P(\kappa, \lambda', M)| + |P(\kappa', \lambda, M) \mp P(\kappa', \lambda', M)| \le 2, \tag{5.63}$$

again because of (5.60). Furthermore, of all things in the town κ, only some are important for measuring the quantity A, such as polarizer orientations. Call them a, while the others (irrelevant) are $\tilde{\kappa}$:

$$\kappa = (a, \tilde{\kappa}). \tag{5.64}$$

We have to verify that averaging over the irrelevant variables will not spoil the inequality (5.63), before we can focus on the relevant ones. That is easy, because the absolute value of an average cannot exceed the average of an absolute value:

$$|\bar{P}(a, b, M) \pm \bar{P}(a, b', M)| \le \overline{|P((a, \tilde{\kappa}), (b, \tilde{\lambda}), M) \pm P((a, \tilde{\kappa}), (b', \tilde{\lambda}'), M)|}. \tag{5.65}$$

Thus, the inequality (5.63) becomes

$$|P(a, b, M) \pm P(a, b', M)| + |P(a', b, M) \mp P(a', b', M)| \le 2, \tag{5.66}$$

where we have removed the bars denoting averages from P to lighten the notation. The expression (5.66) is called Bell's inequality. We show now that the quantum probability (5.54) violates it significantly, which enables experimental comparisons of the two predictions. Decades of ever more refined measurements have shown both that Bell's inequality is violated, and that the predictions of quantum mechanics are borne out in all such cases.

Bell's inequality is usually checked by measurements of polarized photons, not fermions. For them, tilting the polarizer in real space amounts to changing the polarization state by the same angle, so we shall replace half-angles in Eq. (5.54) with angles from now on, but that does not affect the argument. The probability that both detectors measuring the state (5.53) will register a photon is accordingly

$$p_{++} = x \cdot \frac{\Omega}{4\pi} \cdot \frac{1}{2} \sin^2(\alpha - \beta), \tag{5.67}$$

where x is the probability that the photon pair (5.53) was successfully produced, and Ω is the space angle subtended by the detectors at the source. The probability that one photon will be detected, and the other not, amounts to tilting the polarization of one detector by 90°:

$$p_{+-} = p_{-+} = \frac{x\Omega}{4\pi} \cdot \frac{1}{2} \cos^2(\alpha - \beta), \tag{5.68}$$

and similarly

$$p_{--} = \frac{x\Omega}{4\pi} \cdot \frac{1}{2} \sin^2(\alpha - \beta) + x\left(1 - \frac{\Omega}{4\pi}\right) + (1 - x), \tag{5.69}$$

where we have added the probabilities that both photons miss the detectors, and that production was not successful. Introduce the variables $A = \pm 1$ for detection (or not) on one detector, and $B = \pm 1$ on the other. Then the quantum calculation of the expectation value of the product AB, analogous to the local one (5.59), reads

$$P(\alpha, \beta) = \sum_{A,B} AB\, p_{AB} = 1 - \frac{x\Omega}{4\pi}\left[1 + \cos 2(\alpha - \beta)\right], \tag{5.70}$$

which should be compared with Bell's inequality (5.66), inserting $a = \alpha$ etc., while $M = (x, \Omega)$ refer to the common history. One obtains

$$\begin{aligned} |P(\alpha, \beta) - P(\alpha, \beta')| + P(\alpha', \beta) + P(\alpha', \beta') - 2 = \\ = \frac{x\Omega}{4\pi}\big\{|\cos 2(\alpha - \beta) - \cos 2(\alpha - \beta')| \\ - \cos 2(\alpha' - \beta) - \cos 2(\alpha' - \beta') - 2\big\}. \end{aligned} \tag{5.71}$$

The expression in curly braces can be positive, not by a little. Choosing $\alpha = 0$, $2\beta = \pi/4$, $2\alpha' = -\pi/2$ and $2\beta' = 3\pi/4$, we get

$$\{\cdots\} = \frac{1}{\sqrt{2}} + \frac{1}{\sqrt{2}} + \frac{1}{\sqrt{2}} + \frac{1}{\sqrt{2}} - 2 = 2(\sqrt{2} - 1) = 0.828\ldots \tag{5.72}$$

Therefore, quantum mechanics violates Bell's inequality significantly, and observation of such behavior means that local theories are in contradiction with measurements.

5.8.3 The Logic of Bell's Inequality

If x is a random variable, the probability $P_{\mathcal{A}}$ of the event $x \in \mathcal{A}$ can be understood as the normalized size of the subset $\mathcal{A}$ of the set of all possible events $\mathcal{U}$:

$$P_{\mathcal{A}} = \frac{|\mathcal{A}|}{|\mathcal{U}|}, \tag{5.73}$$

where the notation $|\cdot|$ means set size. Denote the statement $x \in \mathcal{A}$ by a. Then the expectation value $P(a, b, M) =: P(a, b)$ in Bell's inequality (5.66) is the probability of the overlap $\mathcal{A} \cap \mathcal{B}$, the simultaneous occurrence of independent events a and b. With that interpretation, $P(a, a) \equiv 1$, so the inequality (5.66) can be rearranged:

$$|P(a,b) - P(a,b')| \leq P(b,b) - P(a',b) + P(a',a') - P(a',b'), \tag{5.74}$$

or in set notation

$$|\mathcal{A} \cap \mathcal{B} \setminus \mathcal{A} \cap \mathcal{B}'| \leq |\mathcal{B} \setminus \mathcal{A}'| + |\mathcal{A}' \setminus \mathcal{B}'|, \tag{5.75}$$

where $\setminus$ is the difference of sets. Figure 5.5 depicts this expression. We see that Bell's inequality reduces to the trivial statement that a subset is not bigger than the set. If memberships of the various sets in Fig. 5.5 are independent events, nothing can violate it. This assumption is encoded graphically by equal surfaces meaning

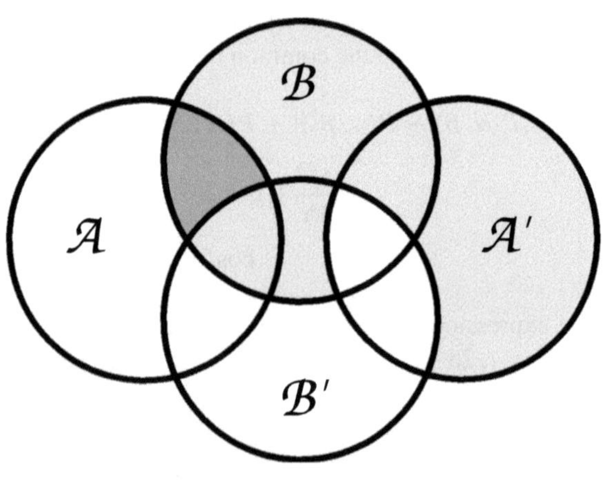

Fig. 5.5 The logic of Bell's inequality. The dark surface corresponds to the left-hand side of the inequality (5.75), while the dark and light surfaces together correspond to the right-hand side

equal probabilities. If measurements are correlated, we cannot replace $P(\mathcal{A}|\mathcal{B})$ with $P(\mathcal{A})$ in the general expression for a joint probability,

$$P(\mathcal{A} \cap \mathcal{B}) = P(\mathcal{A}|\mathcal{B})P(\mathcal{B}), \tag{5.76}$$

which means that events on one detector depend on events on the other, precisely what EPR disputed. In that case, some surfaces in the figure can have more or less weight, so Bell's inequality can be violated.

It is easier to understand the physical meaning of Bell's inequality if we put $\mathcal{A} = \mathcal{B}$. Then, the expression above becomes

$$|\mathcal{B} \setminus \mathcal{B}'| \leq |\mathcal{B} \setminus \mathcal{A}'| + |\mathcal{A}' \setminus \mathcal{B}'|. \tag{5.77}$$

Inserting the set $\mathcal{A}'$ can only increase the left-hand side. The reason is that it forces an additional determination: Either $x \in \mathcal{A}'$, or not. If we find a printing error whose probability is ϵ in a book that was printed in two towns, it occurred either in one town, or the other, with total probability 2ϵ. Bell's inequality then reads $\epsilon < 2\epsilon$. If the printing presses are quantum-entangled, the same error always occurs in both, with probability ϵ^2, even when the books are printed simultaneously, in presses that are so far away that one cannot communicate even by the speed of light where the error occurred in one, before it appears in the other. The violation of Bell's inequality then reads $\epsilon > \epsilon^2$. It is no wonder that Einstein thought that impossible, it is a wonder that it happens.

5.9 Logical Entropy

Correlations between distant detectors, which violate Bell's inequality, are evidence that entangled particles are indistinguishable in principle. The observation of such an effect between photons irrefutably proves that it is kinematic, because there is no interaction between photons. In the discussion accompanying Fig. 5.1, we obtained constructive interference (5.4) directly from indistinguishability, by counting on fingers, without wave functions. We also know that quantum probabilities like (5.54) are also probabilities, in principle subject to the definition (5.73). The question therefore arises whether they could be calculated in general like any other probability, by counting something, rather than squaring wave functions.

As long as the equations of motion of quantum mechanics are written in terms of wave functions, this is not entirely possible. However, classical statistical physics already shows that detailed knowledge of dynamics is rarely necessary. Thermodynamic potentials arise when deterministic trajectories are permanently destroyed by static friction. Entropy is in the denominator of the statistical formula for the probability of fluctuation. It is a quantity a posteriori, the product of the past in which all trajectories that could have been destroyed have already been destroyed.

We need dynamics to calculate the probability of fluctuations in the numerator, or response function, by which we test our understanding of measurements.

Quantum mechanics does not change the essence of this insight, only instead of destruction of trajectories, decoherence appears, i.e., measurement of the wave function. In the sample under observation, static friction has already measured all the wave functions that it could measure, in other words, reduced them to probabilities. Therefore, the quantum formula for entropy is formally identical to the classical one, except that it is summed over quantum states. It preserves quantum coherence to the extent that indistinguishability is taken into account when counting the states, which should reflect the real history of the sample. If thermal fluctuations can access all possible states of indistinguishable particles within the energy range kT, then the standard formula for entropy is sufficient to describe the resulting equilibrium state.

Bell's inequalities are violated by special sample preparations, when this assumption is not true. Specifically, the state $\Psi_{\alpha\beta}$ (5.52) is a vector in four-dimensional space:

$$\begin{aligned}\Psi_{\alpha\beta} &= \left(\cos\frac{\alpha}{2}\begin{bmatrix}1\\0\end{bmatrix}_1 + \sin\frac{\alpha}{2}\begin{bmatrix}0\\1\end{bmatrix}_1\right)\left(\cos\frac{\beta}{2}\begin{bmatrix}1\\0\end{bmatrix}_2 + \sin\frac{\beta}{2}\begin{bmatrix}0\\1\end{bmatrix}_2\right)\\ &= \cos\frac{\alpha}{2}\cos\frac{\beta}{2}\begin{bmatrix}1\\0\end{bmatrix}_1\begin{bmatrix}1\\0\end{bmatrix}_2 + \cos\frac{\alpha}{2}\sin\frac{\beta}{2}\begin{bmatrix}1\\0\end{bmatrix}_1\begin{bmatrix}0\\1\end{bmatrix}_2\\ &\quad + \sin\frac{\alpha}{2}\cos\frac{\beta}{2}\begin{bmatrix}0\\1\end{bmatrix}_1\begin{bmatrix}1\\0\end{bmatrix}_2 + \sin\frac{\alpha}{2}\sin\frac{\beta}{2}\begin{bmatrix}0\\1\end{bmatrix}_1\begin{bmatrix}0\\1\end{bmatrix}_2,\end{aligned} \tag{5.78}$$

and the entangled state (5.53) is just a linear combination of these four basis vectors, antisymmetric on the interchange of labels $1 \leftrightarrow 2$. The remaining three (symmetric) are shown in Fig. 5.1. If each state that falls on the detector is a random choice among the four, the measurements will not violate Bell's inequality. It is violated by the expression (5.53) because it previously selects one state out of four[3] by symmetry. The symmetry restriction has led to kinematic correlations that violate Bell's inequality.

Symmetry constraints are violated by collisions with walls, which can easily change the angular momentum of the state (5.53). Because of the high efficiency of thermalization, Bell's inequality has historically been verified in systems of entangled photons, whose interaction with the environment is weak. Interest in entangled states of material particles has grown in the context of quantum computing, since it was shown that using multiple two-level entangled systems, similar to the ammonia molecule (5.48), one could realize logic circuits that perform many computational operations simultaneously—or rather timelessly—that would take an

---

[3] Symmetric spin combinations of two fermions are possible if the wave function of their relative motion is antisymmetric, so the Pauli principle (5.2) is satisfied overall.

impractically long time for classical sequential computers. Protecting such states from decoherence is an important contemporary area of applied research.

If we want to describe the preparation of entangled states, the standard expression for entropy should be replaced by another one that takes into account additionally given kinematic constraints, above those arising from indistinguishability itself. Such an expression is called logical entropy.

5.9.1 Counting Distinctions

The expression for the relevant entropy (3.28) is a measure of the size of the space of individual events with equal a priori probability. Logical entropy, on the other hand, is the mathematical probability of *pairs of distinct events*. Specifically, if we draw two marbles in succession from a bag (with replacement), what is the probability that they will be of different colors? If p_i is the proportion of the ith color, and there are $\mathcal{N}$ colors, the answer is obviously

$$\sum_{i=1}^{\mathcal{N}} p_i(1-p_i) = 1 - \sum_{i=1}^{\mathcal{N}} p_i^2 =: h, \tag{5.79}$$

where logical entropy is denoted by h. This expression can be compared with the relevant entropy:

$$\langle 1-p\rangle \leftrightarrow \left\langle \ln \frac{1}{p} \right\rangle. \tag{5.80}$$

Both expressions are zero when $p = 1$, which corresponds to the intuitive understanding of entropy as a measure of disorder. Both are also maximal when all $p_i = 1/\mathcal{N}$, with the restriction $\sum_i p_i = 1$. However, the value at the maximum reflects the difference in their meaning:

$$1 - \frac{1}{\mathcal{N}} \leftrightarrow \ln \mathcal{N}. \tag{5.81}$$

When the number of colors is large, the probability of drawing two different ones tends to unity, but it cannot be greater than that. On the other hand, if a number itself grows without limit, its logarithm also grows without limit. The relevant entropy is a measure of the size of the space of individual events, while the logical entropy is the probability of a particular fluctuation in the space of pairs of events. This fluctuation is technically called a *distinction*, or dit. A dit is an ordered pair of events such that one member of the pair is different from the other.

Both formulas (5.80) imply a certain structure of the set on which they are defined, namely its division into disjoint subsets (*partition*):

$$\mathcal{U} = \bigcup_{i=1}^{N} \mathcal{A}_i, \quad p_i = \frac{|\mathcal{A}_i|}{|\mathcal{U}|}. \tag{5.82}$$

Unlike relevant entropy, logical entropy can itself be written as a probability:

$$h[\mathcal{A}] = \frac{|\mathcal{U} \times \mathcal{U}| - \sum_i |\mathcal{A}_i \times \mathcal{A}_i|}{|\mathcal{U} \times \mathcal{U}|} = 1 - \sum_{i=1}^{N} p_i^2. \tag{5.83}$$

Here $\times$ means the direct product of sets, i.e., the set of ordered pairs. Since $|\mathcal{U} \times \mathcal{U}| = |\mathcal{U}| \cdot |\mathcal{U}|$, the formula (5.79) counts all dits for a given partition of the set.

For example, the length of a sequence of coin flips can be thought of as the number of binary decisions (bits) required to determine one element of the set of all such sequences. Thus, three flips generate eight sequences. If we repeat these three flips twice, we have $64 - 8 = 56$ ways to obtain two different sequences. While the number of bits measures the size of the information space, the number of dits measures the amount of information in that space. Repeating the same sequence is not informative, while each change can carry meaning. Similarly, words in natural languages are represented by different combinations of letters. A sequence of letters is efficiently encoded when it is statistically indistinguishable from a random one, i.e., one in which the number of dits is close to the maximum.

5.9.2 *Subsets and Partitions*

Bits and dits encode qualitatively different information. In Fig. 5.6 on the left, we can label all subsets with a sequence of three bits, where the first bit is '1' if the element a is in the subset, and '0' otherwise. Since a really either is or is not in every subset, the bits encode this objective determination (reality), whether we see it or not. Therefore, the logic of subsets necessarily leads to Bell's inequalities.

In the figure on the right we see the ordering of partitions by *refinement*, which is the operation of splitting a subset, i.e. inserting a vertical line into an already existing partition. Before we put in the next line, it is really not determined which elements will remain unseparated. The dits encode this objective indeterminacy (potentiality), which also does not depend on whether we are looking. Refinement therefore naturally describes the process of quantum measurement, and the logic of partitions can describe entangled states.

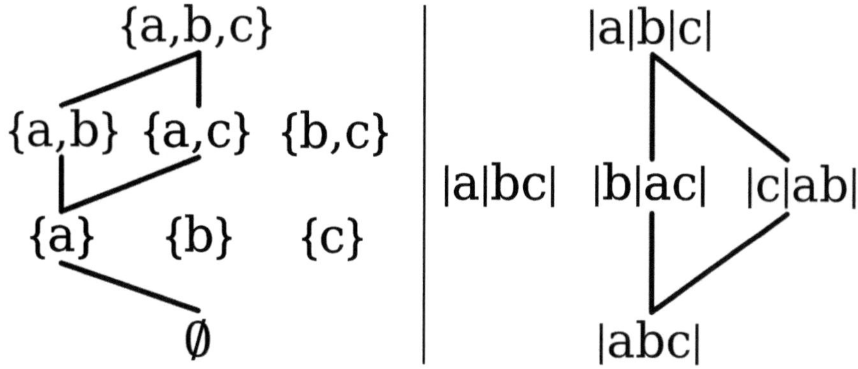

Fig. 5.6 Bits and dits. Left: Subsets ordered by inclusion. Lines indicate branches that always contain the element a. Right: Partitions ordered by refinement. Lines indicate branches along which the element a is resolved last

5.9.3 Logical Entropy of Quantum States. Density Matrices

The states of indistinguishable particles do not have to be entangled. For example, if we denote the spin projection "up" by a and "down" by b, then the four basis vectors in the expression (5.78) can be written as the set $\{(a, a), (a, b), (b, a), (b, b)\}$. The entangled state (5.53) corresponds to the occurrence of the subset $\{(a, b), (b, a)\}$ in some of the 15 partitions of this 4-element set.[4] This subset encodes a certain correlation of measurement outcomes: If the first detector measured a, the second one will surely measure b, or has already measured it. Conversely, the partition $|(a, a)|(a, b)|(b, a)|(b, b)|$ into unit subsets means a random mixture, so that a measurement on one detector means nothing to the other: The probabilities of both measurements are probabilities of independent events.

A probabilistic description of entangled states should therefore distinguish whether the elements (a, b) and (b, a) appear in the same or different subsets of some partition. The event space in quantum mechanics is the space of wave functions, but the usual notation for wave functions, which represents them as vectors or complex numbers, is not general enough to describe everything that can happen in that space equally well. A complete statistical description of quantum mechanics is possible using so-called *density matrices*, which are obtained by

[4] The notation must preserve the information that two-particle states consist of single-particle states a and b.

multiplying the wave function, represented as a column vector, by itself, represented as a row vector. Specifically, for the state Ψ_A (5.53) we have

$$\begin{bmatrix} 0 \\ 1/\sqrt{2} \\ -1/\sqrt{2} \\ 0 \end{bmatrix} \left[0\ 1/\sqrt{2}\ -1/\sqrt{2}\ 0\right] = \begin{bmatrix} 0 & 0 & 0 & 0 \\ 0 & 1/2 & -1/2 & 0 \\ 0 & -1/2 & 1/2 & 0 \\ 0 & 0 & 0 & 0 \end{bmatrix} =: \rho(\Psi_A). \tag{5.84}$$

The general formula is $\rho(\psi) = \psi\psi^\dagger$, where $\dagger$ is the conjugation operation, which exchanges rows with columns and complex-conjugates all elements. The multiplication is matrix multiplication, so when a column $d \times 1$ is multiplied by a row $1 \times d$, the result is a matrix $d \times d$. The density matrix is normalized,

$$\operatorname{Tr} \rho(\psi) = \operatorname{Tr} \psi\psi^\dagger = \operatorname{Tr} \psi^\dagger\psi = 1, \tag{5.85}$$

because the wave function is normalized, $\psi^\dagger\psi = 1$, and for the trace (the sum of the diagonal elements) we have $\operatorname{Tr} AB = \operatorname{Tr} BA$. This matrix is also idempotent, for the same reason:

$$\rho(\psi)^2 = \psi\psi^\dagger\psi\psi^\dagger = \psi\psi^\dagger = \rho(\psi). \tag{5.86}$$

If a density matrix is idempotent, we say that it represents a *pure state*, as can be described by a single wave function.

The main advantage of density matrices over wave functions is that they can be used to represent random mixtures of states equally well. Specifically, if we delete the off-diagonal elements in the matrix $\rho(\Psi_A)$, we obtain a description of the mixture

$$\frac{1}{2}\rho(\psi_{ab}) + \frac{1}{2}\rho(\psi_{ba}), \tag{5.87}$$

where $\psi_{ab} = [0, 1, 0, 0]^\dagger$ and $\psi_{ba} = [0, 0, 1, 0]^\dagger$. In such a mixture of states, (a, b) and (b, a) are found with probability $1/2$ just as if they were black and white balls, i.e., without quantum correlations. The linear combination of density matrices that we use to describe it cannot be written as $\psi\psi^\dagger$ for any wave function (vector) ψ, and thus really extends the notion of density matrix to all linear combinations of the form

$$\rho[p] = \sum_i p_i \psi_i \psi_i^\dagger, \quad \sum_i p_i = 1, \tag{5.88}$$

where ψ_i is the basis in the given state space. Random mixtures are the rule in nature, due to the efficiency of thermalization. Static friction erases off-diagonal elements in density matrices, unless the sample is prepared in such a way that this is impossible, or at least slowed down. When it erases them all, the remaining diagonal elements of the density matrix of the thermalized system are given by the canonical distribution, $p_i = \exp(-E_i/kT)/Z_C$.

Using the density matrix, we can naturally define the logical entropy of a quantum system:

$$h(\rho) = \operatorname{Tr} \rho(1-\rho) = 1 - \operatorname{Tr} \rho^2 = 1 - \sum_{i,j} |\rho_{ij}|^2. \tag{5.89}$$

This logical entropy is called *quantum logical entropy*.[5] We see immediately that it is equal to zero if ρ is idempotent, i.e., represents a pure state. Such is the case of the matrix $\rho(\Psi_A)$. On the other hand, for the mixture above it is easy to see that $h = 1-1/4-1/4 = 1/2$. Deleting off-diagonal elements, i.e., the effect of measurement, is precisely splitting the partition: $|(a,b)(b,a)| \to |(a,b)|(b,a)|$. The increase in logical entropy measures the information gain from the measurement. This increase is obviously equal to the sum of the squares of the deleted off-diagonal elements. Since the logical entropy of a pure state is zero, for it the information gain is equal to the logical entropy of the mixture resulting from the measurement:

$$h(\hat{\rho}) - h(\rho) = 1 - \sum_i |\rho_{ii}|^2 - 0 = 2\sum_{i<j} |\rho_{ij}|^2 > 0, \tag{5.90}$$

where $\hat{\rho}$ denotes the matrix ρ in which the off-diagonal elements have been deleted, and the last equality follows from idempotence.

The density matrix can therefore be read as a pictorial representation of a partition of a set of quantum states. If the matrix element $\rho_{ij} \neq 0$, then the i-th and j-th states are in the same partition, that is, they are parts of some pure state. Otherwise, they are a random mixture. Now we can ask the question: Do all pure states violate Bell's inequality? This is the topic of the next section.

5.9.4 Entangled States and Bell's Inequality

The question of violation of Bell's inequality can be posed as a question of the compatibility of the subset and partition schemes in Fig. 5.6. Technically, we ask whether some subset of two-particle states can be obtained as a direct product of subsets of one-particle states. In our example, the set of one-particle states is $\{a, b\}$, so all its (non-empty) subsets are $\{a\}$, $\{b\}$, and $\{a, b\}$. There are nine subsets of the

[5] The analogous quantum version of the relevant entropy, $-\operatorname{Tr} \rho \ln \rho$, is called the von Neumann entropy.

set of two-particle states that can be obtained as a direct product of these:

$$
\begin{aligned}
\{\{a\}, \{b\}, \{a, b\}\} \times \{\{a\}, \{b\}, \{a, b\}\} \\
&= \{\{(a, a)\}, \{(a, b)\}, \{(a, a), (a, b)\}, \\
&\quad \{(b, a)\}, \{(b, b)\}, \{(b, a), (b, b)\}, \\
&\quad \{(a, a), (b, a)\}, \{(a, b), (b, b)\}, \\
&\quad \{(a, a), (a, b), (b, a), (b, b)\}\}.
\end{aligned}
\tag{5.91}
$$

Statistically, the direct product corresponds to random combinations, as in the thermal contact model. We expect that the partitions in which these subsets appear will satisfy Bell's inequality, that is, the corresponding pure states will behave statistically like random mixtures. However, the set of two-particle states $\{(a, a), (a, b), (b, a), (b, b)\}$ has fifteen subsets, so the list above is missing six of them:

$$
\begin{aligned}
&\{(a, a), (b, a), (a, b)\}, \{(a, a), (b, a), (b, b)\}, \{(a, a), (a, b), (b, b)\}, \\
&\{(b, a), (a, b), (b, b)\}, \{(a, b), (b, a)\}, \{(a, a), (b, b)\}.
\end{aligned}
\tag{5.92}
$$

We suspect that the states in which these subsets appear could exhibit quantum correlations, i.e., violate Bell's inequality.

The structure of the density matrix allows us to quantify the entanglement of a state without having to deal with specific amplitudes of wave functions. First, we replace every non-zero element with unity. (This operation defines the *incidence matrix* of a given matrix.) For the state Ψ_A, we get

	(a, a)	(a, b)	(b, a)	(b, b)
(a, a)	0	0	0	0
(a, b)	0	1	1	0
(b, a)	0	1	1	0
(b, b)	0	0	0	0

(5.93)

The only information remaining here is about the partitions of the set of states that appear in the wave function, specifically $|(a, b)(b, a)|$.

Next, we calculate the correlations of measurements on the first and second detectors, assuming equal probabilities a priori. For the set $\{(a, b), (b, a)\}$, the probabilities of measuring a on the first detector and b on the second, and vice versa, are obviously equal:

$$
P(a, b) = P(b, a) = \frac{1}{2}. \tag{5.94}
$$

On the other hand, $P(a, a) = P(b, b) = 0$, because these states do not appear. To calculate the correlation, we need the probability $P_1(a)P_2(b)$, where $P_i(x)$ are the probabilities of measuring $x = a, b$ at the i-th detector, regardless of what happened at the other one. In statistics, these are called marginal probabilities:

$$P_1(a) = \sum_{x=a,b} P(a, x) = \frac{1}{2}, \quad P_2(b) = \sum_{x=a,b} P(x, b) = \frac{1}{2}. \tag{5.95}$$

The correlation is therefore

$$P(a, b) - P_1(a)P_2(b) = \frac{1}{4} \neq 0. \tag{5.96}$$

We conclude that the state Ψ_A is entangled. If we perform the same calculation on one of the factorized states (5.91), the correlation will be zero, as readers can see for themselves, or notice that it must be so by construction. These states are not entangled.

For subsets of the three states in the list (5.92), which cannot be factored, we obtain a correlation again. Take the first one, $\{(a, a), (b, a), (a, b)\}$. We have $P(a, a) = P(b, a) = P(a, b) = 1/3$ and $P(b, b) = 0$, while the marginal probabilities are $P_1(a) = 2/3$ and $P_2(b) = 1/3$. The correlation is thus

$$P(a, b) - P_1(a)P_2(b) = \frac{1}{3} - \frac{2}{9} = \frac{1}{9}. \tag{5.97}$$

This state is entangled, but it cannot violate Bell's inequality, more precisely: Measuring this correlation will not exclude a local explanation. Namely, if we measure a on the first detector, it means that we have split the partition $|(a, a), (b, a), (a, b)| \rightarrow |(a, a), (a, b)|(b, a)|$, and in the subset $\{(a, a), (a, b)\}$ the second detector can choose a or b independently of the first. It is similar if we measure a on the second detector. On the other hand, the first measurement of b on one detector imposes a on the second, but it is still independent in that case, because it is the first. Accordingly, the measurement of the outcome b is always independent of the other detector, so it can be interpreted locally.

The structural reason why measuring correlations on the state Ψ_A can violate Bell's inequality is that any measurement on one detector uniquely determines a measurement on the other. Among our sets of two-particle states, only $\{(a, b), (b, a)\}$ and $\{(a, a), (b, b)\}$ have this property. They can be elegantly characterized: They are all possible bijections on the set of one-particle states $\{a, b\}$, respectively the inversion $a \mapsto b$, $b \mapsto a$ and the identity $a \mapsto a$, $b \mapsto b$. We conclude that Bell's inequality can be violated only by those entangled states whose content of two-particle states can be read as a bijection in the space of one-particle states.

Finally, note that our identification of candidates for violation of Bell's inequality does not mean that every measurement on such states will actually violate it. The expression (5.71) shows that this depends on preparing states with carefully chosen amplitudes of the wave functions. Our statistical description by reducing the density matrix to its incidence matrix gave the best that can be expected from counting on fingers.

5.10 Coherence in Society

Every even-numbered year, a game of chess with live pieces is held in Marostica. As long as people dressed as chess pieces move according to the commands of two players, it is the same game of chess as one in which the pieces are wooden.

The question arises whether humans can also simulate quantum computation with entangled particles, acting in accordance with the rules of partition logic. Obviously they can, because otherwise such simulations would not be possible even on digital (Boolean) computers. Splitting a partition of the set in Fig 5.6 by tossing a coin has the same effect within those rules as a quantum measurement, even if tossing a coin is not physically the same as measuring entangled photons. The real question is whether this actually happens in human society. We give a few examples that indicate that the answer is yes, in fact, that partition logic evolved early in the service of community cohesion, where coherence is expressed as a shared system of values.

5.10.1 The Rug-Seller's Story

A rug-seller in Turkey told the following story:

> There are villages in eastern Anatolia where only women weave rugs. When a girl is old enough to be married, she weaves a rug with a special pattern that means "a marriageable girl lives here," and hangs it on the balcony of her parents' house. A boy who wants to marry will knock on the door of such a house and be invited to coffee, which the girl will prepare alone. If she likes him, she will put sugar in it, if not, salt. In any case, he will drink the coffee and take his leave. If there was sugar in the coffee and he likes her as well, he will visit a second time and propose to her.

This story describes a quantum logical AND circuit (conjunction), which is minimally realized with two atoms. Let a denote "yes" and b denote "no." The circuit splits the partition $|(a,a)(b,a)(a,b)(b,b)| \rightarrow |(a,a)|(b,a)(a,b)(b,b)|$. According to the previous analysis, the state $|(b,a)(a,b)(b,b)|$ is entangled, so it is not known who said "no." The source of randomness is the variability of human beings. The key to the coherence of the circuit is that an atom can return to its ground state without breaking the entanglement. For people, preserving entanglement means that they can go on living without damage to their reputations. The irreversible gain of information, whether these two are right for each other or not, becomes part of the real history of the community.

5.10.2 Prisoner's Dilemma

The prisoner's dilemma is a classic game theory problem. If two burglars are placed in separate cells and told that the first one to betray the other will be released and the second one will be convicted, it is easy to predict that both will betray. Controlled experiments on subjects, as well as police practice, do not agree with this prediction. The probability of cooperation, i.e., silence, is significantly higher than expected from this rational analysis. Several counterintuitive phenomena have been observed in different variants of the experiment, of which we describe the so-called disjunction effect here.

In this variant, one subject (Alice) is first told that the other (Bob) has betrayed her, and by repeating this on many volunteers, the conditional probability $P(A_C|B_B)$ that Alice will cooperate (keep quiet) even if Bob has betrayed her is established. Then, Alice is told that Bob is silent, i.e., cooperating with her, and the probability $P(A_C|B_C)$ that she will cooperate in that case is similarly established. In the third part of the experiment, Alice is not told what Bob did, so the absolute probability $P(A_C)$ that Alice will cooperate is measured. Classical (Boolean) probability theory predicts a relation between these numbers:

$$P(A_C) = p \cdot P(A_C|B_C) + (1-p) \cdot P(A_C|B_B), \tag{5.98}$$

where p is Alice's estimate of the probability that Bob will cooperate. It is in the nature of Boolean logic that Bob either cooperates or not, so the estimate p that he does necessarily implies the estimate $1-p$ that he does not. Since $0 \leq p \leq 1$, linear interpolation means that the absolute probability $P(A_C)$ is bracketed by the two conditional ones:

$$P(A_C|B_B) \leq P(A_C) \leq P(A_C|B_C). \tag{5.99}$$

Observations do not agree with this prediction. In two independent experiments, a significant deviation was observed:

$$P(A_C|B_B) = 3\% < P(A_C|B_C) = 16\% < P(A_C) = 37\%, \tag{5.100}$$

where the numbers are from the original study. We can understand the first inequality. Clearly, Alice is less inclined to cooperate when Bob betrayed her than when he did not. However, the second inequality is puzzling. Why is it *twice* more likely for Alice to cooperate if she does not know what Bob is doing, than if she knows that he is cooperating?

Not forgetting that it is rational to betray in this context, we can imagine that Alice feels freer to decide if she knows that Bob has already decided, i.e., that the partition in which they were together has already split. Indeed, every partition of a set is an equivalence relation, so we conjecture that the two burglars will be maximally solidary as long as they feel like "one soul in two bodies."

Maximal solidarity without calculation protects the equivalence that defines the community, while if that is absent, one is left with rationally optimal solidarity with an individual.

Be that as it may, a number of attempts to model the disjunction effect statistically use a density matrix, i.e., they model human behavior with quantum statistics. Such modelling is not arbitrary. As long as the event space is modelled as a vector space (of dimension > 2), which is standard in statistics, so-called Gleason's theorem holds: Squaring vectors, i.e., quantum probability calculus, is the only probability measure that is additive over subspaces. Solidarity appears in these models as constructive interference, and the inequality (5.100) corresponds to (5.4).

5.10.3 Ostracism

In ancient Athens, the council of free citizens determined once a year whether someone should be exiled from the city for ten years. The process had two stages. In the first stage, the citizens were asked whether someone should be exiled. If the majority voted in the affirmative, in the second stage, two months later, a secret ballot was held to determine who would be exiled. Whoever received a relative majority of votes was exiled, provided that at least 6000 citizens participated in the second vote.

The key point for our consideration is that no one was obliged to explain why he thought someone should be exiled. Ostracism is an implementation of the logic of partitions, not Boolean logic. It answers the question "who is not one of us," that is, who has already separated himself from the coherence of the *polis*. Rules about who should be exiled cannot be introduced, because such rules themselves introduce divisions. Introducing them abandons the logic of partitions in favor of the Boolean logic of a legal process that determines the commission, yes or no, of a previously defined act. The logic of partitions is reversed: If a citizen does not understand why he was exiled, that in itself justifies the ostracism, and it also justifies it if he does.

Ostracism was reversible, and the exiled man could return after ten years without losing his social position. The obvious aim was to restore disrupted social coherence, similarly as in the rug-seller's story.

If Boolean logic is "yes–no," partition logic is "we'll see–I won't say." Hindsight is the privilege of the stronger. Conversely, forcing someone to declare is an act of domination, isolating the individual from a coherent group. While such declarations under pressure from ambitious individuals increase political tensions, there is, on the contrary, historical evidence that ostracism under pressure from the community calmed them down.

5.10.4 Heresy of the Second Kind

In the introductory chapter, we met heretics of the first kind. They willfully arrange accepted observations into an alternative narrative. Because truth is objective and knowable, such heretics are ultimately either right or wrong, even if that is not always easy to determine, because truth is also invisible. Heresy of the first kind threatens the Boolean structure of society, the lattice of yes–no switches that determine how the community interprets observations.

The heretic of the *second kind* introduces division by demanding declarations and rules, or, technically, enforces measurement of the wave function of the current state. Heresy of the second kind threatens the partition structure of society, the lattice of accepted equivalences, which corresponds to the original meaning of the word *haíresis*: Choice. It replaces ineffable values with computable rules, i.e., attempts to replace a partition lattice with a Boolean one, which is impossible because they are structurally incompatible, as observed in Fig. 5.6.

While the consequences of heresy of the first kind are in principle reversible, the second kind disrupts the coherence of society, which is an irreversible process. If we have made a specific mistake, we can try to correct it, but once a collective value system is disrupted, it is difficult to reestablish it, because every attempt to make it explicit is just another cause for discord.

5.11 Quantum Mechanics and Free Will

The discovery of quantum indeterminacy, especially the real indeterminacy that manifests itself in the violation of Bell's inequality, has renewed the debate on the physical conditionality of free will. In light of the insights above, we can see that such a debate is narrower than it seems. First of all, Newtonian determinism fails already in the classical ideal gas, as soon as the physical nature of thermalization is understood, so we do not need quantum mechanics to liberate us from the idea that our actions are physically predetermined.

Furthermore, the human brain is a material object, and is subject to the laws of quantum mechanics like everything else. In this sense, there is no debate, unless we are to claim that the brain is a supernatural phenomenon, because that is the only thing left if it does not obey natural laws. Thus, the question whether free will depends on quantum phenomena is rather specific: Whether *entangled states* are essential for free will.

The will is free if our decision is not determined by the past. This is the case with entangled states that choose, for example, the orientation of the spin at the very moment of measurement. The problem with this low-level (physiological) interpretation of free will is that a living cell is generally more similar to a steam engine than to an entangled state of photons. Cellular and tissue functions are controlled by deterministic feedback loops that have a resilience to thermal noise that presently known entangled states do not. Several mechanisms have been proposed by which the brain could nevertheless be a quantum computer, but none have yet been shown to actually work.

Alternatively, we can postulate that free will arises at a high (functional) level, independently of the physiological implementation. In that case, it would be manifested by the brain's ability to simulate both Boolean and quantum logic. The examples from sociology above suggest that quantum logic could have evolutionary advantages over Boolean.[6] It is possible that the brain has a multitude of evolved logics, only a small part of which are known to mathematicians, among which it freely chooses in given circumstances, thereby changing the criterion of purposeful behavior from case to case.

At the functional level, both past and future are spiritual rather than physical categories. The problem of past conditioning can then be resolved either apodictically, by claiming that free will manifests itself by including hope, i.e., the future, in decision-making, or pragmatically, by claiming that no one knows us well enough to predict our behavior. If we were to prove that a given physical past necessarily leads to a particular spiritual experience of the past and future, i.e., particular personality, we would be proving that free will is an illusion. At the present level of understanding of physiology and quantum mechanics, free will is real because of the variability of human beings, rather than because of details of biochemical processes.

5.12 Timeless Strangeness

Reducing Bell's inequalities, and their violation, to logical phenomena reopens the question of what was strange about quantum mechanics in the first place. We might reason as follows: Blaise Pascal analyzed games of chance, so he naturally defined expectation values equivalently to the probability of drawing one ball. If he had been a marriage counselor, he would have defined them by drawing two balls, so he would have invented logical entropy. Then quantum statistics would not seem to us to be fundamentally different from such classical statistics. So, all we need is a change of attitude for quantum mechanics to cease to be strange. Indeed, quantum correlations of two detectors are logically clear: Two sequential measurements of the same state cannot contradict each other, if material reality is not in contradiction with itself.

On the contrary, there is something really strange about quantum mechanics, which cannot be compensated by a change of attitude: The quantum correlations that violate Bell's inequality occur outside of time. Specifically, we can change the state of the polarizer by rapid oscillations, of such high frequency that the change of

[6] We may conjecture that it is older, noting how children accept their parents' value system uncritically until adolescence, when they begin challenging it with impeccable Boolean logic.

state cannot be communicated to the other detector even at the speed of light. The accompanying electronics, however, is fast enough to record when something was measured on both detectors. When the records are subsequently synchronized and compared, it is shown that the correlations existed all along, exactly as predicted by quantum mechanics. So, the logical sequence of observations is established through synchronization with the clock on the wall, which determines which detector is first and which is second, but the measurement process seems to happen instantaneously, better say timelessly.

We have already noted that time does not exist in quantum mechanics, and the possibility remains open that laboratory time is emergent, like temperature in thermodynamics. Some may find the reverse possibility, offered by the theory of relativity, more elegant, namely that time is as fundamental a coordinate as position. However, then the inverse problem to the emergence of time appears, which is such a reduction of the theory of relativity in which time disappears, and the time-*independent* Schrödinger equation remains. Correlations of entangled states, which cannot be realized sequentially even at the speed of light, indicate that this problem is not formal, but is just as real a physical problem as emergence in the previous case. In other words, until we understand what happens physically with time in entangled states, we will not know which of these possibilities is true. It is improbable that formal manipulation of equations can solve such a problem without a clear idea of its physical cause.

Chapter 6
Fermions

> Как известно, в целом ряде случаев для рассмотрения системы ферми-частиц применяется модель ферми-газа, несмотря на то, что взаимодействие между частицами отнюдь не является слабым. [...] Такое состояние теории не является удовлетворительным, так как остается неясным, какие свойства газовой модели отвечают действительности, а какие присущи только газу.
>
> *Lev Landau*

The fermionic ideal gas is an unexpectedly good model for understanding electrons in metals. One would expect that the strong electrostatic forces between electrons, and between electrons and ions in the crystal lattice, would significantly change their behavior compared to a gas without interaction. The success of the ideal gas model in describing real electrons is a consequence of the Pauli principle, which kinematically restricts the possible states of electron motion at low temperature so much that very little phase space remains for dynamical effects. To understand how this is possible, one needs to understand the ground state of the fermionic ideal gas. It is the lowest possible energy state, the only one that remains in the absence of thermal fluctuations at zero temperature.

6.1 Ground State. Fermi Level

The wave function of a particle in a one-dimensional container (rectangular potential well) of width ℓ is

$$\psi_n(x) \sim \sin k_n x, \quad k_n = \frac{n\pi}{\ell}. \tag{6.1}$$

The associated kinetic energy is $\varepsilon_n = \hbar^2 k_n^2/2M$, where M is the mass of the particle. If we want to construct a state of the lowest possible energy for N fermions of spin

D. Sunko, *Conceptual Thermodynamics*,
https://doi.org/10.1007/978-3-032-20877-4_6

1/2 while respecting the Pauli principle, we can put exactly 2 fermions in each state of wave number k_n, where $1 \leq n \leq N/2$. (The calculation in three dimensions is essentially the same.) Such a state is unique, because fermions are assumed to be indistinguishable particles, so its entropy is $S = k \ln 1 = 0$. Its internal energy,

$$U|_{S=0} = 2 \sum_{n=1}^{N/2} \frac{\hbar^2 k_n^2}{2M} \sim N \left(\frac{N}{\ell} \right)^2, \tag{6.2}$$

is extensive and quadratic in density. It is called the ground-state energy or the zero-motion energy of the Fermi gas. To estimate it, we can reason as follows: The wavelength of the highest occupied state $\psi_{N/2}$ is $\lambda = 2\pi/k_{N/2} \sim \ell/N$. A sodium cube has an edge of $\ell \approx 3$ cm when N^3 is of the order of Avogadro's number, so the wavelength of the highest occupied electron state is of the order of the interatomic distance (a few Ångstroms). In simple terms, to make all the electrons' wave functions orthogonal, we have to put as many nodes into the last one as there are electrons. This means that the kinetic energy ϵ_F of that state, the so-called *Fermi level*, is large compared to the thermal fluctuation energy kT at room temperature. Due to the high electron density, the electron gas in metals is in the quantum regime $kT \ll \epsilon_F$ even above the melting temperature, because ϵ_F/k in metals is of the order of 10–100,000 K, while melting temperatures are usually 1–2000 K.

6.2 Fermions at Low Temperatures

The Fermi-Dirac (Fermi) distribution does not differ from the Maxwell-Boltzmann (Maxwell) distribution at high temperatures, as shown in Fig. 5.4. When the chemical potential is negative, only its exponentially decreasing part appears in the physical range $\varepsilon > 0$, which coincides with the Maxwell distribution. However, this cannot remain so at low temperatures, because the step of the Fermi distribution becomes sharper, as shown in Fig. 6.1, and the expression (5.35) tends to the Heaviside function in the limit $T = 0$,

$$f_\mu(\varepsilon)\big|_{T=0} = \begin{cases} 1, & \varepsilon < \mu, \\ 1/2, & \varepsilon = \mu, \\ 0, & \varepsilon > \mu. \end{cases} \tag{6.3}$$

If we set a negative chemical potential and lower the temperature, the concentration of particles with positive kinetic energy will tend to zero, i.e., all will flow into the ground. Thus, upon lowering the temperature, the equation of state for a given number of particles,

$$N = \langle N \rangle = \sum_i f_\mu(\varepsilon_i), \tag{6.4}$$

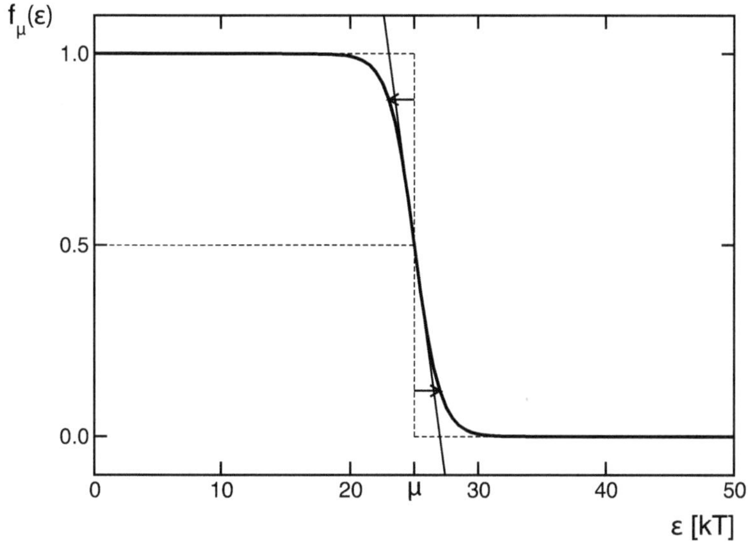

Fig. 6.1 Fermi-Dirac distribution with chemical potential in the quantum regime, $\mu = 25\,kT$. The curve is centrally symmetric around the point $(\mu, 1/2)$. The arrows show the step width estimate $\mu \pm 2kT$. The thin solid line is the tangent through the inflection point

becomes solvable only for positive values of the chemical potential, $\mu = \mu^\star(N) > 0$, because the constant part $f_\mu(\varepsilon < \mu) = 1$ then appears in the physical range of energies $\varepsilon > 0$.

When solving the equation of state (6.4), it is natural to consider the right-hand side as a function of the chemical potential for a given spectrum ε_i. Then the Fermi distribution in Fig. 6.1 is mirrored about the vertical axis, i.e., becomes an ascending step. [Formally, $f_\varepsilon(\mu) = 1 - f_\mu(\varepsilon)$.] The dependence of the expected number of particles on the chemical potential is shown in Fig. 6.2. The low-temperature regime here is $kT \ll \Delta\varepsilon$, with $\Delta\varepsilon$ the smallest difference of the energies ε_i. It corresponds to atomic and nuclear spectra, where all energy levels are resolved and grouped into shells between which gaps appear, such as between states 1 and 2, 5 and 6, etc., in the figure. For electrons in crystals, typically $kT \gg \Delta\varepsilon$, and the shells become bands within which individual levels cannot be resolved. Between the bands, however, gaps remain whose width is $\Delta \gg kT$, and in them the function $\mu^\star(N)$ is almost vertical, as can be seen from the figure by exchanging the axes. This observation is important for the technology of semiconductors, which have one band filled with electrons and the next one empty, so the position of the chemical potential in the gap is a singularly sensitive function of small concentrations of impurities, whose contribution to the right-hand side of Eq. (6.4) would otherwise be negligible. Manipulation of the electronic properties of semiconductors by impurities is a fundamental technology of the electronics industry.

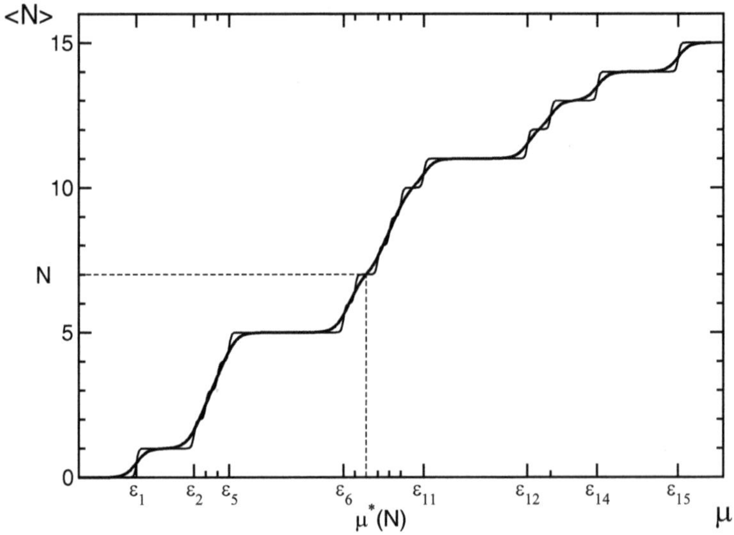

Fig. 6.2 Graphical representation of the dependence $\langle N\rangle\ (\mu)$, for fifteen unequally spaced levels, at low temperature. Thick line: $\Delta\varepsilon < kT \ll \Delta$. Thin line: $kT < \Delta\varepsilon \ll \Delta$. Dashed line: Solution of the equation of state for given $N = 7$, $\mu^\star(7) = (\varepsilon_8 - \varepsilon_7)/2$. The right-hand side of the equation of state (6.4) is integer-valued at the points of horizontal inflection, which are at the energy midpoints due to the central symmetry of the Fermi distribution

We have previously justified the Legendre transform (4.63) by appealing to negative values of the chemical potential. The same asymptotic estimate,

$$Z_G(\mu, \beta) = \sum_N Z_C(N, \beta)e^{\beta\mu N} \simeq Z_C(N^\star, \beta)e^{\beta\mu N^\star}, \tag{6.5}$$

can be justified for $\mu > 0$ by noting that the assumption that $Z_C(N, \beta)$ increases as an N-th power without limit is no longer valid. Namely, $Z_C(N, \beta)$ increases with the number of particles only as long as the number of states they can occupy itself increases with N. This is not the case at low temperatures, specifically, at absolute zero only single-particle states whose energy does not exceed a given value μ can be occupied, and $Z_C(N, \beta)$ drops sharply to zero as soon as the number of particles exceeds the number of such states. At finite but not too high temperatures $0 < kT < \mu$, this drop is milder, but still exponential, as can be seen in Fig. 6.1. Thus, at positive chemical potential thermodynamic equilibrium with the particle reservoir is achieved by Pauli blocking, due to which the system cannot draw an arbitrarily large number of particles of positive energy $0 < \varepsilon < \mu$ from the reservoir, but finds the equilibrium number again as the maximum of a product of rapidly increasing and rapidly decreasing functions. Finally, for temperatures $kT \gg \epsilon_F$ the chemical potential becomes negative, and the classical limit described by (4.63) is valid again.

Upon lowering the temperature, the part of the Fermi distribution that dominates the equilibrium equation of state (6.4) shifts from the exponential (Maxwellian) tail $f_\mu(\varepsilon) \ll 1$ to the constant $f_\mu(\varepsilon) = 1$. The consequence is a dramatic change in the behavior of the Fermi gas. As can be seen in Fig. 6.1, thermal fluctuations of size kT cannot affect the constant part of the distribution, so fermions with energies lower than approximately $\mu - 2kT$ move deterministically, just as if the temperature were zero. This means that only a small fraction of fermions can interact with their environment at low temperatures, which reduces the phase space for mutual dynamics, because in thermodynamic equilibrium the particles are each other's environment. Therefore, fermions at temperatures $kT \ll \epsilon_F$ behave similarly to an ideal gas even when they are subject to strong interactions.

In order to pass from such qualitative considerations to effective computation with quantum gases in the thermodynamic limit, a technical step by which sums over discrete states are converted into integrals is essential. That is the topic of the next section.

6.3 Level Density

Instead of considering sums by state index (i.e., by the energy-eigenstate ordinal number), we can consider them as integrals in energy:

$$\sum_i f(\varepsilon_i) = \int_{-\infty}^{\infty} f(\varepsilon) \left[\sum_i \delta(\varepsilon - \varepsilon_i) \right] d\varepsilon, \tag{6.6}$$

where $f(\varepsilon)$ is any function, and $\delta(x)$ is the Dirac delta function. The expression above is a formal identity, by definition of the delta function. The quantity

$$g(\varepsilon) := \sum_i \delta(\varepsilon - \varepsilon_i) \tag{6.7}$$

is called the *level density*. It is practically useful whenever the change in energy with a change in index is so small that we can claim that the energies change continuously, so that this sum can be replaced by an integral, making the level density a real function of energy.

The most common case in practice is when the energy indices refer to the wavenumbers of free particles in a box of size ℓ,[1]

$$\varepsilon_m = \varepsilon(k_m), \quad k_m = \frac{2\pi m}{\ell}.$$

[1] We put 2π here instead of π as in (6.1), because we imagine travelling waves (sines and cosines of integer periods, cyclic boundary conditions), not standing waves (sines of half-integer periods, fixed boundary conditions).

To pass to the integral, it is incidental that a shift of the index m by $\Delta m = 1$ gives a very small shift of the wavenumber k, by $\Delta k = 2\pi/\ell$, when ℓ is large ("dense" levels). In three dimensions, we can write the density of levels per unit volume as

$$\frac{1}{\ell^3} \sum_{m_x,m_y,m_z} \delta(\varepsilon - \varepsilon_{m_x,m_y,m_z}) = \left(\frac{1}{2\pi}\right)^3 \sum_{k_x,k_y,k_z} \delta(\varepsilon - \varepsilon(\boldsymbol{k}))\Delta k_x \Delta k_y \Delta k_z$$

$$\rightarrow \frac{1}{(2\pi)^3} \int_{-\infty}^{\infty} dk_x dk_y dk_z \delta(\varepsilon - \varepsilon(\boldsymbol{k})), \tag{6.8}$$

when we let $\ell \rightarrow \infty$. This integral is of the form[2]

$$\int \delta(f_0 - f(x))dx = \int \delta(f_0 - f) \left|\frac{dx}{df}\right| df = \sum_{\{x_i : f(x_i) = f_0\}} \left|\frac{\partial f}{\partial x}\right|_{x=x_i}^{-1}. \tag{6.9}$$

In our case, there is a triple integral, so the sum over zeros of the argument of the delta function becomes an integral over the surface defined by the energy condition $\varepsilon(\boldsymbol{k}) = \varepsilon$, where the argument vanishes:

$$\int_{-\infty}^{\infty} dk_x dk_y dk_z \delta(\varepsilon - \varepsilon(\boldsymbol{k})) = \int_{\varepsilon(\boldsymbol{k})=\varepsilon} \frac{dS}{|\nabla\varepsilon(\boldsymbol{k})|}. \tag{6.10}$$

When we add the spin degeneracy factor $2s + 1 = 2$ and put $\ell^3 = V$, the level density becomes

$$g(\varepsilon) = \frac{V}{4\pi^3} \int_{\varepsilon(\boldsymbol{k})=\varepsilon} \frac{dS}{|\nabla\varepsilon(\boldsymbol{k})|} = \frac{V}{4\pi^3} \int_{\varepsilon(\boldsymbol{k})=\varepsilon} \frac{dS}{|\hbar\, \boldsymbol{v}|}, \tag{6.11}$$

where we remembered that the derivative of kinetic energy with respect to momentum is the group velocity.

6.3.1 The Case of Free Particles

For free particles, we have

$$\varepsilon(\boldsymbol{k}) = \frac{\hbar^2}{2M}(k_x^2 + k_y^2 + k_z^2), \tag{6.12}$$

[2] The absolute value of the Jacobian appears because we count all points $f = f_0$ equally, whether f increases or decreases at the intersection.

whence

$$|\nabla\varepsilon(\boldsymbol{k})|_{\varepsilon(\boldsymbol{k})=\varepsilon} = \frac{\hbar^2}{M}\sqrt{k_x^2+k_y^2+k_z^2}\Big|_{\varepsilon(\boldsymbol{k})=\varepsilon} = \frac{\hbar}{M}\sqrt{2M\varepsilon}, \tag{6.13}$$

so the absolute value of the gradient under the integral is constant, and the area of the isoenergic surface in momentum space is the area of a sphere,

$$\int_{\varepsilon(\boldsymbol{k})=\varepsilon} dS = 4\pi k^2\Big|_{\varepsilon(\boldsymbol{k})=\varepsilon} = 4\pi\frac{2M\varepsilon}{\hbar^2}, \tag{6.14}$$

so the outcome is

$$g(\varepsilon) = V\frac{4\pi}{h^3}(2M)^{3/2}\sqrt{\varepsilon} \sim V\sqrt{\varepsilon}. \tag{6.15}$$

This calculation is simple because for free particles $|\nabla\varepsilon(\boldsymbol{k})|$ is a constant on the isoenergic surface. This is the same as saying that for free particles it is easy to invert the dispersion $\varepsilon(\boldsymbol{k})$. But then we can easily check it directly, as a change of variable in momentum space. Namely, the number of single-particle states in a phase space volume of radius p is

$$\mathcal{N}(\varepsilon) := 2\frac{VV_p}{h^3} = \frac{8\pi Vp(\varepsilon)^3}{3h^3}, \tag{6.16}$$

and their increase with the change in energy is

$$d\mathcal{N}(\varepsilon) = \frac{8\pi V}{h^3}p(\varepsilon)^2dp(\varepsilon) =: g(\varepsilon)d\varepsilon, \tag{6.17}$$

which agrees with (6.15) when we insert $p(\varepsilon) = \sqrt{2M\varepsilon}$.

6.4 Number Fluctuation

As can be seen in Fig. 6.2, the expected number of particles grows faster when the chemical potential passes through regions where the single-particle states are closer together in energy, i.e., "denser." This suggests a thermodynamic definition of the density of levels, denoted by a tilde:

$$\tilde{g}(\mu) = \frac{\partial\langle N\rangle}{\partial\mu}, \tag{6.18}$$

in which the expected number of particles appears instead of the cumulative number of states as in (6.17). Using Eq. (5.34),

$$\widetilde{g}(\mu) = \sum_i \frac{\partial f_\mu(\varepsilon_i)}{\partial \mu} = \sum_i \beta f_\mu(\varepsilon_i)\left[1 - f_\mu(\varepsilon_i)\right], \tag{6.19}$$

The Fermi function is a constant, so its derivative is zero, everywhere except in the step region, where it changes rapidly, the faster the temperature is lower (β larger). Then the derivative is a narrow, high peak around the value $\mu = \varepsilon_i$. But since the integral of that peak is the step from 0 to 1, the area below it is always equal to unity, regardless of the value of β. For large β, this spike under the integral is equivalent to a rectangle of height $\beta/4$ and width $4/\beta$, centered about ε_i, which gives the estimate of the width of the step in Fig. 6.1. It follows that

$$\lim_{\beta\to\infty} \int_{-\infty}^{\infty} h(\mu)\beta f_\mu(\varepsilon_i)\left[1 - f_\mu(\varepsilon_i)\right] d\mu = h(\varepsilon_i), \tag{6.20}$$

where $h(x)$ is any continuous function of bounded variation. In other words, the derivative of the Fermi function tends to the delta function when the temperature is low, so

$$\lim_{\beta\to\infty} \frac{\partial \langle N\rangle}{\partial \mu} = \sum_i \delta(\mu - \varepsilon_i) = g(\mu), \tag{6.21}$$

and the level density, defined as the derivative of the expected number of particles, coincides with the microscopic definition (6.7) at absolute zero. This limit is easy to visualize in Fig. 6.2: The derivatives in the sum (6.19) diverge on the vertical parts of the staircase, and are equal to zero on the horizontal ones.[3] At zero temperature, there is no difference between the functions of the expected number of particles $\langle N\rangle(\mu)$ and of the cumulative number of states $\mathcal{N}(\mu)$, because each new state found by shifting the chemical potential is filled by exactly one fermion from the reservoir, with probability 1.

Our thermodynamic level density $\widetilde{g}(\mu)$ does not have a common name, because a related dimensionless quantity, the number fluctuation (4.67), is used instead:

$$\Delta N^2 = \frac{1}{\beta}\frac{\partial \langle N\rangle}{\partial \mu} = \sum_i f_\mu(\varepsilon_i)\left[1 - f_\mu(\varepsilon_i)\right]. \tag{6.22}$$

This expression indicates that at low temperature the fluctuation of the number of fermions in contact with a particle reservoir (ground) is strongly suppressed

[3] Caution: The expression (6.6) is exact and valid at all temperatures.

compared to the classical ideal gas, where we had $\Delta N^2 = \langle N \rangle$. Indeed,

$$\Delta N^2(T, V, \mu) \xrightarrow{kT \ll \mu} kTg(\mu) \sim TV\sqrt{\epsilon_F}, \tag{6.23}$$

so the number fluctuation vanishes at $T = 0$. When $0 < kT \ll \epsilon_F$, only fermions whose energies are close to the chemical potential at zero temperature, $\varepsilon_i \approx \epsilon_F \pm 2kT$, contribute to it. Their number can be estimated as $N_{\mathit{eff}} \sim NkT/\epsilon_F$.

6.5 Entropy and Heat Capacity

We can obtain an operational expression for the entropy of the Fermi gas by inverting the Legendre transform (2.49):

$$\begin{aligned}\sigma &= \Psi_G - \beta\mu \langle N \rangle + \beta \langle E \rangle = \Psi_G + \sum_i \beta(\varepsilon_i - \mu) f_\mu(\varepsilon_i) \\ &= \sum_i \left[-f_\mu(\varepsilon_i) \ln f_\mu(\varepsilon_i) - (1 - f_\mu(\varepsilon_i)) \ln(1 - f_\mu(\varepsilon_i)) \right],\end{aligned} \tag{6.24}$$

where we replaced the thermodynamic solutions of the equations of state N and U with the statistical expected values $\langle N \rangle$ and $\langle E \rangle$ first, and expressed Ψ_G and $\beta(\varepsilon_i - \mu)$ in terms of the Fermi distribution (5.35) next. We can understand this expression as repeating the formula (3.28) with $\mathcal{N} = 2$ for each state. Namely, for each state, there is a probability $f_\mu(\varepsilon_i)$ that it will be occupied and $1 - f_\mu(\varepsilon_i)$ that it will not, and statistically the entropy is the average of the logarithms of these probabilities, with a negative sign. Since there is no interaction between states, the total entropy is the sum of the entropies of all states. (The number of states is like the number of independent coin flips.)

The heat capacity is defined as

$$c_V = \left(\frac{\partial U}{\partial T}\right)_{N,V} = T\left(\frac{\partial S}{\partial T}\right)_{N,V}, \tag{6.25}$$

according to Eq. (2.59). One usually introduces

$$\gamma(T) := \frac{c_V}{T} = \left(\frac{\partial S}{\partial T}\right)_{N,V}, \tag{6.26}$$

where $\gamma(T)$ is called the *linear specific-heat coefficient*. For fermions, $\gamma(0) > 0$, so the specific heat of a Fermi gas increases linearly near absolute zero. Intuitively, in the classical expression $U = 3NkT/2$ the number N is replaced by the effective number $N_{\mathit{eff}} \sim kT$, so the total is $U \sim (kT)^2$, or $c_V \sim T$. This rough estimate is based on the property that the Fermi distribution (Fig. 6.1) is a constant, i.e., a rectangle of fixed height, so the area under the step (the number of particles absorbing heat) is related to the total area as the width of the step kT is related to the total length of the side ϵ_F.

The precise calculation is slightly complicated by the fact that the chemical potential is fixed in the grand canonical expression (6.24) for the entropy, rather than the number of particles. The corresponding derivatives are connected by a change of variables from (N, V, T) to (μ, V, T):

$$\left(\frac{\partial S}{\partial T}\right)_{N,V} = \left(\frac{\partial S}{\partial T}\right)_{\mu,V} + \left(\frac{\partial S}{\partial \mu}\right)_{V,T} \left(\frac{\partial \mu}{\partial T}\right)_{N,V}. \tag{6.27}$$

The part containing $(\partial\mu/\partial T)_{N,V}$ does not need to be calculated. Namely, it is multiplied by

$$\left(\frac{\partial S}{\partial \mu}\right)_{V,T} = \sum_i \frac{\partial f_\mu(\varepsilon_i)}{\partial \mu} \ln \frac{1 - f_\mu(\varepsilon_i)}{f_\mu(\varepsilon_i)}, \tag{6.28}$$

and when the temperature tends to zero, the derivative of the Fermi function tends to the Dirac function $\delta(\mu - \varepsilon_i)$. But at the points $\mu = \varepsilon_i$, the Fermi function is equal to $1/2$, so the logarithm is zero. Therefore,

$$\gamma(0) = \left(\frac{\partial S}{\partial T}\right)_{\mu,V}\Bigg|_{T=0}. \tag{6.29}$$

It follows from the expression for entropy that

$$\left(\frac{\partial S}{\partial T}\right)_{\mu,V} = k^2 \int_{-\infty}^{\infty} \beta^2(\mu - \varepsilon)^2 \cdot \beta f_\mu(\varepsilon)[1 - f_\mu(\varepsilon)] \cdot g(\varepsilon) d\varepsilon, \tag{6.30}$$

where we have introduced the density of states. By changing the variable $x = f_\mu(\varepsilon)$, the infinite energy range becomes the finite range of the Fermi step:

$$\left(\frac{\partial S}{\partial T}\right)_{\mu,V} = k^2 \int_0^1 g\left(\mu + kT \ln \frac{1-x}{x}\right) \left(\ln \frac{1-x}{x}\right)^2 dx. \tag{6.31}$$

Hence,

$$\gamma(0) = k^2 g(\epsilon_F) \int_0^1 \left(\ln \frac{1-x}{x}\right)^2 dx = \frac{k^2\pi^2}{3} g(\epsilon_F), \tag{6.32}$$

where we have substituted the value ϵ_F of the chemical potential for a given number of particles at zero temperature. The slope of the line by which the heat capacity increases from zero is given by the density of states around the Fermi level. That is clear, because thermal fluctuations can only affect states whose occupancy changes with temperature, and these are on the slope of the Fermi step, according to Fig. 6.1. The more are found there, the faster the heat capacity increases.

6.6 Practical Determination of the Fermi Energy

When the number of fermions is given, it is customary to rename the function

$$\mu^{\star}(N) =: \varepsilon_F(T), \tag{6.33}$$

and call the chemical potential *Fermi energy*, so that the Fermi energy at zero temperature is equal to the energy of the Fermi level, $\varepsilon_F(0) = \epsilon_F$. Thus, the term "chemical potential" remains restricted to the grand canonical ensemble, when it is an independent variable. The equation of state (6.4) is rewritten in terms of the density of states,

$$N = \int_{-\infty}^{\infty} f_{\varepsilon_F(T)}(\varepsilon) g(\varepsilon) d\varepsilon. \tag{6.34}$$

In this form, it can be solved operationally in terms of $\varepsilon_F(T)$.

6.6.1 *Fermi Energy at T = 0*

At absolute zero, the Fermi distribution is a step, so we get

$$N = \int_{-\infty}^{\varepsilon_F(0)} g(\varepsilon) d\varepsilon, \tag{6.35}$$

which can usually only be solved numerically, if the level density is even a mildly complicated function. The only simple case is that of free particles, where $g(\varepsilon)$ is given by (6.15) for $\varepsilon > 0$, and is zero otherwise. Then the integral gives

$$\varepsilon_F(0) = \frac{\hbar^2}{2m}\left(3\pi^2 n\right)^{2/3} =: \frac{\hbar^2 k_F^2}{2m}, \tag{6.36}$$

where $n = N/V$ is the concentration. The expression $p_F = \hbar k_F = \hbar(3\pi^2 n)^{1/3}$ is called the Fermi momentum; it corresponds to the momentum of a fermion in the highest occupied energy state. The Fermi wavenumber $k_F = p_F/\hbar$ and the Fermi temperature $T_F = \varepsilon_F(0)/k$ are defined analogously. As we have already said, it is very high for electrons in metals compared to room temperature, $T \ll T_F$, so the electron gas is said to be *degenerate*. The kinetic energy of quantum fluctuations (zero-point motion) of the highest occupied state of the electron gas at absolute zero is hundreds of times greater than the kinetic energy of thermal fluctuations at room temperature.

6.6.2 Fermi Energy at $T > 0$. Sommerfeld Expansion

Even if the electron distribution in metals is almost indistinguishable from that at absolute zero, the shift of the Fermi energy with temperature cannot generally be neglected. Electrons deep below the Fermi energy cannot participate in many effects, because their excitation to unoccupied states requires too much energy. Thus, often the electrons in the immediate vicinity of the Fermi energy are the only ones that participate in a measurement, and their distribution changes significantly even with small changes in temperature.

Corrections arising from the smearing of the Fermi distribution at finite temperatures can be calculated for many quantities by an elegant mathematical trick. It can be applied whenever the quantity of interest is expressed as an integral over the Fermi function:

$$H(\mu) = \int_{-\infty}^{\infty} h(\varepsilon) f_\mu(\varepsilon) d\varepsilon. \tag{6.37}$$

For example, if $h = g$, the level density, then $H = N$, the number of particles. First we notice that the Fermi function can be written as a function of one variable, $f_\mu(\varepsilon) =: f(\varepsilon - \mu)$. We introduce $H_0(\varepsilon)$ as the indefinite integral of the function h:

$$h(\varepsilon)d\varepsilon =: dH_0(\varepsilon). \tag{6.38}$$

$H_0(\mu)$ is the functional dependence of the quantity H on the chemical potential at absolute zero. Namely, the Fermi distribution is step-like there, so that

$$H(\mu)|_{T=0} = \int_{-\infty}^{\mu} h(\varepsilon) d\varepsilon, \tag{6.39}$$

is a function of the upper limit of integration, whose derivative is $h(\mu)$. Then one partial integration gives

$$H(\mu) = -\int_{-\infty}^{\infty} H_0(\varepsilon) f'(\varepsilon - \mu) d\varepsilon, \tag{6.40}$$

where we assumed that the integrated part vanishes. That is not an important assumption, but it facilitates the notation, and it is satisfied when h contains the level density. By changing the variable, $x = \varepsilon - \mu$,

$$H(\mu) = -\int_{-\infty}^{\infty} H_0(\mu + x) f'(x) dx, \tag{6.41}$$

and now comes the main step: The function H_0 is expanded in a Taylor series, which is expressed as an *operator*:

$$H_0(\mu + x) = \sum_{i=0}^{\infty} \frac{1}{i!} \left(x \frac{\partial}{\partial \mu} \right)^i H_0(\mu) =: e^{x\delta} H_0(\mu), \tag{6.42}$$

where δ is a shorthand for the derivative by μ, $\delta := \partial/\partial\mu$. Now we have simply

$$H(\mu) = O(\beta) H_0(\mu), \tag{6.43}$$

where the operator O can be calculated:

$$O(\beta) = \beta \int_{-\infty}^{\infty} \frac{e^{x(\beta+\delta)}}{(e^{\beta x} + 1)^2} dx = \frac{\pi k T \delta}{\sin \pi k T \delta}, \tag{6.44}$$

which should be understood in terms of its Taylor expansion. Specifically,

$$H(\mu) = \left[1 + \frac{(\pi k T)^2}{6} \frac{\partial^2}{\partial \mu^2} + \frac{7(\pi k T)^4}{360} \frac{\partial^4}{\partial \mu^4} + \dots \right] H_0(\mu). \tag{6.45}$$

This expression is called the *Sommerfeld expansion*.

For example, the functional dependence of the number of particles on the chemical potential is

$$N(\mu) = N_0(\mu) + \frac{(\pi k T)^2}{6} g'(\mu) + \dots, \tag{6.46}$$

if $N_0(\mu)$ is the same dependence at absolute zero. For free particles, $N_0(\mu) = C\mu^{3/2}$, $g(\mu) = 3C\mu^{1/2}/2$, so

$$N(\mu) = C\mu^{3/2} + \frac{(\pi k T)^2}{8} C\mu^{-1/2} + \dots \tag{6.47}$$

If the number of particles is given, $N(\mu) = N = N_0(\mu) = C\varepsilon_F(0)^{3/2}$, where we have expressed N using the known functional dependence at $T = 0$. We have

$$\varepsilon_F(0)^{3/2} = \varepsilon_F(T)^{3/2} \left[1 + \frac{\pi^2}{8} \left(\frac{kT}{\varepsilon_F(T)} \right)^2 + \dots \right], \tag{6.48}$$

which is easy to invert to second order in temperature:

$$\begin{aligned}\varepsilon_F(T) &\approx \varepsilon_F(0)\left[1+\frac{\pi^2}{8}\left(\frac{kT}{\varepsilon_F(0)}\right)^2\right]^{-2/3}\\ &= \epsilon_F\left[1-\frac{\pi^2}{12}\left(\frac{kT}{\epsilon_F}\right)^2+\dots\right].\end{aligned} \tag{6.49}$$

As the temperature increases, the Fermi energy decreases slightly, shifting towards the negative values corresponding to high temperature.

6.6.3 Behavior of the Chemical Potential

Previously, we explained in Fig. 4.2 why the chemical potential of a classical ideal gas is negative. To explain why it is positive for a degenerate Fermi gas, we recall the expression (4.19),

$$\mu = \left(\frac{\partial U}{\partial N}\right)_{S,V}.$$

Again, we ask ourselves how much the entropy will change when a particle is added. At $T = 0$ all fermions are in the lowest possible states, and there is only one such configuration: The entropy of the ground state is zero. If a fermion is now added, it will occupy the lowest empty state, so the entropy is still zero. Thus, when a particle is added, the entropy does not change at all, so we can read directly how much the energy has changed, by ϵ_F, thus, $\mu(T = 0) = \epsilon_F$. When the temperature is higher than zero, but not much higher, it is no longer known which level within $\pm 2kT$ around ϵ_F a new electron will occupy, so the entropy increases somewhat. The system needs to be cooled a little to compensate, but not as much as the energy increased by adding electrons; the overall effect is that the chemical potential is still positive but has decreased slightly, just as we calculated in the previous section.

Hidden in the above consideration is the assumption that entropy increases with internal energy. That is true for free particles, whose single-particle spectrum is unbounded from above, and new particles find an increasing number of new states as the chemical potential increases. But if the fermion spectrum is grouped into shells or bands, an increase in the internal energy with the number of particles can lead to a decrease in the number of available states. Specifically, when the last fermion fills a band at low temperature, such that the gap between the bands $\Delta \gg kT$, the entropy drops to zero. In such a case, the entropy is maximal when the band is half-filled, and decreases with the number of particles thereafter. Therefore, the chemical potential in that filling range will increase when the temperature starts to rise from zero.

6.7 Landau Fermi Liquid

In 1956, Landau described a Fermi liquid motivated by liquid ^{3}He, a fermionic quantum liquid directly analogous to the bosonic ^{4}He, which was of paramount scientific interest at the time. However, his arguments are so general that its presentation is usually cited in textbooks as a theory of any Fermi liquid. His basic idea is that quantum effects, specifically indistinguishability, require corrections to the Boltzmann expression for the entropy, whose maximum value is determined by the equation of state (4.76) that looks like the microscopic Maxwell distribution (4.71). Therefore, it is reasonable (says Landau!) to add a term that reflects Pauli blocking to the Boltzmann entropy, patterned after the microscopic expression (6.24):

$$\sigma(\boldsymbol{r}) = 2V \int \left[-n_{\boldsymbol{p}}(\boldsymbol{r}) \ln n_{\boldsymbol{p}}(\boldsymbol{r}) - (1 - n_{\boldsymbol{p}}(\boldsymbol{r})) \ln(1 - n_{\boldsymbol{p}}(\boldsymbol{r}))\right] \frac{d^3 p}{h^3}. \tag{6.50}$$

Here, $n_{\boldsymbol{p}}(\boldsymbol{r}) = N(\boldsymbol{p}, \boldsymbol{r})/N$ is the number density of particles of momentum $\boldsymbol{p}$ in the *macroscopic* cell centered at position $\boldsymbol{r}$, and the factor 2 comes from spin degeneracy. Entropy maximization, which previously led to the equation of state (4.76), now gives the equation of state of the *Fermi liquid*

$$n_{\boldsymbol{p}}(\boldsymbol{r}) = \frac{1}{e^{(\varepsilon(\boldsymbol{p},\boldsymbol{r})-\mu)/kT} - 1}, \tag{6.51}$$

where we wrote the internal energy of the cell as

$$U(\boldsymbol{r}) = 2V \int \varepsilon(\boldsymbol{p}, \boldsymbol{r}) n_{\boldsymbol{p}}(\boldsymbol{r}) \frac{d^3 p}{h^3}. \tag{6.52}$$

The equation of state (6.51) looks like the microscopic Fermi distribution (5.35), however, the dispersion may depend on temperature here. Because of this similarity, Landau called its self-consistent solutions, $n_{\boldsymbol{p}}(\boldsymbol{r})$ and $\varepsilon(\boldsymbol{p}, \boldsymbol{r})$, the density and dispersion of *quasiparticles*. A quasiparticle is a collective state of many particles, characterized by the same quantum numbers as a single particle.

Since thermodynamic equations of state correspond to measurements directly, the observation of a Fermi distribution for electrons in metals is immediate evidence that they are a Fermi liquid. This justifies their similarity to the ideal Fermi gas phenomenologically, despite strong Coulomb interactions.

6.7.1 Evolution of One-Particle States with the Interaction

The equation of state of the Fermi liquid (6.51) was derived from a crude macroscopic correction of the Boltzmann entropy. As always in thermodynamics, its justification is in agreement with measurements. However, Landau also gave a plausible microscopic argument for why such an approach would work.

It is clear that it cannot always work. For example, if deuterium atoms D bind into molecules D_2, the fermions have turned into bosons, so the interaction has changed the statistical distribution, which means that liquid D_2 is not a Fermi liquid. We must assume something about the interaction that will exclude such a possibility.

Imagine two states of free particles with kinetic energies $\varepsilon_{1,2}$. If they scatter off each other by some interaction $\mathcal{V}$, the Schrödinger equation for such a system is

$$\begin{pmatrix} \varepsilon_1 & \mathcal{V} \\ \mathcal{V} & \varepsilon_2 \end{pmatrix} \begin{pmatrix} \psi_1 \\ \psi_2 \end{pmatrix} = E \begin{pmatrix} \psi_1 \\ \psi_2 \end{pmatrix}, \tag{6.53}$$

whose well-known solutions are

$$E_\pm = \frac{\varepsilon_1 + \varepsilon_2}{2} \pm \sqrt{\left(\frac{\varepsilon_1 - \varepsilon_2}{2}\right)^2 + |\mathcal{V}|^2}. \tag{6.54}$$

Obviously, $|E_+ - E_-| > |\varepsilon_1 - \varepsilon_2|$ for any $\mathcal{V}$. The solutions are said to avoid, or repel, each other.

Landau's condition for the formation of a Fermi liquid is that all single-particle levels behave in the same way when the interaction is slowly (adiabatically) increased from zero to its physical value. If the levels do not intersect, the structure of the ground state will remain the same: The lowest N states in energy will be filled, and the rest will be empty. In fact, the quantum numbers of all single-particle states will remain the same, simply because each energy level is continuously connected to its value at the point $\mathcal{V} = 0$, and we know the quantum numbers there: They are the momentum and spin in an ideal Fermi gas.

Landau's condition is sufficient for the formation of a Fermi liquid from a gas. It played a significant role in the development of many-body theory, where the stability of perturbation calculations against variations in the interaction strength is a standard assumption. Based on this assumption, it can be shown that the density of the interacting system satisfies the Landau equation of state (6.51), which justifies the perturbation approach physically. However, Landau's condition is not necessary. Observation of the equation of state (6.51) means that the physical system is a Fermi liquid regardless of whether Landau's microscopic explanation can be applied to the case at hand.

6.7.2 Effective Mass

As we have already said, the quasi-particle spectrum depends on temperature. If we put $\delta n_{\boldsymbol{p}}(\boldsymbol{r}) = n_{\boldsymbol{p}}(\boldsymbol{r}) - n_{\boldsymbol{p}}(\boldsymbol{r})\big|_{T=0}$, we have up to first order

$$\varepsilon(\boldsymbol{p}, \boldsymbol{r}) = \varepsilon_0(\boldsymbol{p}, \boldsymbol{r}) + 2 \int f(\boldsymbol{p}, \boldsymbol{p}')\delta n_{\boldsymbol{p}'}(\boldsymbol{r}) \frac{d^3 \boldsymbol{p}'}{h^3}. \tag{6.55}$$

Recall that $\varepsilon(\boldsymbol{p}, \boldsymbol{r})$ was the first variation of the total energy, so $f(\boldsymbol{p}, \boldsymbol{p}')$ is the second. Thermodynamically, it is the susceptibility to variations in the momentum distribution that originates from the interaction between quasiparticles, so it is sometimes called the scattering integral. It is the principal phenomenological parameter of Landau's theory.

For small deviations from the Fermi energy, one can write the Taylor expansion

$$\varepsilon(\boldsymbol{p}, \boldsymbol{r}) - \epsilon_F = \varepsilon(\boldsymbol{p}, \boldsymbol{r}) - \varepsilon(\boldsymbol{p}_F, \boldsymbol{r}) \approx \boldsymbol{v} \cdot (\boldsymbol{p} - \boldsymbol{p}_F), \tag{6.56}$$

where the velocity is

$$v = \left|\nabla\varepsilon(\boldsymbol{p}_F, \boldsymbol{r})\right| =: \frac{p_F}{m^*}, \tag{6.57}$$

but the mass m^* is not the mass of fermions that make up the liquid itself, because it was introduced macroscopically, so it includes all reversible collective effects. It is called the *effective mass*. "Reversible" means that the effective mass does not include dissipative responses. If we imagine the quasiparticle as a multitude of particles connected by springs, we should imagine these springs oscillating without dissipation.

Landau showed that on the Fermi surface of an isotropic liquid, the difference between the physical (bare) and effective mass is given by the scattering integral over that surface:

$$\frac{1}{m} - \frac{1}{m^*} = \frac{2p_F}{h^3} \int G(\cos\theta)\cos\theta d\Omega, \tag{6.58}$$

where $G(\cos\theta) := f(\boldsymbol{p}_F, \boldsymbol{p}'_F)$, and θ is the angle between $\boldsymbol{p}_F$ and $\boldsymbol{p}'_F$. In general, we would expect the effective mass to be greater than the bare one, because moving a single particle drags a cloud of particles that are following it self-consistently. This is indeed the case for atoms in Fermi liquids such as ^{3}He, but in crystals it is not uncommon for the effective mass of the conduction electrons to be smaller than the mass of the free electron. The reason is that electron dispersions in crystals originate from the strong Coulomb (chemical) forces that caused the crystal to form, so the quasiparticle velocity reflects the nature of the chemical bond between atoms, not only the interaction among electrons, as in Landau's theory. This merely confirms that Landau's equation of state (6.51) is more general than the context in which it originally appeared.

6.8 Superconductivity

Superconductivity exposes a loophole in Landau's argument. Namely, what if the energies ε_i in the construction (6.53) are all equal? The question is not idle, because at low temperatures $kT \ll \epsilon_F$ all thermodynamically observable fermions have energies within a thin shell around the Fermi surface, in the range $\epsilon_F \pm 2kT$. We are led to consider a degenerate version of Landau's argument, where a large number n of scatterers all have the same energy and interact with the same attractive interaction $\mathcal{V} = -1$, so the Hamiltonian matrix is

$$H_{sc} = -\begin{pmatrix} 0\,1 & \cdots & 1\,1 \\ 1\,0 & \cdots & 1\,1 \\ \vdots\,\vdots & \ddots & \vdots\,\vdots \\ 1\,1 & \cdots & 0\,1 \\ 1\,1 & \cdots & 1\,0 \end{pmatrix}. \tag{6.59}$$

The trace of the matrix is an invariant, so the eigenvalues should add to zero. They do, but in the worst possible way (for Landau): $n - 1$ eigenvalues with the value 1, and one with the value $-n + 1$. The degeneracy is broken in a maximally coherent way, such that one collective state appears whose energy gain is *extensive*. That is the superconducting state. The fermions are still fermions, but the superconducting state does not look like anything in the ideal gas.

A complication in the argument above is that the maximally coherent eigenstate is symmetric in the scatterers (it is the column vector $[1, \ldots, 1]$), so one wonders how they can be fermions. They are not. Each scatterer is a *pair* of fermions, colliding head-on, so the momentum of their center of mass is zero ($\boldsymbol{p}_{CM} = 0$). If the Fermi surface is pictured as a sphere, such a pair appears as two antipodal dots. These pairs are called Cooper pairs, and the scattering above, Cooper scattering. The wave function of a collective of such pairs is a symmetric superposition of them, called the BCS wave function, after Bardeen, Cooper, and Schrieffer, who developed the now-standard theory of superconductivity. In their approach, the attractive interaction between electrons is mediated by sound waves, so it is very weak relative to the kinetic energy ϵ_F at the Fermi surface.

The characteristic energies of sound waves are in the hundred-Kelvin range, while ϵ_F is of the order of tens of thousands Kelvin, as noted already. Thus, Cooper pairs are not bound. The coherence consists in one scattering with $\boldsymbol{p}_{CM} = 0$ being followed by another with overwhelming probability. If a thermodynamically large fraction of fermions spends its time in such head-on collisions, i.e., as Cooper pairs, they constitute the superconducting ordered state, so-called condensate.

Notably, the condensate does not preserve particle number, because a collision with $\boldsymbol{p}_{CM} = 0$ can be followed by another with $\boldsymbol{p}_{CM} \neq 0$, or vice versa, upon which the condensate will lose, or gain, a pair. At equilibrium, the quantum amplitude for the two events must be the same. This amplitude controls the average number of pairs in the condensate, like a chemical potential for Cooper pairs. It is the order parameter of the superconducting state, just like the magnetization is the order parameter for ferromagnetism.

Historically, when BCS published their microscopic theory, there was already a macroscopic description of superconductivity developed by Landau and Ginzburg. In their approach, the thermodynamic order parameter for superconductivity was a complex number (some "effective" wavefunction, in their words). The connection between the two was worked out by Gor'kov a few years later, who showed that the microscopic BCS order parameter became the macroscopic Landau-Ginzburg order parameter upon suitable averaging near the transition point. Given that the Landau-Ginzburg approach connected directly to measurements by equations of state, Gor'kov's derivation provided the macroscopic physical validation of BCS theory.

6.8.1 *Superfluidity in* ^{3}*He*

The paradigmatic Fermi liquid ^{3}He also manages to avoid Landau's argument at extremely low temperatures, by a mechanism similar to superconductivity. It turns out that the scattering integral $f(\boldsymbol{p}, \boldsymbol{p}')$ in ^{3}He is spin-dependent, in such a way that scattering in the triplet channel (both spins parallel, total spin $S = 1$) is preferred relative to the singlet channel (spins antiparallel, $S = 0$). At temperatures below 2 mK, atoms of ^{3}He spend so much time in the triplet channel that the pairs become coherent, which causes the liquid to exhibit bulk superfluidity.

The detailed picture is different than superconductivity, however. The triplet spin state is symmetric, so the orbital (space) part must be antisymmetric, which means that the wave function of relative motion is dominated by the spherical harmonic Y_{LM} with angular momentum $L = 1$. The orbital and spin angular momentum can couple in different ways, accounting for two distinct superfluid phases, which can be further affected even by relatively weak magnetic fields (below 1 T).

The superconducting instability is generic in Fermi liquids, because it is triggered by the degeneracy of thermodynamically accessible states that necessarily occurs near zero temperature. However, the condensation energy is small for the same reason, namely, only the small fraction of fermions at the Fermi surface can participate in Cooper scattering. Thus, it tends to occur at low temperatures, for the ordered state not to be destabilized by thermal fluctuations. The better questions are why it sometimes does not occur, and why it sometimes occurs at unexpectedly high temperatures, up to 150 K in cuprate perovskites. As of this writing, some of these questions are still open.

Chapter 7
Bosons

In der von mir verwendeten Bose'schen Statistik werden die Quanten bzw. Moleküle nicht als voneinander unabhängig behandelt. [...] Ich verabsäumte es, deutlich hervorzuheben, dass hier eine besondere Statistik angewendet ist, die durch nichts anderes als durch den Erfolg vorläufig begründet werden kann.

Albert Einstein

In 1924, Bose sent his rederivation of Planck's blackbody formula to Einstein, who translated it into German and had it published in the *Zeitschrift für Physik*. However, he did not stop there. Bose's main idea was to apply Gibbs' ensemble construction to an assembly of indistinguishable photons, by a purely combinatorial argument. Einstein realized that the same idea could be applied to a gas of material particles, which he proceeded to do himself, in two papers he published within a year in the *Sitzungsberichte der Preussischen Akademie der Wissenschaften zu Berlin*. In the second of these, he predicted the phenomenon of Bose-Einstein condensation, which was experimentally observed only in 1995.

It was noted by contemporaries that the statistics employed by Bose and Einstein was incompatible with the assumption of independent particles. Indistinguishability causes kinematic correlations even in the absence of interactions, a point not made sufficiently clear in Einstein's first exposition, as he admitted later. Bose's innovation was to treat states of motion as independent, rather than particles. He thus inverted standard physical reasoning, which begins with observable particles, in favor of an abstract approach in which states of motion are the primary quantities. This inversion was present already in Gibbs' phase-space ensembles, but it was latent there, because one could (mostly) leave it undetermined, whether $\boldsymbol{p}$ and $\boldsymbol{q}$ in some expression referred to laboratory or phase space. It was arguably Einstein's reframing of Bose's work that brought the abstract approach to its first radical conclusion in the realm of quantum mechanics.

D. Sunko, *Conceptual Thermodynamics*,
https://doi.org/10.1007/978-3-032-20877-4_7

7.1 Bose-Einstein Condensation

The chemical potential of an ideal gas of bosons cannot be greater than zero, because that implies a thermodynamic instability in which the system draws particles from the reservoir until it is drained. However, thermodynamic equilibrium is achieved at low temperatures in a significantly different way than at high temperatures, where the equation of state (5.37) is dominated by the Maxwell distribution, as shown in Fig. 5.4. We can see this if we try to pass from the sum to the integral in the limit of a large system,

$$N = \langle N \rangle = \sum_i n_\mu(\varepsilon_i) \to \int_0^\infty n_\mu(\varepsilon) g(\varepsilon) d\varepsilon \quad \text{(wrong!)}. \tag{7.1}$$

This integral vanishes in the low-temperature limit, indeed, there is a certain finite temperature below which the equation of state is unsolvable as written. This is easily proven: When $\mu < 0$, $n_\mu(\varepsilon) < n_0(\varepsilon)$, and the change of variable $x = \varepsilon/kT$ gives

$$\int_0^\infty n_0(\varepsilon) g(\varepsilon) d\varepsilon = CVT^{3/2}, \tag{7.2}$$

without calculating the constant C. Thus, if we define some temperature T_0 by the total number of particles,

$$N =: CVT_0^{3/2}, \tag{7.3}$$

the equation of state turns out to be unsolvable for temperatures $T < T_0$. Where did we go wrong?

Mathematically, the transition from sum to integral is allowed if the function $n_\mu(\varepsilon)$ converges *uniformly* on the integration interval $[0, \infty)$. This is not the case, because the singularity at $\varepsilon = \mu$ can approach zero from the left at will, thus spoiling any would-be uniform convergence evaluation for $\varepsilon \geq 0$. But if we replace the integration interval by $[\delta, \infty)$, with an arbitrary $\delta > 0$, then we can take any point in the interval $(0, \delta)$ for a uniform evaluation of convergence in $[\delta, \infty)$, including for $\mu = 0$. Even better, the preceding analysis bounded the integral from above by $CVT^{3/2}$ for $\delta = 0$ as well, which allows us to replace the interval $[0, \infty)$ with $(0, \infty)$ at the cost of keeping only the first term of the sum outside the integral, so the correct equation of state is

$$N = \langle N \rangle = n_\mu(0) + \int_0^\infty n_\mu(\varepsilon) g(\varepsilon) d\varepsilon. \tag{7.4}$$

It is clearly solvable, because the first term balances what the second cannot, $n_\mu(0) = N_0$. At temperatures $T < T_0$, a macroscopic number N_0 of particles is at rest. This phenomenon is called *Bose-Einstein condensation* (BEC).

7.2 Chemical Potential and Transition Temperature

At temperatures below T_0, the chemical potential is easily determined from the equation

$$n_\mu(0) = N_0 \implies \mu = -kT/N_0, \tag{7.5}$$

because we do not have to go to higher orders in $1/N_0$ when N_0 is of the order of Avogadro's number. Figure 7.1 shows that this approximation is good already for $N_0 \approx 100$. Thus, in the limit of a large system, $\mu = 0$ as soon as $T < T_0$. Then $CVT^{3/2}$ is the exact value of the integral in this temperature range, so we can write

$$CVT_0^{3/2} = N_0 + CVT^{3/2} \implies \frac{N_0}{N} = 1 - \left(\frac{T}{T_0}\right)^{3/2}. \tag{7.6}$$

As soon as we know the chemical potential, dimensional analysis of the equation of state gives the dependence of the fraction of condensed particles on the temperature. The number of particles in the condensate is called the *order parameter* of the low-temperature (condensed) phase.

If we want to determine the temperature T_0, we need to calculate the constant C. For this, it is convenient to introduce the *fugacity* of the distribution,

$$\xi := e^{\beta\mu}. \tag{7.7}$$

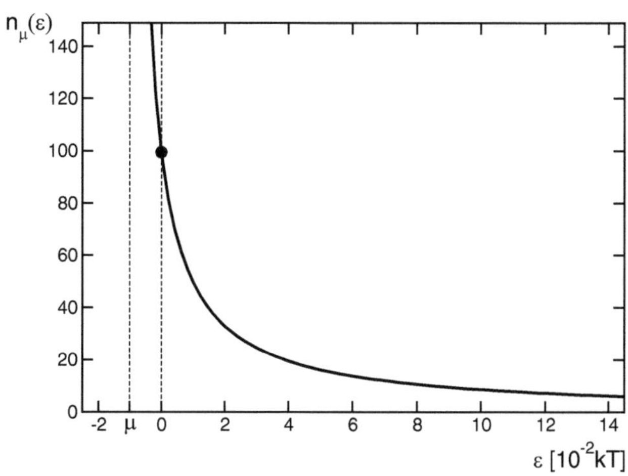

Fig. 7.1 Bose-Einstein distribution with chemical potential in the quantum regime, $\mu = -kT/100$. The closed circle indicates the expected number of particles at rest, $N_0 \approx -kT/\mu = 100$

Expanding in a series and integrating term by term,

$$\int_0^\infty \frac{g(\varepsilon)d\varepsilon}{e^{\beta\varepsilon}/\xi - 1} = \frac{V}{\lambda_T^3} F(\xi), \quad F(\xi) = \sum_{m=1}^{\infty} \frac{\xi^m}{m^{3/2}}. \tag{7.8}$$

The function $F(\xi)$ is called London's function. To determine the constant C we need[1] $F(1) = 2.6124\ldots$, and we obtain the condition on the transition temperature:

$$F(1)\frac{V/N}{\lambda_{T_0}^3} = 1 \Longrightarrow kT_0 = 3.3125\ldots \frac{\hbar^2}{M}\left(\frac{N}{V}\right)^{2/3}. \tag{7.9}$$

When the mass of ^{4}He and its density at the liquefaction temperature $T = 4.2$ K, $N/V = 27.6$ mol/cm^3, are inserted, one obtains $T_0 \approx 3.2$ K, which is not far from the temperature of the superfluid transition, $T_c = 2.17$ K. As we shall see in Sect. 7.5, the question whether this is a coincidence is subtle and took decades to be settled.

7.2.1 *The Meaning of the Chemical Potential*

We can repeat the analysis of Sect. 6.6.3 to understand the behavior of the boson chemical potential as well. If the temperature is $T < T_0$, the new boson will enter the condensate, whose entropy is zero, no matter how many particles are in it, so the chemical potential of the boson system is zero as long as the condensate exists. As soon as it is gone, the new boson increases the entropy of the system, so we need to decrease the internal energy to remain on the adiabate. Thus, the chemical potential of the bosons becomes negative as soon as the temperature increases above T_0.

7.3 Entropy and Heat Capacity

The inversion of the Legendre transform (2.49) gives the operational expression for the entropy of a Bose gas:

$$\begin{aligned}\sigma &= \Psi_G - \beta\mu\langle N\rangle + \beta\langle E\rangle = \Psi_G + \sum_i \beta(\varepsilon_i - \mu)n_\mu(\varepsilon_i) \\ &= \sum_i \left[-n_\mu(\varepsilon_i)\ln n_\mu(\varepsilon_i) + (1+n_\mu(\varepsilon_i))\ln(1+n_\mu(\varepsilon_i))\right],\end{aligned} \tag{7.10}$$

[1] $F(1) = \zeta(3/2)$, where $\zeta(s)$ is the Riemann zeta-function.

which we could have guessed by comparison with the fermion expression (6.24). Along the same lines,

$$\begin{aligned}\left(\frac{\partial S}{\partial T}\right)_{\mu,V} &= k^2 \int_{-\infty}^{\infty} \beta^2(\mu-\varepsilon)^2 \cdot \beta n_\mu(\varepsilon)[1+n_\mu(\varepsilon)] \cdot g(\varepsilon)d\varepsilon \\ &= k^2 \int_0^{\infty} g\left(\mu + kT\ln\frac{1+x}{x}\right)\left(\ln\frac{1+x}{x}\right)^2 dx,\end{aligned} \tag{7.11}$$

where we changed the variable, $x = n_\mu(\varepsilon)$. The calculation of the heat capacity in the condensed phase is facilitated by the fact that the chemical potential is constant, $\mu = 0$ for $T < T_0$, so the correction in the change of variables (6.27) disappears. We see immediately that $c_V \sim Tg(kT) \sim T^{3/2}$, or more precisely:

$$\begin{aligned}c_V &= T\left(\frac{\partial S}{\partial T}\right)_{N,V} = T\left(\frac{\partial S}{\partial T}\right)_{\mu,V} \qquad (T < T_0) \\ &= k\frac{V}{\lambda_T^3}\left[\frac{2}{\sqrt{\pi}}\int_0^{\infty}\left(\ln\frac{1+x}{x}\right)^{5/2} dx\right] = aNk\left(\frac{T}{T_0}\right)^{3/2}.\end{aligned} \tag{7.12}$$

Note that $c_V \sim V/\lambda_T^3$, which is the size of the phase space of one particle. The constant $a = 1.925... > 3/2$, so the heat capacity at the transition point is higher than that of a classical ideal gas. At higher temperatures, it decreases asymptotically towards that value, thus, it has a sharp spike at $T = T_0$ (discontinuity of the derivative).

7.4 Continuous Phase Transitions

Bose-Einstein condensation is the only phase transition in nature that is caused exclusively by kinematic correlations, without interactions. Despite this uniqueness, it shares some general properties of all second-order phase transitions. These are most easily seen in the canonical picture. From the expression (5.32) for the grand canonical sum over states,

$$Z_G(\mu,\beta) = \prod_i \left(1 + e^{\beta(\mu-\varepsilon_i)} + e^{2\beta(\mu-\varepsilon_i)} + e^{3\beta(\mu-\varepsilon_i)} + \ldots\right),$$

it can be seen that terms of the form $e^{N\beta(\mu-\varepsilon_i)}$ appear in it exactly once for a given N, namely, when we choose that term in the i-th factor of the product $Z_G(\mu,\beta)$, and unity in all others. Thus, in the canonical sum over states for bosons, there appears a contribution $\sum_i e^{-N\beta\varepsilon_i} = Z_C(1, N\beta)$ from configurations where all bosons are in the same single-particle state. On the other hand, there also appears a classical contribution, so we write the end terms in Faà di Bruno's formula (5.27) by inspection,

$$Z_C(N,\beta) = \frac{1}{N!} Z_C(1,\beta)^N + Z_C(1, N\beta) + \ldots, \tag{7.13}$$

where the remaining terms give quantum corrections to the contribution of the first term, in which there are more than one and less than N bosons in the same state. [The second term should be multiplied by a factor of $(1 - 1/N!)$ here, because one contribution of that form is already in the first term, but that is clearly negligible.]

The single-particle canonical sum over states is $Z_C(1,\beta) = V/\lambda_T^3$, so we get

$$Z_C(N,\beta) = \left(\frac{eV/N}{\lambda_T^3}\right)^N + \frac{1}{\sqrt{N}} \frac{V/N}{\lambda_T^3} + \ldots. \tag{7.14}$$

This expression exhibits the main competition between classical and quantum effects. All terms in this expansion are dominated by the first one when the ratio in parentheses is greater than 1, and by the second when it is smaller. The transition occurs when the ratio[2]

$$e \frac{V/N}{\lambda_T^3} = 1. \tag{7.15}$$

The expression (7.14) indicates that the higher the number of particles N, the sharper the transition will be in temperature. The transition is formally continuous, but if N is of the order of Avogadro's number, the Nth power of the ratio practically has only two values, zero and infinity. This explains why thermodynamic phase transitions are discontinuous, even if the sum over states is a sum of exponential functions, which are analytic everywhere. For a small number of particles, it is observed both in simulations and in measurements that the phase transition is smeared over a finite range in temperature.

However, when the ratio passes through unity, the subdominant terms of the expansion (7.14) cannot be ignored. They represent *critical fluctuations*, states of motion whose probability is negligible both above and below the transition temperature. We can estimate the width of the temperature range over which they can be seen. Critical fluctuations are significant whenever the magnitude of the order parameter is less than its standard deviation, i.e., consistent with zero:

$$N_0 < \sqrt{\Delta N_0^2}. \tag{7.16}$$

[2] This condition is similar to the one before, only with $e = 2.71\ldots$ instead of $F(1) = 2.61\ldots$, but the criterion "vanishing of the classical contribution" is not measurable.

This estimate is called the *Ginzburg criterion*. It can be applied to all *continuous phase transitions*, where the order parameter tends to zero continuously as the temperature approaches the transition temperature from below. For a Bose gas, the number fluctuation is

$$\frac{1}{\beta}\frac{\partial n_\mu(\varepsilon)}{\partial \mu} = n_\mu(\varepsilon)[1 + n_\mu(\varepsilon)], \tag{7.17}$$

so the Ginzburg criterion for N_0 gives

$$0 < 1 - \left(\frac{T}{T_0}\right)^{3/2}, \tag{7.18}$$

which is satisfied for all temperatures $0 \leq T < T_0$. The entire ordered phase is a region of critical fluctuations! Given the previous discussion, we would expect them to disappear below T_0 as quickly as the classical contribution begins to dominate above T_0. This is indeed the case in most second-order phase transitions, such as magnetic transitions, where the susceptibility diverges only within a small range around the transition temperature, and is otherwise finite in the ordered phase as well, where the order parameter behaves in accordance with classical expectations. Bose-Einstein condensation is unique in that the order parameter N_0 itself has fluctuations of order N_0, instead of $\sqrt{N_0}$ as in the classical ideal gas. (If the magnetization changed incessantly like that due to quantum effects, electric motors would not work.) The formal reason for such behavior is that in the expression (7.13), due to indistinguishability, the probability that we will see all bosons in *any* same state is $N!$ times higher than the classical expectation. Physically, we may imagine that the center of mass of the whole condensate fluctuates microscopically.

7.5 Superfluidity

Helium is the only atomic gas that does not solidify under normal conditions. This is because the attractive van der Waals force between the ^{4}He atoms is very weak, so the potential well in the interatomic potential is only about 9 K deep, while the momentum uncertainty at the corresponding interatomic distance is about 5 K, so the solid melts spontaneously by quantum fluctuations. A pressure of about 2.5 MPa is required to obtain a solid. Thus, liquid helium is a macroscopic quantum liquid. It liquefies at 4.2 K, and at 2.17 K it passes into another thermodynamic state, in which it exhibits a number of unexpected properties, the most easily observable one being flow without internal friction. It was observed in 1938 as a sudden drop in viscosity (by a factor of at least 1500) in flow through a narrow capillary with an imposed pressure gradient, as soon as the temperature dropped below 2.17 K.

F. London immediately attributed this behavior (*superfluidity*) to a macroscopic Bose-Einstein condensate (BEC) of ^{4}He atoms, which are obviously bosons, which was ignored by Landau in his quantum hydrodynamic description (cf. Sect. 9.3). Thermodynamically, BEC is a gas state, while a liquid is a bound state of lower energy, so it apparently makes no sense to refer to a gas below 2 K when it vanishes already above 4 K. A specific objection is that the heat capacity of the BEC increases as $T^{3/2}$ at low temperatures, while that of liquid ^{4}He increases as T^3, as expected from the analysis of isotropic sound waves in Chap. 9 below. The BEC has served as an aid to physical intuition in understanding superfluidity for decades, without the theoretical predictions of its presence in superfluid helium being confirmed experimentally. Finally, two classes of experiments in the 1990s, neutron scattering and sound wave evaporation, demonstrated convincingly that a BEC fraction exists in superfluid helium, of about 8% when the data are extrapolated to absolute zero. The resolution of the apparent thermodynamic paradox is that the liquid is very weakly bound, so that due to quantum fluctuations it spends part of its time in an unbound (gaseous) state even at absolute zero. (Such behavior is not unheard of; for example, it can be shown that a neutron and a proton in deuterium spend approximately $1/3$ of their time at a distance greater than the range of the attractive nuclear force.)

7.5.1 BEC in a Superfluid

A Bose-Einstein condensate (BEC) in an ideal gas should be distinguished from a superfluid, such as occurs in ^{4}He at 2.17 K. BEC means that N_0, the number of particles at rest, is thermodynamically large. The expectation value (first derivative of the Massieu potential) appears as an order parameter of the system that vanishes at the transition, according to Eq. (7.6). The number of particles in the ground state therefore behaves directly analogously to the spontaneous magnetization in ferromagnets, which also disappears above the transition temperature. It is also the value of the susceptibility, that is, the number fluctuation, which also diverges in the ordered phase, as we saw above.

On the other hand, superfluidity means that if we have ^{4}He in a pipe below the transition temperature, and we tug the pipe a little along the axis, due to surface roughness it will drag along only a part of the liquid that is *not* superfluid, that is, if we transform to the system where the pipe is at rest, we will see that only the superfluid moves through the liquid as a single block. At the level of the free energy, such a state can be described by

$$F = F_0 + N_s \frac{Mv^2}{2}, \tag{7.19}$$

where N_s is the number of atoms of mass M in the superfluid fraction, and v is the pipe velocity. Obviously $N_s \sim \partial^2 F/\partial v^2$, so N_s is the second derivative of the thermodynamic potential, and therefore the divergence (macroscopic number) N_s physically represents the divergence of a susceptibility. The imaginary experiment with which we identified it is precisely the definition of susceptibility: If we push the system a little, and it tilts a lot, the susceptibility is large. Specifically, if we tug the pipe a little, and all the ^{4}He stays still, we have $N_s/N = 100\%$, all the helium is superfluid. This is the case at absolute zero, while the BEC fraction is only around 8%. The superfluid fraction is also the order parameter. Thus, identical thermodynamic reasoning can be applied to the superfluid fraction as previously to the BEC, but they are not the same physical objects.

The role of the BEC in the superfluid state can be understood by considering the single-particle density matrix constructed by projection from the many-particle coherent wave function Ψ of the ground state of the bosonic liquid:

$$\begin{aligned}\rho(\boldsymbol{r},t;\boldsymbol{r}',t') &= \int \Psi^*(\boldsymbol{r},\boldsymbol{r}_2,\ldots,\boldsymbol{r}_N,t)\Psi(\boldsymbol{r}',\boldsymbol{r}_2,\ldots,\boldsymbol{r}_N,t')d^3r_2\ldots d^3r_N \\ &= \sum_i \lambda_i \chi_i^*(\boldsymbol{r},t)\chi_i(\boldsymbol{r}',t'),\end{aligned} \tag{7.20}$$

where we diagonalized it in the second step. Now, let us think what happens at one point in space, $\boldsymbol{r} = \boldsymbol{r}'$, i.e., what $\rho(\boldsymbol{r},0;\boldsymbol{r},t)$ looks like after a not too short time t. Individual functions χ_i change in both amplitude and phase over time, while a not too short time is one in which all phases have made a full circle several times, each at its own pace, so that there are no correlations between them. If all the eigenvalues of λ_i are of order 1, and N is the number of particles, the one-particle density matrix is the sum of many random contributions, whose total phase behaves as a random variable at each point of space, while its amplitude is of the order of the standard deviation of all λ_i. This scenario describes an incoherent quantum system. On the other hand, if one of the values λ_i is of order N_A while the others are of order 1, the overall density matrix will be dominated by that one:

$$\rho(\boldsymbol{r},0;\boldsymbol{r},t) = \lambda_0|\chi_0^*(\boldsymbol{r},0)\chi_0(\boldsymbol{r},t)|e^{i[\phi(\boldsymbol{r},t)-\phi(\boldsymbol{r},0)]}. \tag{7.21}$$

Numerically, say λ_0 is 0.1% of N, amounting to 10^{21} for $N = 10^{24}$, while the other $\lambda_i \sim 1$ add with random phases, so their sum is of the order $\sqrt{N} \sim 10^{12}$, which justifies the equality.

In other words, as soon as a macroscopic BEC component appears, no matter how small it is observationally, the density matrix will have a phase as if it all consisted of that component. By comparing (7.20) and (7.21), we see that we can write

$$\arg\Psi(\boldsymbol{r},\ldots,t) - \arg\Psi(\boldsymbol{r},\ldots,0) = i(\phi(\boldsymbol{r},t) - \phi(\boldsymbol{r},0)), \tag{7.22}$$

which means that the macroscopic component "slaves" the entire many-body function to behave as if it had its phase alone. Since the current is proportional to the phase gradient, the BEC plays a similar role as the vector potential in the

Meissner effect: It provides an external reference point for the phase to all particles, thus enabling a macroscopic coherent current without dissipation.

7.5.2 Long-Range Order in Space

Instead of observing correlations at a single point in space for long periods of time, we can observe them at a given instant over large distances. Due to translational invariance, the density matrix depends only on the distance $\boldsymbol{r} - \boldsymbol{r}'$, and for $T < T_c$ it is found by numerical calculation that, even in the limit of strong interactions,

$$\lim_{|\boldsymbol{r}-\boldsymbol{r}'|\to\infty} \rho(\boldsymbol{r}, t; \boldsymbol{r}', t) = n_0 + \frac{const.}{|\boldsymbol{r} - \boldsymbol{r}'|}, \tag{7.23}$$

where n_0 is the BEC fraction, which is finite, but can be much smaller than 100%. The phenomenon that the single-particle density matrix does not vanish over large distances is called off-diagonal long-range order (ODLRO); "off-diagonal" means $\boldsymbol{r} \neq \boldsymbol{r}'$.

For this reason, it is believed that superfluidity is not possible without a finite BEC fraction for $T < T_c$. Namely, correlations over large spatial distances are the essential property of superfluidity, and one can see above that the survival of these correlations at infinite range is due to the BEC. Interestingly, however, it does not matter how much of it there is. The phase of the entire ground state wave function is fixed over large distances, although the BEC is only a part of it. This can be intuitively understood as an effect of indistinguishability: Every atom can be found in the BEC at any time with a fixed probability. The momentum of the BEC is zero, so each of its atoms is delocalized with a constant wave function. One way that any atom anywhere can switch places with one in a BEC is obviously that they have the same phase, so the exchange occurs without interference effects. (The numerical calculation above suggests that this is the only way, although that has not been proven.) Thus, it turns out that the BEC provides a reference point for the phase of the entire superfluid, however small its fraction may be, and the presence of the BEC makes that phase a physical order parameter, rather than an arbitrary choice of origin of the phase. In more scholarly language, the BEC is a sign of gauge-symmetry breaking, i.e., disappearance of arbitrariness in the choice of the phase of the wave function, and such a physically fixed collective phase is essential for superfluidity.

Chapter 8
Light

Will man sich also mit der Thermodynamik, und auch mit der Erfahrung, in Uebereinstimmung setzen, so ist das nur möglich auf Grund einer neuen, von den Maxwell'schen Gleichungen unabhängigen Hypothese.

Max Planck

When we walk down a street in summer, the open windows on the upper floors appear black. Light entering the window scatters on the interior before it finds a way out, by which time it is in thermal equilibrium with the room. Electromagnetic radiation at room temperature is mostly in the infrared part of the spectrum, which is why we see it as black. One can reproduce this arrangement in a laboratory, by drilling a small hole in a polished metal box and measuring the radiant energy emitted by the hole as a function of frequency and temperature, the so-called *blackbody radiation*. Based on earlier measurements by Dulong and Petit, Josef Stefan postulated the empirical equation of state for the energy density

$$\frac{U}{V} = \frac{4\sigma}{c} T^4, \tag{8.1}$$

with a measured value of σ of the order of $10^{-8}\,\mathrm{Wm^{-2}\,K^{-4}}$. Ludwig Boltzmann later established theoretically that $U \propto VT^4$, but attempts to calculate the constant ran into a serious problem, that the value appeared to be infinite, a result known as the Rayleigh-Jeans ultraviolet catastrophe. Using the formalism developed so far, we first give a short technical derivation of this equation of state, and then discuss the physics at length.

D. Sunko, *Conceptual Thermodynamics*,
https://doi.org/10.1007/978-3-032-20877-4_8

8.1 Stefan-Boltzmann Equation of State

We take light to be a gas of non-interacting bosons called photons, an attitude evidently based on hindsight. The internal energy of the photon gas is the average over all one-body energies,

$$U = \int_0^\infty \varepsilon n_\mu(\varepsilon) g(\varepsilon) d\varepsilon. \tag{8.2}$$

Because photons are bosons, we use the Bose-Einstein distribution

$$n_\mu(\varepsilon) = \frac{1}{e^{\beta(\varepsilon-\mu)} - 1},$$

which raises the question, what is the chemical potential of light. One cannot control how many photons of a given frequency the gas will contain, so the only unprejudiced approach is a sum of all sums over states with fixed photon number with equal a priori probabilities:

$$\sum_N Z_C(N, \beta) = \sum_N e^{\beta \cdot 0 \cdot N} Z_C(N, \beta) = Z_G(\mu, \beta)|_{\mu=0} \,. \tag{8.3}$$

This sum is formally the *grand* canonical sum over states at $\mu = 0$, which fixes the chemical potential.

The expression for the level density must be derived anew, because the dispersion of light is different than for free material particles. We can obtain it following the same reasoning as before. The cumulative number of one-particle states in a spherical volume of radius p in momentum space is

$$\mathcal{N} = 2\frac{V V_p}{h^3} = \frac{8\pi}{3}\frac{V p^3}{h^3}, \tag{8.4}$$

where V is the volume of the container as usual, and the factor 2 refers to the two polarizations of light. The dispersion is $\varepsilon = cp$, which can be easily inverted to write the cumulative number of states as a function of energy:

$$\mathcal{N}(\varepsilon) = \frac{8\pi}{3}\frac{V\varepsilon^3}{c^3 h^3}, \tag{8.5}$$

so that the level density is

$$g(\varepsilon)d\varepsilon = \frac{d\mathcal{N}}{d\varepsilon}d\varepsilon = 8\pi V \frac{\varepsilon^2}{(hc)^3}d\varepsilon. \tag{8.6}$$

Following Planck, we set $\varepsilon = h\nu$, whence the density of states becomes a function of frequency:

$$g(\varepsilon)d\varepsilon = V\frac{8\pi\nu^2}{c^3}d\nu =: V \cdot G(\nu)d\nu. \tag{8.7}$$

Notably, h has dropped out, so the single-mode level density is the same as obtained by a derivation within Maxwell's electromagnetism. Putting everything together, the total energy per unit volume is

$$\frac{U}{V} = \int_0^\infty u(\nu, T)d\nu, \tag{8.8}$$

where

$$u(\nu, T) = G(\nu) \cdot \frac{h\nu}{e^{h\nu/kT} - 1} = \frac{8\pi h}{c^3}\frac{\nu^3}{e^{h\nu/kT} - 1} \tag{8.9}$$

is the famous *Planck distribution*, the first correct quantum formula in the history of physics. It fits the observed spectrum of blackbody radiation over all frequencies at any given temperature.

Now we can calculate the constant σ. Rescaling with temperature, $s = h\nu/kT$, simple dimensional analysis of the integral gives $U \propto VT^4$:

$$\frac{U}{V} = \int_0^\infty u(\nu, T)d\nu = \frac{8\pi h}{c^3}\left(\frac{kT}{h}\right)^4 \int_0^\infty \frac{s^3 ds}{e^s - 1}. \tag{8.10}$$

The now-dimensionless integral evaluates to $\pi^4/15$, so finally

$$\frac{U}{V} = \frac{4\sigma}{c}T^4, \quad \sigma = \frac{2\pi^5 k^4}{15c^2h^3} = 5.6704\ldots \times 10^{-8}\,\mathrm{Wm^{-2}K^{-4}}. \tag{8.11}$$

Measuring this constant is not easy. The first reported measured value consistent with theory within one standard deviation dates from 1971.

8.2 Comparison with the Classical Approach

The efficiency of the derivation above is a testimony to the formalism, but makes it difficult to understand what is happening. One should first understand where classical reasoning so failed *physically* in the particular case of light that it produced a nonsensical result. Notably, it does not help understanding that Planck's constant h is in the denominator of the expression for σ, because h is classically the unit of measure for the volume of phase space, so it makes no sense to take $h = 0$ as the classical limit.

Physically, the classical limit is the high-temperature limit, here $kT \gg h\nu$. Then the average energy of a mode becomes

$$h\nu \cdot \frac{1}{e^{h\nu/kT} - 1} \to h\nu \cdot \frac{kT}{h\nu} = kT, \tag{8.12}$$

so that we recover the equipartition theorem. Now we can get an infinity without setting h to zero:

$$\frac{U}{V} = \int_0^\infty u(\nu, T) d\nu \to \frac{8\pi h}{c^3} \left(\frac{kT}{h}\right)^4 \int_0^\infty s^2 ds \to \infty, \tag{8.13}$$

because the *dimensionless* integral is infinite, while h retains its classical role. This is the ultraviolet catastrophe. The classical prediction fails because it assigns a fixed energy kT to all vibration modes of light, irrespective of frequency. One needs to understand why it is wrong to do so.

8.3 Planck's Insight, Modern Version

The problem in the classical expression for the radiant energy is that the Stefan-Boltzmann constant diverges when the equipartition theorem is applied. But that theorem is itself the result of an integral, the average energy of a degree of freedom with quadratic dispersion $\varepsilon(s) = s^2$:

$$\langle \varepsilon \rangle = \frac{\int_{-\infty}^{\infty} s^2 e^{-s^2/kT} ds}{\int_{-\infty}^{\infty} e^{-s^2/kT} ds} = \frac{kT}{2}, \tag{4.79}$$

or kT for the elementary oscillator, which has two such terms. In effect, Planck discovered that this integral does not describe the thermalization of light correctly. He made a toy model of the interaction of light with the walls of the container, in other words, a simplified picture of thermalization. The model accounted for thermalization by "counting on fingers," meaning that it only allowed *discrete* packets of energy to be exchanged between light and the walls. In such a model, sums appear instead of the integrals in the equipartition theorem. Planck's original approach was to calculate the entropy of such a system of packets in terms of their energy, and connect the resulting expression with the temperature via the equation of state $\partial S/\partial U = 1/T$. We replace it here by a direct calculation of the average energy of a mode in the same model.

The substitute for the equipartition theorem is easy to calculate in the discrete-packet model of thermalization, assuming that the energy of the exchange at frequency ν is a whole multiple n of a fixed value $h\nu$. The arbitrary constant h fixes the size of packets in the model. Then an exchange of size $nh\nu$ happens with the probability $\exp(-nh\nu/kT)$ by Gibbs, so the average energy of the thermalized mode of frequency ν is

$$\langle\varepsilon(\nu)\rangle = \frac{h\nu e^{-h\nu/kT} + 2h\nu e^{-2h\nu/kT} + \dots}{1 + e^{-h\nu/kT} + e^{-2h\nu/kT} + \dots}, \tag{8.14}$$

where we recognize the derivative of the geometric series in the numerator:

$$q(1 + 2q + \dots) = q\frac{d}{dq}(1 + q + q^2 + \dots) = \frac{q}{(1-q)^2}, \tag{8.15}$$

with $q := \exp(-h\nu/kT)$. The same geometric series appears in the (discretized) Gibbs norm in the denominator, so the model replaces the equipartition $\langle\varepsilon\rangle$ for light with

$$\langle\varepsilon(\nu)\rangle = h\nu \cdot \frac{q/(1-q)^2}{1/(1-q)} = \frac{h\nu}{e^{h\nu/kT} - 1}. \tag{8.16}$$

The same result can be obtained even more compactly by the standard statistical expression

$$\langle\varepsilon(\nu)\rangle = -\frac{\partial}{\partial\beta}\ln Z,$$

where Z is the sum over states for the discrete-packet model,

$$Z = 1 + e^{-\beta h\nu} + e^{-2\beta h\nu} + \dots = \frac{1}{1 - e^{-\beta h\nu}}. \tag{8.17}$$

Whatever the technicalities, one cannot set $h = 0$ to remove quantization, because the model then reverts to the classical equipartition theorem, Eq. (8.12). The correct expression avoids the ultraviolet catastrophe because the discreteness of the packets introduces a new low-temperature physical regime, $kT \ll h\nu$, in which the contribution of the high-energy modes is exponentially small.

Thus, it turns out that Planck's counting on fingers is how light thermalizes in physical reality: Exchange of energy between light of frequency ν and the walls happens only in discrete packets of size $h\nu$. Planck's result, which fitted the whole measured spectrum, showed that h was a fundamental natural constant, like the speed of light, called Planck's constant ever since.

8.4 Planck's Insight, Original Version

Planck's original approach combines an extensively reasoned determination of the main problem, a physical adaptation of kinetic gas theory to light, and a statistical approach following Boltzmann. Here, only the physics is recounted, in so far as it is missed by the formal application of Gibbs' method above.

Light appears very differently from the point of view of Maxwell's equations and from the thermodynamic one. This observation is the same as in gases, only with Newton's equations replaced by Maxwell's. A physical treatment is even more difficult here, because a single ray of light is unimaginably more complex than a particle trajectory: Two time-dependent vectors at each point in space, complicated further because no ray is truly monochromatic, so we have to deal with a frequency distribution at the same time. Yet no trace of such complexity is observed in blackbody radiation, which empirically depends only on temperature. The main problem is: How does light thermalize?

Planck's starting point is Kirchoff's claim that the spectrum of light in thermal equilibrium with matter does not depend on what the matter is: It is always the same and determined solely by the temperature. Therefore, it is legitimate to model matter as simply as possible. He chooses a system of "resonators," imagined as dipoles with opposing charges connected by springs. They are an idealized representation of the material thermostat, not of light itself. The resonators' damping is purely by emission of light, because a mechanical friction term introduces Joule heating, which would make the problem intractable. Modes of different frequencies are independent by Maxwell's equations, so it is sufficient to equilibrate light of one fixed frequency with the resonators.

The essential new hypothesis, independent of Maxwell's equations, is that light in the container is not just any solution of these equations. It is "natural radiation" (*natürliche Strahlung*), namely radiation whose available energy is spread randomly over all oscillation modes compatible with its boundary conditions.

With this hypothesis, the resonator model reduces the physical thermalization problem to a technical one: Find the entropy of the resonators as a function of their energy, $S = S(U)$. The equation of state $1/T = \partial S/\partial U$ can then be solved to give $U(T)$ at one frequency, which should give Stefan-Boltzmann's equation when integrated over all frequencies. The distribution of energy in the field must be the same as among the resonators, otherwise they would not be in equilibrium with it.

Enter Boltzmann's statistical method: The entropy is given by the logarithm of the number of "complexions," or ways one can distribute a given amount of energy among the resonators. Now, however, the method encounters an obstacle: This number is uncountably large. To make it countable, one must (says Planck!) allow each resonator to absorb or emit energy only in discrete packets—quanta—of a fixed size ϵ. A large number N of resonators then have the energy

$$U_N = P\epsilon, \tag{8.18}$$

where P is another large number. The resonators are taken to be independent, so N resonators have total energy and entropy, respectively,

$$U_N = N\mathcal{U}, \quad S_N = N\mathcal{S}, \tag{8.19}$$

where $\mathcal{U}$ and $\mathcal{S}$ are the equilibrium energy and entropy of a single resonator.

The number of complexions R is obtained combinatorially, crucially by treating the N resonators as distinguishable, but the P energy packets distributed among them as indistinguishable:

$$R = \binom{N+P-1}{N-1} \simeq \frac{(N+P)^{(N+P)}}{N^N P^P}. \tag{8.20}$$

A brief calculation then yields

$$\mathcal{S} = \frac{1}{N} k \ln R = k\left\{-\frac{\mathcal{U}}{\epsilon}\ln\frac{\mathcal{U}}{\epsilon} + \left(1+\frac{\mathcal{U}}{\epsilon}\right)\ln\left(1+\frac{\mathcal{U}}{\epsilon}\right)\right\}, \tag{8.21}$$

which gives the equation of state for light in terms of $\mathcal{U}$ and ϵ:

$$\frac{1}{T} = \frac{\partial\mathcal{S}}{\partial\mathcal{U}} \Longrightarrow \mathcal{U} = \frac{\epsilon}{e^{\epsilon/kT}-1}. \tag{8.22}$$

We are not done. One must connect this $\mathcal{U}$ to the observed energy density $u(\nu, T)$, and interpret ϵ in terms of the same parameters, $\epsilon = \epsilon(\nu, T)$.

Planck achieves the first by kinetic theory, in a lengthy argument which does not bear repeating here. The idea is to find the flux K emitted by a surface element of the blackbody, which can be related both to the energy density of the radiation it emits into a cavity, and to the resonator illuminating it from the cavity. The two calculations must give the same flux, because the surface is in equilibrium with both the radiation and the resonators, like a thermometer. The result is

$$u(\nu, T) = \frac{8\pi}{c}K \quad \& \quad \mathcal{U} = \frac{c^2}{\nu^2}K \Longrightarrow u(\nu, T) = \frac{8\pi\nu^2}{c^3}\mathcal{U}. \tag{8.23}$$

The net outcome is that the factor connecting u and $\mathcal{U}$ is just the density of states of the radiation field, as found very easily in Eq. (8.9) by Gibbs' method, which was applied to light for the first time by Bose. His success shows why phase-space ensembles have replaced real-space kinetic theory wherever possible.

The final determination $\epsilon = h\nu$ rests on a scaling argument. Wien found that the energy-density of the field as a function of wavelength must have the form

$$\mathcal{E}(\lambda, T)d\lambda = T^5\psi(\lambda T)d\lambda, \tag{8.24}$$

where ψ is a universal function of a single variable. This so-called Wien's displacement law can be proven kinetically, by considering the compression of a radiation field by an infinitely slowly moving mirror. Converting to frequency, $u(\nu, T)d\nu := \mathcal{E}(\lambda, T)d\lambda$, Planck finds

$$u(\nu, T)d\nu = T^5\psi\left(c\frac{T}{\nu}\right)d\left(\frac{c}{\nu}\right) = \nu^3 f\left(\frac{T}{\nu}\right)d\nu, \tag{8.25}$$

with f universal. Eliminating $\mathcal{U}$ between Eqs. (8.22) and (8.23) and comparing with Eq. (8.25), it is clear that ϵ must be linear in ν and independent of temperature.

8.5 Photons

The year 1905 has been called Einstein's *annus mirabilis* for the three foundational articles he published in it. All three can be summarized in a single statement: It is really happening. Time dilation is real, hence relativity. Thermalization is real, hence Brownian motion. Planck's quanta are real, hence photons.

Significantly, Einstein's argument for the reality of photons is valid only in the high-frequency limit $h\nu \gg kT$, where Planck's distribution reduces to the previously known Wien distribution:

$$u(\nu, T) = \frac{U(\nu, T)}{V} = \frac{8\pi h}{c^3}\frac{\nu^3}{e^{h\nu/kT} - 1} \rightarrow \frac{8\pi h}{c^3}\nu^3 e^{-h\nu/kT}. \tag{8.26}$$

Inverting, we find an explicit expression for $1/T$ which can be inserted into the equation of state:

$$\frac{1}{T} = -\frac{k}{h\nu}\ln\frac{c^3 U}{8\pi h\nu^3 V} = \frac{\partial S}{\partial U}, \tag{8.27}$$

where S is the entropy of Planck's resonators corresponding to their energy U in the high-frequency limit. Integrating with respect to U, one finds

$$S(V) = -\frac{kU}{h\nu}\left(\ln\frac{c^3 U}{8\pi h\nu^3 V} - 1\right) + const., \tag{8.28}$$

from which it follows that the difference in entropy between two volumes is

$$S(V) - S(V') = k\ln\left(\frac{V}{V'}\right)^{U/h\nu}. \tag{8.29}$$

This difference is just what the ideal gas equation would give for $U/h\nu$ particles. This number was called P in Eq. (8.18)—it is the number of quanta, not resonators.

Hence Planck's quanta are like molecules of some gas, randomly scattered in a container.

The photoelectric effect is observed in the ultraviolet, where experiment had already amply confirmed Wien's asymptotic formula. Hence, Einstein claimed that light causing the emission of electrons was like a gas of particles impinging on the metal, each of a fixed energy $h\nu$ and separated from the others. He thus resolved the main puzzle of the photoelectric effect, that increasing the intensity of radiation only increased the number of emitted electrons, while increasing the frequency increased their energy.

This claim was a significant pivot from Planck's point of view, who introduced the resonators as a simplified model of the material thermostat. According to Einstein, photons are thermalized light, just like randomly moving molecules are a thermalized gas. The resonators have dissolved into photons: Saying "this resonator has absorbed five quanta" is the same as saying "there are five photons of this frequency." Thus, thermodynamics can be applied directly to empty space filled with radiation, also known as the vacuum, inside a container.

In the long-wavelength limit $kT \gg h\nu$, the equipartition of energy takes over, the equation of state is no longer Wien's, and Einstein's analogy to a gas disappears. The question whether it makes sense to speak of photons as elementary particles at all wavelengths requires the more refined discussion in Sect. 8.8 below.

8.6 Atoms

Having settled that Planck's resonators are not a model of matter, Einstein turned to the interaction of light with atoms in the container some ten years later, after Bohr had established the principle of quantization of atomic levels. In this context, blackbody radiation is a powerful constraint on possible theories of the atom, powerful because it rests on quite mild assumptions.

First, assume that atoms absorb and emit radiation by changing their discrete excitation energies, say $E_n < E_m$, so the energy balance for an absorption process is

$$E_m = E_n + \Delta, \quad \Delta > 0, \tag{8.30}$$

and vice versa for emission. Second, the absorbed (emitted) energy is carried by a photon of the requisite frequency,

$$\Delta = h\nu. \tag{8.31}$$

Third, the absorption rate is of the form

$$P_{n\to m} = B_{nm} N_n N(\nu), \tag{8.32}$$

proportional to both the number of atoms in the lower level N_n and the number of photons $N(\nu)$ surrounding them. The factor of proportionality B_{nm} is to be provided by some microscopic theory. The emission rate is analogously

$$P_{m\to n} = A_{mn}N_m + B_{mn}N_m N(\nu), \tag{8.33}$$

where the first term is the rate of spontaneous emission, because an excited atom has the required energy. The other term is called induced emission. It is written by symmetry with absorption ("induced excitation"), but note that there is no a priori claim even that $B_{mn} \neq 0$. Finally, the ratio of excited atoms is given by the ratio of Gibbs probabilities, because they do not interact:

$$\frac{N_m}{N_n} = \frac{e^{-E_m/kT}}{e^{-E_n/kT}} = e^{-\Delta/kT}, \tag{8.34}$$

a relation known as the Nernst equation.

The two transition rates are equal at equilibrium, so we can eliminate the number of atoms to obtain the number of photons:

$$P_{n\to m} = P_{m\to n} \implies N(\nu) = \frac{A_{mn}}{B_{nm}e^{h\nu/kT} - B_{mn}} = \frac{U(\nu,T)}{h\nu}. \tag{8.35}$$

Comparing with the experimentally verified Planck distribution (8.26), one finds that

$$B_{nm} = B_{mn}, \quad \frac{A_{mn}}{B_{mn}} = V\frac{8\pi\nu^2}{c^3} = VG(\nu). \tag{8.36}$$

We have discovered two things. First, the rates of absorption and induced emission are equal. Second, the ratio of spontaneous and induced emission rates is just the (extensive) density of states of radiation. Both results are stringent tests of any microscopic theory of atoms and light.

In his article, Einstein assumed a priori that $B_{nm} = B_{mn}$. His interpretation of his own analysis was that he derived Planck's distribution from first principles. One of them was that resonance processes are naturally symmetric between emission and absorption, which had the same effect as Planck's neglecting dissipation in the resonators as a matter of expediency. Without introducing that assumption, we can see above that experiment *implies* that photons are emitted and absorbed reversibly, in other words, that the responsible microscopic process is not dissipative. Far from being an assumption, either expedient or plausible, this direct observation means that neither atoms nor photons have internal "moving parts" which absorb heat.

8.7 Fermi's Golden Rule

Einstein's B-coefficient expresses the microscopic reversibility of the atom-light interaction. In contrast, the A-coefficient expresses the irreversible nature of spontaneous emission. To see that, consider the expression

$$A = B \cdot VG(\nu). \tag{8.37}$$

Evidently, $A = B$ if there is just one photon of the required frequency, $VG = 1$. This situation can be realized experimentally in small cavities, where one can observe significant suppression of spontaneous emission in favor of undamped oscillations between the ground and excited states. When there are many photon states of the same frequency, i.e., the room is large, $A \gg B$ and the process is irreversible for one atom. In that limit, excited atoms minimize their energy by radiation: The emitted photon has a very low probability to find the atom in the room again. Blackbody equilibrium depends on there being many atoms as well, so the emitted photon eventually finds some other atom to excite in return.

The formula (8.37) contains the *equilibrium* density of states of the photon gas, which means that the individual atomic transitions always occur against a background of light in thermal equilibrium. Therefore, the time ΔT between any two successive atomic transitions is much longer than the thermalization time of light τ. The adiabatic condition $\Delta T \gg \tau$ is so often fulfilled that Fermi has called the relation (8.37) between macroscopic (total) and microscopic (individual) transition rates a "golden rule." Like Ohm's law, Fermi's golden rule is not a fundamental law, but has a wide range of applicability because of the efficiency of thermalization processes in general.

Dirac has expressed the microscopic meaning of the B-coefficient in terms of quantum perturbation theory. In the limit of weak coupling and under adiabatic conditions, it is simply the square of the matrix element of the interaction $\mathcal{V}$ causing the transition:

$$B_{nm} = |\langle n|\mathcal{V}|m\rangle|^2 \,. \tag{8.38}$$

Microscopic time-reversibility is expressed in quantum theory by the hermiticity of matrix elements, so the experimental constraint $B_{nm} = B_{mn}$ is satisfied.

Fermi's golden rule applies to banking as well. Suppose that investors are quick to adapt to market conditions, while a bank's lending rules change slowly. Then the bank's total credit emission is simply the probability that an individual investor obtains credit, multiplied by the density of such investors in the bank's surroundings. On the other hand, if many investors obtain credit in a short time, they may deplete the bank's resources, affecting the conditions faced by the next one. Adiabatic factorization no longer applies, because an effective correlation has appeared among the investors, even if they don't know about each other. Similarly, laser radiation is not blackbody radiation, because kinematic correlations appear among photons in a laser beam, even if there is no interaction between them, when many atoms deexcite before the emitted light can thermalize.

8.8 Bosons

The derivation of the sum over states for an ideal boson gas in Chap. 5 is quite different than Planck's derivation of the blackbody spectrum. In Planck's approach, energy is just a number whose unit summands are obviously indistinguishable *mathematically*, e.g., $3 = 1 + 1 + 1$ can be written in only one way. Thus, the material resonators are distinguishable, while the energy packets which determined their states of motion are not. Bose imagined, on the contrary, that light consisted of indistinguishable particles occupying distinguishable quantum states of motion, implemented as discrete cells in phase space, each of size h^3. Einstein was able to exploit Bose's innovation in his theory of the Bose gas, as discussed in Chap. 7. This difference in approach opens the question: Are photons *physically* bosonic particles like helium atoms, or is that just a formal analogy, because the distribution in Eq. (8.20) becomes symmetric in N and P when both are large?

Bose's derivation treats photons as elementary particles of light. The Bose photon distribution is a one-body thermal distribution of particles with a microscopic spectrum $\varepsilon_n = nh\nu$, the *solution of the quantum Hamiltonian* for a single photon. Crucially, it is a distribution in quantum-state space. By contrast, Planck obtained his distribution as a *thermodynamic equation of state* for all the light in the real-space container, in which the relation $\epsilon = h\nu$ was derived from a macroscopic scaling argument. In principle, ϵ could have turned out to depend on temperature as well, which is impossible for the spectrum of a microscopic Hamiltonian.

This distinction is exactly parallel to the one between microscopic particles and collective quasiparticles in Chap. 4. Bose's photons are elementary particles distributed in quantum-state space, while Einstein's photons are their quasiparticles distributed in real laboratory space. They are the physical realization of Planck's natural radiation, which was historically the first intimation of the quasiparticle concept. Because the microscopic and effective (macroscopic) dispersions are the same for light, the distinction is less prominent than for massive particles, where quasiparticles typically have a different effective mass than the microscopic ("bare") one. Nevertheless, it is real: Laser radiation is not natural radiation. In infrared and even microwave lasers (masers), photons behave like indistinguishable elementary particles deep in the thermal regime $h\nu \ll kT$, where Einstein's photons lose their similarity to a particle gas and revert to classical electromagnetic waves. This continuous transition from particles to waves is physically expected for quasiparticles, because they are by construction the lowest-lying thermalized fluctuations of the equilibrium state at any given temperature. The same transition in a Bose gas is between individual atoms at high energy and sound waves at low energy.

Energy in the field is not just a number. It has real physical existence in the form of photons, which are elementary particles in their own right. Einstein's unification of classical electromagnetism with gravity, confirmed by the bending of starlight, extends into quantum mechanics. Photons gravitate just like neutrons, and both phenomena have been observed directly by Mössbauer spectroscopy and neutron diffraction, respectively. Thus, Bose's attitude has been vindicated to the point that there is nothing left to say. Once we call something "elementary," we have accepted it as given: Light is rationalized in terms of the molecular hypothesis.[1]

Not only are photons bosons, strictly speaking they are the only free bosons. The only other massless bosons are gluons, which are never observed as free particles. Massive conserved bosons, such as helium atoms, are composed of fermions. There are no conserved elementary bosons. The uniqueness of photons is further underlined by their heat capacity, the only one in nature which increases without bound with temperature, $c_V \propto T^3$ by the Stefan-Boltzmann law. It does so because light has an infinite number of degrees of freedom (wave numbers) in a finite volume, so raising the temperature continuously increases the *number* of thermalized degrees of freedom, not only their *energy* as in a material gas. If one heats the oven enough, gamma-rays will come out! No one knows how light does that. In that sense, the luminiferous aether still retains some miraculous properties.

8.9 Vacuum Energy

There is a puzzling aspect to the distinction between microscopic and effective spectra of light. When one quantizes the electromagnetic field, there appears a zero-point energy $h\nu/2$ in the exact expression $\varepsilon_n = (n + 1/2)h\nu$. This energy is theoretically present in all modes, even those that are not excited according to the Planck distribution, so it appears that the energy of the field is again infinite at all temperatures. However, the quantization procedure assumes that the field is uncoupled from the surroundings. Conversely, the zero-point energy of the field coupled to the container is already counted in the total ground-state energy, which includes the chemical condensation energy of the box material on an equal footing with light. It is conventionally set to zero for a blackbody, but is certainly finite as long as the box is finite.[2]

To visualize this argument, imagine a mechanical ball-and-spring oscillator welded to the ceiling. The theoretical zero-point energy is present in the oscillator only if it is perfectly isolated from the surroundings, as assumed in standard quantization.

---

[1] Unlike gravity, which has resisted all attempts at microscopic rationalization so far.

[2] This argument is of the same kind as Bethe gave in his calculation of Lamb's shift. The shift comes out infinite if one calculates the correction to the binding energy of an interacting electron relative to a non-interacting electron. But the difference in energy of a bound and a free electron, *both* of which interact fully with the electromagnetic field, is finite, and that was what Lamb had measured.

But if the oscillator is able to excite sound waves in the box through the weld, it is thermalized, so only the probability of *exchange* of packets of size $nh\nu$ is included in its thermodynamic average energy, like in Sect. 8.3.

In his Nobel lecture, Planck recounts how he originally misunderstood the equilibrium of his resonators with the field, long before the $h\nu/2$ term was known:

> Furthermore, my hypothesis that the resonator could exercise a unilateral, i.e., irreversible, effect upon the energy in the surrounding radiation field, was strongly contested by Ludwig Boltzmann, who, with his riper experience in these problems, proved that according to the laws of classical dynamics each of the processes observed by me can proceed in exactly the opposite direction, ...

In other words, it is impossible for a Bose photon (elementary particle) to retain its vacuum energy irreversibly in contact with a thermostat. It must share all its energy reversibly, which turns it into an Einstein photon (quasiparticle). Generally, no measurement can observe a non-interacting particle, hence all observations are necessarily of quasiparticles, which may approach the ideal ("bare") elementary particles more or less, depending on the experimental situation.

Chapter 9
Sound

Models in physics come in two varieties, those that don't work even where they should, and those that work even where they shouldn't.

Igal Talmi

A macroscopic system in thermodynamic equilibrium is in its most probable state, i.e., the extremum of the corresponding thermodynamic potential. Because potentials are generally quadratic in the vicinity of extrema, small deviations of the volume from equilibrium will be met with a restitutive force that is linear in the deviation, i.e., harmonic. If the damping is not too strong, such a deviation is relaxed by oscillations, which we call sound.

The energy of typical sound waves tends continuously to zero as their wavelength increases, which can be visualized by an infinite wavelength corresponding to a sample at rest. Hence sound is easy to excite, and sound waves are a regular manifestation of thermal fluctuations at equilibrium. Measuring the dispersion of sound as a function of temperature is a standard method for studying structural phase transitions in crystals, which manifest themselves as a softening, or disappearance, of the restitutive force for sound waves (deformations) of the symmetry that corresponds to the new stable structure.

More generally, any harmonic excitation of the equilibrium state can be called sound. Thus, in magnetic materials there are spin waves, oscillations of local magnetization, and in superfluid helium there are elastic waves of temperature, so-called second sound. Due to its most general possible thermodynamic origin, sound is an unavoidable mode of interaction between macroscopic systems, and research into sound encompasses broad areas of physics, technology, biology, medicine, sociology, and art. The description of the heat capacity of sound in crystals has historically played a significant role in the development of quantum statistical physics.

D. Sunko, *Conceptual Thermodynamics*,
https://doi.org/10.1007/978-3-032-20877-4_9

The basic division of sound is according to the mechanism of dissipation. In general, long waves excite shorter ones, which eventually disappear in thermal noise. In more detail, we can divide the system into small cells, each of which is a large thermodynamic system in itself, with a defined temperature and pressure. In the classical limit, there is no coherence between particles, and the sound wave propagates in such a way that each cell affects the neighboring one macroscopically, due to the difference in thermodynamic parameters between them. Thermalization of the system, or the equalization of pressures and temperatures, appears as dissipation, i.e., increase in entropy due to the return of the system to equilibrium.[1] Here we will ignore this internal friction, but it is important to understand that it is assumed, as much as is necessary to stabilize macroscopic oscillations hierarchically in the continuum limit, on time scales that are neither too long nor too short, such that short waves dissipate faster than longer ones.

In the quantum limit, we can divide the system into cells again, and establish macroscopic equations of motion for their thermodynamic parameters. However, the stability of this motion is no longer guaranteed by internal friction, but by quantum coherence over large distances, larger than a single cell. The quantum regime appears at low temperatures, where, due to indistinguishability, the number of states of motion becomes smaller than the number of particles, which means that particles in different cells cannot move independently, as in the classical limit. This kinematic restriction closes dissipation channels, short-wave excitations in particular, but also establishes long-lived collective states, coherent quasi-particles, which ensure the stability of sound waves on intermediate time scales by deterministic motion. Their damping is again dissipative, but at the level of these collective excitations, not individual particles.

9.1 Heat Capacity of Crystals

At high temperatures, the heat capacity of a monoatomic crystal is $3Nk$, which is consistent with the equipartition theorem, if we imagine that each of the N atoms is bound in place by a harmonic force in all three directions. This result is called the Dulong-Petit rule. At low temperatures, a drop in heat capacity is observed, which vanishes near absolute zero with the third power of temperature. This drop obviously contradicts the equipartition theorem, while on the other hand it is consistent with Carnot's interpretation of temperature, according to which the heat capacities of all bodies vanish at absolute zero. In the harmonic approximation, sound waves of different wavenumbers are independent states of motion, and form an ideal gas whose heat capacity is the heat capacity of the crystal lattice.

[1] This is also called "entropy production," e.g., when air is heated by speech.

9.1.1 Sound and Light

We have already seen that the heat capacity of light increases as $c_V \sim T^3$, where the power 3 came from the dispersion $\varepsilon = cp$. The dispersion of sound is the same, only the velocity is different, which means that the level density is the same, except that sound in a solid has three polarizations, because in addition to the two transverse ones it has a third in the direction of motion, which light does not. Quantized sound waves, called phonons, are bosons like photons, and are also not conserved particles, so their chemical potential is zero. The same calculation as for light will give the same result, except for the aforementioned differences in constants. For someone who believes in quantum mechanics, it is not a problem how to get the dependence T^3 for sound as well, but how to reach the classical limit of Dulong and Petit. Why does the heat capacity of sound not increase without limit, as of light?

In the expression for the internal energy of light (8.2) we summed the energies of modes of all frequencies up to infinity, that is, all wavelengths down to zero. This was correct for light, which can have an arbitrarily small wavelength, which means that it has an infinite number of degrees of freedom in a finite volume. Its heat capacity increases without limit because new independent states of motion are excited with increasing temperature, unlike in a gas, where the existing degrees of freedom (atoms) gain energy, but their number does not increase. Obviously, the latter is the case in a crystal. The difference between light and sound is that sound is a wave in granular matter, so it cannot have an arbitrarily short wavelength, because there is nothing to vibrate between atoms.

The shortest wavelength in a crystal is the one in which neighboring atoms vibrate opposed in phase, so if one goes up, the other goes down. The next goes up again, giving the wavelength $\lambda_{min} = 2a$, if the atomic spacing is a. For a cubic lattice, this calculation is performed independently in all three directions, so the range of allowed (physically different) wavenumbers $k = 2\pi/\lambda$ in each direction is

$$-\frac{\pi}{a} < k \leq \frac{\pi}{a}, \tag{9.1}$$

where both signs appear because "up" and "down" are different vibrational states, except for $k = \pm\pi/a$, which represent the same state with cyclic boundary conditions. This range of wavenumbers is called the *first Brillouin zone*. The continuum of wavenumbers in it reflects the thermodynamic limit of an infinitely large crystal, while the restriction to the first Brillouin zone reflects its granular structure. Readers can draw waves of wavenumber k and $k + 2\pi/a$, and notice that they always intersect at the integer coordinates spaced by a, thus describing the same physical elongations of atoms at their positions in the lattice.

[The same situation has already been encountered by anyone who has studied Fourier series using the example of a vibrating string. The string is continuous and of finite length ℓ in real space, while its Fourier coefficients are functions of integer multiples of $2\pi/\ell$ in wavenumber space, thus describing an infinite crystal in inverse space. Here it is the other way around: We have an infinite crystal in real space, and the first Brillouin zone is a finite continuous string in inverse space.]

9.1.2 Debye Model

We can further simplify the representation of the crystal by ignoring the crystal structure and replacing the cube of the first Brillouin zone with a sphere. Granularity is taken into account by the fact that the sphere is of finite size, such that it contains the same number of degrees of freedom as N atoms, $\mathcal{N}(\varepsilon_{max}) = 3N$, where

$$\mathcal{N}(\varepsilon) = 3\frac{V V_p}{h^3} = 3V\frac{4\pi}{3h^3}\left(\frac{\varepsilon}{c}\right)^3 \tag{9.2}$$

is the number of sound states (wavenumbers) in phase space, with three polarizations. This model of crystal oscillations is called the *Debye model*. It is customary to express the maximum energy in temperature units, $\varepsilon_{max} = k\Theta_D$, where Θ_D is called the *Debye temperature*. In real crystals, it is most often 100–500 K. Explicitly,

$$k\Theta_D = c\,\hbar\left(\frac{6\pi^2 N}{V}\right)^{1/3}. \tag{9.3}$$

By changing the variable, $s = \varepsilon/kT$, the energy can be written

$$U = \int_0^{\varepsilon_{max}} \varepsilon n_0(\varepsilon) g(\varepsilon) d\varepsilon = 9Nk\,\Theta_D\left(\frac{T}{\Theta_D}\right)^4 \int_0^{\Theta_D/T} \frac{s^3 ds}{e^s - 1}, \tag{9.4}$$

similarly to photons (8.2), only the integral does not extend to infinity, but to the highest allowed vibrational energy. Here the dimensionless integral is no longer a constant, but retains the scale Θ_D, so we have two physical regimes. At low temperatures, $T \ll \Theta_D$, the upper limit of integration is effectively infinite, so the integral is $\pi^4/15$, whence the heat capacity is

$$c_V = \frac{\partial U}{\partial T} = \frac{12\pi^4}{5} Nk\left(\frac{T}{\Theta_D}\right)^3 \quad (T \ll \Theta_D), \tag{9.5}$$

which gives the experimentally observed behavior, as for light. On the other hand, when $T \gg \Theta_D$, e^s can be replaced by $1 + s$ throughout the integration region, so the integral gives $(\Theta_D/T)^3/3$, or the energy $U = 3NkT$, which brings us back to the Dulong-Petit rule,

$$c_V = 3Nk \quad (T \gg \Theta_D), \tag{9.6}$$

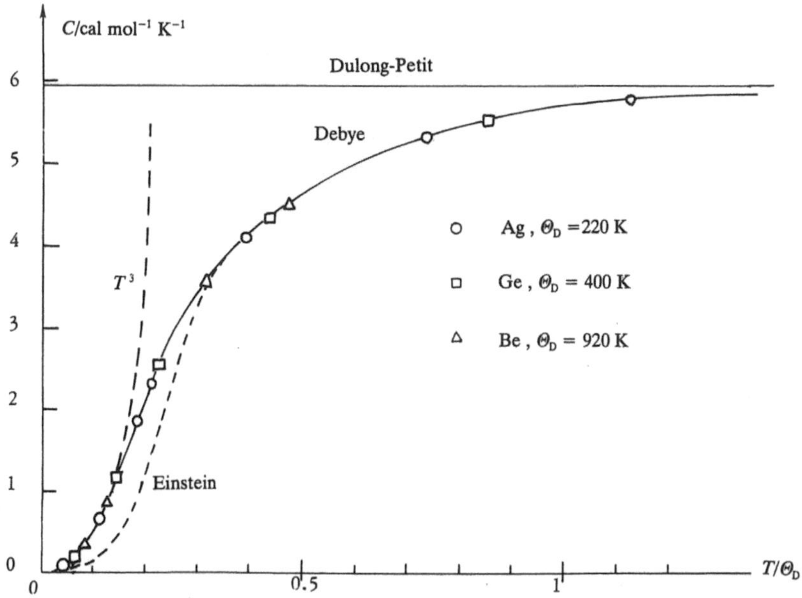

Fig. 9.1 Heat capacity of simple crystals, fitted with Debye's and Einstein's models. [Roger Balian, From Microphysics to Macrophysics, Vol. II. Springer, Berlin 1982. Figure 11.20. Reproduced with permission.]

as it should be. The Debye model agrees well with the measured heat capacity of simple crystals, over a wide range of temperatures, as observed in Fig. 9.1. It explains why the heat capacity of diamond varies significantly around room temperature, owing to its very high value of $\Theta_D \approx 2200\,\mathrm{K}$. The reason for it is that the scales of both terms in the harmonic-oscillator model of the C–C bond are large. The kinetic scale $p^2/2M$ is large because the mass of $^{12}\mathrm{C}$ is small. The potential scale is large because the covalent bond is due to the second ($n = 2$) atomic shell, which is significantly more strongly bound, according to the Balmer formula $E_n = -13.6/n^2\,\mathrm{eV}$, than those with $n \geq 3$, which comprise the bulk of common crystals. The higher atomic energy scales translate into 5–10 times greater values of elasticity constants than usually observed.

9.1.3 Optical Modes: Einstein's Model

A unit cell of a crystal lattice rarely has only one atom. If it has more, relative movements of atoms within the unit cell are possible without moving its center of mass, i.e., without pushing the neighboring cell. Such excitations behave as localized oscillators to first approximation. In ionic crystals, atoms participating in relative motion can have opposite charge, so these oscillations carry an electric dipole moment. It is therefore possible to excite them with infrared light, due to which the dielectric response of the crystal is anomalous in the corresponding frequency range. That is why these modes are called "optical," as opposed to Debye's "acoustic." The simplest model of this behavior was derived by Einstein, by ignoring the connections between the unit cells of the crystal, which reduces the model to a sum of contributions of independent harmonic oscillators. It is customary to present this model as Einstein did originally, normalized to a monoatomic crystal with N atoms, although that is the only case when it is not realistic.[2] The internal energy of the crystal is then

$$U = 3\sum_{i=1}^{N} \langle n \rangle h\nu = 3N \langle n \rangle h\nu, \tag{9.7}$$

because all contributions are equal. The expected value of $\langle n \rangle$ is

$$\langle n \rangle = \frac{1}{Z}\sum_{n=0}^{\infty} n e^{-\beta n h\nu} = -\frac{1}{h\nu}\frac{\partial}{\partial \beta} \ln Z, \tag{9.8}$$

where the sum over states of one (quantized) oscillator is

$$Z = \sum_{n=0}^{\infty} e^{-\beta n h\nu} = \frac{1}{1 - e^{-\beta h\nu}}, \tag{9.9}$$

therefore

$$U = \frac{3Nh\nu}{e^{\beta h\nu} - 1}, \tag{9.10}$$

which corresponds to the Bose-Einstein distribution with chemical potential $\mu = 0$. The heat capacity is

$$c_V = 3Nk\beta^2 \frac{\partial^2}{\partial \beta^2} \ln Z = 3Nk \left(\frac{h\nu}{kT}\right)^2 \frac{e^{h\nu/kT}}{\left(e^{h\nu/kT} - 1\right)^2}. \tag{9.11}$$

This expression has two regimes. At high temperatures, $kT \gg h\nu$, heat capacity tends to a constant: $c_V \rightarrow 3Nk$, which is the result of the equipartition theorem, $c_V = 3(1 + 1)Nk/2$, because we have a three-dimensional harmonic oscillator at each lattice node. At low temperatures, $kT \ll h\nu$, the heat capacity decays

---

[2] Einstein's main point was that sound was quantized similarly to light, while the fit to experiment he published did not include measurements at the very lowest temperatures.

exponentially, as in the spin case (Fig. 10.3), associated with the energy gap $h\nu$ in the excitation spectrum:

$$c_V \to 3Nk \left(\frac{h\nu}{kT}\right)^2 e^{-h\nu/kT}, \tag{9.12}$$

which is inconsistent with measurements at low temperatures, where the Debye acoustic contribution $c_V \sim T^3$ dominates in insulators. (In metals, the contribution of free electrons, which is linear at low temperatures, dominates the cubic contribution of the crystal lattice below $\sim$ 10 K.) This temperature dependence of the heat capacity is universal for all gapped systems in the limit when the temperature is much smaller than the gap, as one can verify in a system of only two levels, to which they reduce.

9.1.4 *Real Crystals*

Both Debye's and Einstein's models of sound grossly simplify the vibration of atoms in real crystals, by neglecting the crystal structure. Namely, since the wavenumber shift $k \to k + 2\pi/a$ describes the same state of motion of atoms in a crystal, all functions of wavenumbers must be periodic with that period, in particular the dispersion $\varepsilon(k)$. Debye's linear dispersion, cut off at the value ε_{max}, does not satisfy this condition. The simplest periodic function that satisfies it is

$$\varepsilon(k) = \frac{2\hbar}{a} c \left|\sin \frac{ka}{2}\right|, \tag{9.13}$$

where we adjusted the constants so that $\varepsilon'(0) = \hbar c$, where c is the speed of sound, and the sine is maximal when $k = \pi/a$, at the shortest wavelength. This expression can be obtained as a solution to the equation of motion of a monoatomic lattice in the harmonic approximation. The corresponding level density (6.7) replaces the expression (8.6) and thus corrects the Debye model by including the crystal structure.

Acoustic modes are always present in polyatomic unit cells, because they correspond to motion of the center of mass of the entire cell. Einstein's constant dispersion for optical modes is a periodic function, but it is not realistic, because an atom vibrating in one unit cell affects the neighboring one, even if the center of mass of the cell does not move. Indeed, if we imagine that the atoms are connected by springs, we see that Einstein's assumption of independent vibrations is realized only for the motion in which neighboring cells are in opposite phase (when one is compressed, the neighboring ones expand), because the spring connecting atoms in neighboring cells is then not engaged. A realistic optical mode is less dispersive

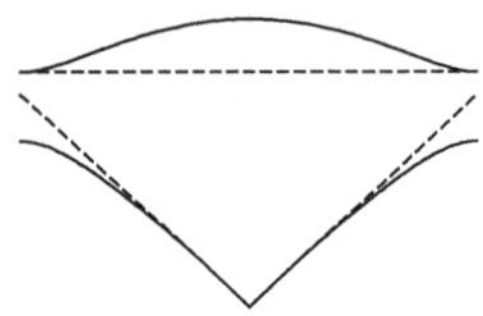

Fig. 9.2 Acoustic and optical vibrations

than an acoustic one, and its energy increases from the edge of the Brillouin zone towards the center ($k = 0$), where all unit cells vibrate in phase (Fig. 9.2).

While Fig. 9.1 shows some best-possible cases, the apparently very good agreement of the Debye model with the heat capacities of most real crystals is too good to be true. First, the value of the parameter Θ_D that achieves the optimal fit does not correspond to the theoretical formula (9.3). Furthermore, one would expect that a satisfactory adjustment of Θ_D in a small range near zero would be maintained at higher temperatures, because only a few points should be sufficient to fix the cubic hyperbola. However, this is not the case. As the range of temperatures over which the Debye expression for the heat capacity is adjusted increases, the optimal Θ_D changes, and even oscillates as a function of temperature in some crystals.

The reason for this strange behavior of the theoretically constant parameter is that the phonon level density in real crystals is a much more complex function than the photon level density, which Debye effectively used in his model. In particular, optical modes are only activated at finite temperatures, which means that the total number of degrees of freedom, which determines the Debye temperature, changes discontinuously with temperature. In a certain range of excitations, the cumulative number of states tracks the photon one more closely, so Θ_D is stationary in the corresponding temperature range, that is, the curve $\Theta_D(T)$ has an extremum where the agreement with measurements improves, but not with the same value of Θ_D as at lower temperatures. The overall agreement of the model with the entire heat capacity curve for a single value of Θ_D is to some extent an optical illusion, which comes from the fact that a rapidly increasing curve can be laterally close to the measured point even if it significantly deviates from it vertically. Having understood that, we are prompted to ask why the measured points nevertheless persistently stay close to the Debye curve with some effective value Θ_D. What attracts them to it?

The Debye model is robust. This means that the structure of the formula used to adjust the measured values is realistic, even if the theoretical values of the parameters are not. An important feature of robust models is that their parameters can be optimally adjusted to measurements without losing their physical meaning. Thus, the best-fit values of Debye temperatures for real crystals reliably estimate how much energy is required to excite all their modes of vibration. Robustness in a model is achieved by taking into account kinematic restrictions, allowed symmetries and available phase space, first, considering the dynamics only in the next step. An improved approach may replace parameters in a robust formula with functions, but it will not change the structure of the formula. On the contrary, if the kinematics is properly captured, the parameters will be measurable directly, such as the speed of sound in the expression (9.13), refining the motivation to calculate them theoretically. The Debye model is a textbook example of the success of such an approach. It is robust because the heat capacity is essentially determined by the number of excited degrees of freedom. The change of that number with temperature is efficiently modelled by the Debye temperature because the model is exact in both asymptotic limits, $T \ll \Theta_D$ and $T \gg \Theta_D$, so there is little room for deviation from measurements in the transition region between them.

9.2 Zero Sound

When the sound frequency ω is low, or the relaxation time τ is short, so that $\omega\tau \ll 1$, a Fermi liquid can transmit ordinary (hydrodynamic) sound just like a Fermi gas, because then one cell of the liquid can influence another by usual thermodynamic collective effects, expressed by the equation of state. Then it can be shown that the attenuation of sound is $\alpha \sim \omega^2/T^2$. On the other hand, as the temperature decreases, the relaxation time increases, so that at $\omega\tau \approx 1$ a single cell can no longer be considered as an incoherent thermodynamic collective, stabilized by diffusion. Any attempt to increase its density then results in the emission of particles into other cells without collisions, because the added particles do not have time to relax thermally by collisions in one period of sound oscillation. Simply put, when $\omega\tau \gg 1$, an increase in density results in an instantaneous (on the scale $1/\omega$) loss of particles from the cell, not an increase of pressure inside it. Note that a change in density causes a change in the radius of the Fermi sphere, $p_F \sim (N/V)^{1/3}$, so that in normal sound the Fermi sphere "breathes" from cell to cell, i.e., its volume changes.

However, in this collisionless (dispersive) regime, the Fermi liquid, unlike a gas, has an alternative way of transmitting pressure waves, *without* a change in density, by modifying the collision integral, or susceptibility, $f(\boldsymbol{p}, \boldsymbol{p}')$. Because the collision integral appears in the variation of energy, its modification appears as a deformation of the shape of the Fermi sphere without a change in volume. This sound is called zero sound. The transition from one regime to the other is seen in both velocity and attenuation, more clearly in attenuation because it is qualitative there, from $\alpha \sim \omega^2/T^2$ at higher to $\alpha \sim T^2$ at lower temperatures.

The attenuation is proportional to T^2 at low temperatures due to indistinguishability, which leads to a kinematic restriction of phase space. Since the dissipation is caused by two-particle scatterings of those quasiparticles that can change states of motion, and there are $N_{\mathit{eff}} \sim kT/\varepsilon_F$ of these at low temperature, the probability that two will scatter is $\sim (kT/\varepsilon_F)^2$. Specifically for ^{3}He, first sound of velocity 180 m/s changes to zero sound of velocity 190 m/s at temperatures around 10 mK, while its Fermi velocity is $v_F \approx 50$ m/s.

9.3 Second Sound

In Sect. 7.5.1 we saw that due to the macroscopic component of the BEC in a superfluid, it behaves as if its wave function had only one phase, determined by the BEC. This motivates us to average the wave function of a macroscopic cell at position $\boldsymbol{r}$ as

$$\Psi_{N_r}(\boldsymbol{r}, t) := \int \Psi(\boldsymbol{r}, \dots, t) = \Psi_{N_r}(\boldsymbol{r}, 0)e^{-iE(\boldsymbol{r})t/\hbar}, \tag{9.14}$$

where the integration is over the remaining coordinates of the particles within one cell, and we used the fact that the phase is related to the energy of the cell. This integration is analogous to one where we would obtain the magnetization as a sum of the microscopic magnetic moments of the atoms. Such sums are called *order parameters* of the system. Physically, the order parameter of a superfluid is a complex number because the wave functions of all particles behave as if they had a single unique phase. This insight is enough for us to obtain the most striking macroscopic property of the superfluid, mass transfer without dissipation. A superfluid screens a chemical potential difference by a noiseless phase shift, just as a superconductor screens an electromagnetic potential difference. In both cases, the phase shift manifests itself as macroscopic transport of particles.

The chemical potential of the cell is $\mu(\boldsymbol{r}) = E(\boldsymbol{r}, N_{\boldsymbol{r}} + 1) - E(\boldsymbol{r}, N_{\boldsymbol{r}})$, so we have

$$\begin{aligned} \mu(\boldsymbol{r}) - \mu(\boldsymbol{r}') &= E(\boldsymbol{r}, N_{\boldsymbol{r}} + 1) - E(\boldsymbol{r}, N_{\boldsymbol{r}}) - E(\boldsymbol{r}', N_{\boldsymbol{r}'} + 1) + E(\boldsymbol{r}', N_{\boldsymbol{r}'}) \\ &= E(\boldsymbol{r}, N_{\boldsymbol{r}} + 1) - E(\boldsymbol{r}', N_{\boldsymbol{r}'} + 1) \\ &= i\hbar\frac{\partial}{\partial t}\left(\arg \Psi_{N_{\boldsymbol{r}}+1}(\boldsymbol{r}) - \arg \Psi_{N_{\boldsymbol{r}'}+1}(\boldsymbol{r}')\right) \\ &= -\hbar\frac{\partial}{\partial t}\left(\phi(\boldsymbol{r}) - \phi(\boldsymbol{r}')\right), \end{aligned} \tag{9.15}$$

because $E(\boldsymbol{r}, N_{\boldsymbol{r}}) = E(\boldsymbol{r}', N_{\boldsymbol{r}'})$ at equilibrium.[3] The last equality follows from (7.22), because the addition of a particle does not change the overall phase: The added particle either has that phase, or some other that is negligible with respect to it, as explained in Sect. 7.5.1. The quantum current $\boldsymbol{j}$ (5.11) is proportional to the phase gradient, so we have

$$\nabla\mu(\boldsymbol{r}) + \hbar\frac{\partial\nabla\phi}{\partial t} = \nabla\mu(\boldsymbol{r}) + m\frac{\partial \boldsymbol{v}_s}{\partial t} = 0, \tag{9.16}$$

[3] The second line is not zero because $N_{\boldsymbol{r}} + 1$ is not the equilibrium number of particles.

which means that the superfluid short-circuits the chemical potential with a noiseless stream of particles, $\boldsymbol{j}_s = \rho_s \boldsymbol{v}_s$. This expression is analogous to Fourier's equation, which similarly relates heat flow to a temperature gradient. Note that the chemical potential is related to the change in the number of all particles, not just those in the superfluid component, so we would naturally expect $\boldsymbol{v}_s + \boldsymbol{v}_n$ instead of $\boldsymbol{v}_s$ in this expression, where $\boldsymbol{v}_n$ is the velocity of the normal component. But this is precisely the special property of superfluidity that is analogous, for the flow of particles, to superconductivity for the flow of charge: If we try to establish a voltage difference in a parallel connection of a normal conductor and a superconductor, the superconductor will short-circuit it, with no current flowing through the normal conductor. Likewise, the gradient of the chemical potential is shorted by a current of the superfluid component, while the normal one remains at rest.

For any fluid in the hydrodynamic limit, the equations linearized in velocities are reduced to three conserved currents:

$$\frac{\partial \rho}{\partial t} + \nabla \cdot \boldsymbol{j} = 0, \tag{9.17}$$

$$\frac{\partial S}{\partial t} + \nabla \cdot \boldsymbol{j}_S = 0, \tag{9.18}$$

$$\frac{\partial \boldsymbol{j}}{\partial t} + \nabla p = 0, \tag{9.19}$$

where $\boldsymbol{j}_S$ is the entropy current, $\boldsymbol{j}$ the mass current, and p the pressure. These three equations in five unknown functions $(p, T, \boldsymbol{v})$ are complemented by two connections between currents and densities:

$$\boldsymbol{j} = \rho \boldsymbol{v}, \tag{9.20}$$

$$\boldsymbol{j}_S = S \boldsymbol{v}, \tag{9.21}$$

which in principle allows us to calculate the evolution of a system from a given spatial distributions of pressure, temperature, and velocity at $t = 0$, while the entropy is obtained from the equation of state. The appearance of the superfluid component implies an additional contribution to the mass current,

$$\boldsymbol{j} = \rho_n \boldsymbol{v}_n + \rho_s \boldsymbol{v}_s, \tag{9.22}$$

which, however, does not contribute to the heat current:

$$\boldsymbol{j}_S = S \boldsymbol{v}_n, \tag{9.23}$$

and now we miss just one equation, which describes the origin of the superfluid current. That is precisely the Eq. (9.16).

Since the chemical potential is rarely under direct control, we can use the Gibbs-Duhem relation to replace

$$\nabla\mu = -\frac{S}{N}\nabla T + \frac{V}{N}\nabla p = -s\frac{m}{\rho}\nabla T - \frac{m}{\rho}\frac{\partial \boldsymbol{j}}{\partial t} \tag{9.24}$$

so in general

$$(\rho_n + \rho_s)\frac{\partial \boldsymbol{v}_s}{\partial t} = -\frac{\rho}{m}\nabla\mu = s\nabla T + \frac{\partial(\rho_n \boldsymbol{v}_n + \rho_s \boldsymbol{v}_s)}{\partial t}, \tag{9.25}$$

where expressions like $\boldsymbol{v}\partial\rho/\partial t$ appear, which are of higher order, because in a state close to equilibrium all temporal changes are considered as small first-order quantities, like velocities themselves. When these are neglected, the linearized equations of quantum hydrodynamics are

$$\rho_n\left(\frac{\partial \boldsymbol{v}_n}{\partial t} - \frac{\partial \boldsymbol{v}_s}{\partial t}\right) = -s\nabla T, \tag{9.26}$$

$$\rho_n\frac{\partial \boldsymbol{v}_n}{\partial t} + \rho_s\frac{\partial \boldsymbol{v}_s}{\partial t} = -\nabla p, \tag{9.27}$$

$$\frac{\partial\rho}{\partial t} + \rho_n\nabla\cdot\boldsymbol{v}_n + \rho_s\nabla\cdot\boldsymbol{v}_s = 0, \tag{9.28}$$

$$\frac{\partial s}{\partial t} + s\nabla\cdot\boldsymbol{v}_n = 0. \tag{9.29}$$

We have eight equations for two vector fields $\boldsymbol{v}_s$ and $\boldsymbol{v}_n$, and five scalar fields ρ_s, ρ_n, s, p and T. Since among the thermodynamic fields, two determine the third by the equation of state, we can consider any two scalar fields in these equations as determined by external conditions ("sources"), while the other two are solutions for the given situation.

Adding and subtracting Eqs. (9.26) and (9.27) we obtain

$$\rho\frac{\partial \boldsymbol{v}_s}{\partial t} = s\nabla T - \nabla p, \tag{9.30}$$

$$\rho\frac{\partial \boldsymbol{v}_n}{\partial t} = -s\frac{\rho_s}{\rho_n}\nabla T - \nabla p. \tag{9.31}$$

The divergence of these expressions together with the time derivative (9.28) and (9.29) allows the elimination of velocities, so a pair of coupled second-order equations in time appear,

$$\frac{\partial^2\rho}{\partial t^2} - \nabla^2 p = 0, \tag{9.32}$$

$$\frac{\partial^2 \rho}{\partial t^2} - \frac{\rho_s}{s}\frac{\partial^2 s}{\partial t^2} + s\frac{\rho_s}{\rho_n}\nabla^2 T = 0. \tag{9.33}$$

Now we need to decide which of these quantities are independent. If we are going to consider density and temperature waves, we can eliminate pressure and entropy using a linearized change of variables:

$$\nabla^2 p = \left(\frac{\partial p}{\partial T}\right)_\rho \nabla^2 T + \left(\frac{\partial p}{\partial \rho}\right)_T \nabla^2 \rho, \tag{9.34}$$

$$\frac{\partial^2 s}{\partial t^2} = \left(\frac{\partial s}{\partial T}\right)_\rho \frac{\partial^2 T}{\partial t^2} + \left(\frac{\partial s}{\partial \rho}\right)_T \frac{\partial^2 \rho}{\partial t^2}, \tag{9.35}$$

Putting it all together, we get two coupled wave equations:

$$\nabla^2 \rho - \frac{1}{c_1^2}\frac{\partial^2 \rho}{\partial t^2} + \gamma_1 \nabla^2 T = 0, \tag{9.36}$$

$$\nabla^2 T - \frac{1}{c_2^2}\frac{\partial^2 T}{\partial t^2} + \gamma_2 \frac{\partial^2 \rho}{\partial t^2} = 0. \tag{9.37}$$

Here, we have identified two velocities,

$$c_1^2 = \left(\frac{\partial p}{\partial \rho}\right)_T^{-1}, \tag{9.38}$$

$$c_2^2 = \frac{s^2 \rho_s}{\rho \rho_n (\partial s/\partial T)_\rho} = \frac{T S^2 \rho_s}{c_1 \rho_n}, \tag{9.39}$$

and the coupling constants

$$\gamma_1 = \frac{(\partial p/\partial T)_\rho}{(\partial p/\partial \rho)_T} = -\left(\frac{\partial \rho}{\partial T}\right)_p, \tag{9.40}$$

$$\gamma_2 = \frac{\rho_n}{s\rho_s}\left[1 - \frac{\rho}{s}\left(\frac{\partial s}{\partial \rho}\right)_T\right]. \tag{9.41}$$

If the coupling constants are negligible, density waves and temperature waves propagate independently through liquid helium. Such behavior is observed: Both density waves at constant temperature (first sound) and temperature waves at constant density (second sound) can appear. This is a consequence of the very small value of the thermal expansion coefficient γ_1, due to which local heating of helium by a temperature wave does not trigger a density wave.

Chapter 10
Applications

Et inde est quod quanto aliqua scientia est abstractior et simpliciora considerans, tanto eius principia sunt magis applicabilia aliis scientiis.

St. Thomas Aquinas

Thermodynamics is the phenomenology of macroscopic phenomena, so any description of such a phenomenon is an application of thermodynamics in the broadest sense. In the narrow sense, we call phenomena thermodynamic when random effects of a multitude of particles, manifested as pressure or heat, cannot be neglected. Here we shall limit ourselves even more, to review only applications that played a significant role in understanding the fundamental concepts of statistical physics and its connection with thermodynamics.

10.1 Barometric Formula

One of the early problems with the molecular hypothesis was that it was not clear whether there was any limit on the size of the particles participating in thermodynamics. Specifically, if we have a colloidal suspension, i.e., of particles large enough to be seen,[1] in water, do these macroscopic particles contribute to the pressure according to the equipartition theorem? Direct measurement of their contribution is impossible because their partial pressure is very small, due to their small number.

Einstein solved this problem in his second paper on Brownian motion. He pointed out that the gravitational potential of the particles should be added to their thermodynamic potential, a simple exercise if one believed that atoms were real.

[1] Their diameter is usually about 1 μm, which is near the limit of light microscopes, but if they are optically active, their concentration can be measured by observing the rotation of the plane of polarization.

D. Sunko, *Conceptual Thermodynamics*,
https://doi.org/10.1007/978-3-032-20877-4_10

Therefore, when the suspension is in a gravitational field, their distribution will not be uniform.

Namely, for a system in thermodynamic equilibrium, the chemical potentials of its various parts must be equal, otherwise there is transport of particles from one part to another. When a variation of the chemical potential is imposed by an external field, thermodynamic equilibrium is established with an uneven distribution of particles, which cancels the imposed variation of the chemical potential so that macroscopic transport stops.

If a beaker is at the given temperature and pressure of the laboratory, the Gibbs potential is extremal, so we describe the equilibrium shift by adding the potential of the external field to it:

$$G = G_0 + U_{external}, \tag{10.1}$$

or, using $G = N\mu$,

$$\mu = \mu_0 + u, \tag{10.2}$$

where u is the external potential per particle. In this case, the external field binds directly to the chemical potential. A gravitational field will thus shift it by the potential energy per particle at height z,

$$\mu = \mu_0 + mgz. \tag{10.3}$$

In this formula, z is not the microscopic position of a single particle, rather mgz is the macroscopic potential energy of a layer of particles at height z, per particle. We recall the equation of state (4.14) of an ideal gas that relates the chemical potential and concentration, without an external field:

$$\mu_0 = kT \ln\left(\frac{N\lambda_T^3}{V}\right). \tag{10.4}$$

At equilibrium, the chemical potential is constant despite the field:

$$kT \ln\left(\frac{N\lambda_T^3}{V}\right) + mgz = const., \tag{10.5}$$

and since only the concentration changes at a given temperature, we get

$$\frac{N}{V} = const.e^{-mgz/kT}, \tag{10.6}$$

from which the same formula follows for the pressure, $p = kTN/V$. It is called the barometric formula, or the exponential atmosphere. Measurements confirm it for the concentration of colloidal particles. Thus, there are no fundamental restrictions on the size of the particles from which we can build ensembles.

We applied the equation of state of an ideal gas for particles in a solution. This is allowed if the solvent is inert, and therefore serves only as a wall (source of randomness) for the dissolved particles. If we were to consider chemical reactions, the same assumption would not hold for colloidal particles in a polar solvent, such as water, because they polarize it significantly. This polarization affects the reaction barriers, which are different in this case than for reactions in the gas phase.

In his celebrated first paper on Brownian motion, Einstein invoked the osmotic pressure of large particles heuristically, claiming that it was obvious that they would also be hitting the walls. He needed the argument to justify the key step in the development, the estimate of the average kinetic energy of a suspended particle by the equipartition theorem, $\langle Mv^2/2\rangle = 3kT/2$. This estimate assumes that the macroscopic particle is thermalized equally with all water molecules, therefore it also participates in thermodynamics. The barometric formula enables an experimental verification of the same assumption, alternative to the direct observation of the Brownian particle's excursion over time. It is historically the first application of Einstein's insight into the physical process of thermalization.

10.2 Diatomic Molecules

A diatomic molecule apparently has seven quadratic contributions to the energy. Three come from the translational motion of the center of mass, two from the rotation of the molecular axis, and two from vibrations along the axis, one from the kinetic and one from the potential energy of a one-dimensional oscillator. By the equipartition theorem, this should correspond to a heat capacity of $c_V = 7R/2$ per mole. In reality, however, most diatomic gases have a heat capacity of $c_V = 5R/2$ at room temperature, which even *decreases* further with a decrease in temperature. The question is, where have some of the degrees of freedom disappeared?

Recall that the equipartition theorem was derived by calculating the mean squared contribution to the energy $\varepsilon(s) = as^2$,

$$\langle \varepsilon \rangle = \frac{\int as^2 e^{-as^2/kT} ds}{\int e^{-as^2/kT} ds} = \frac{kT}{2}, \tag{4.79}$$

which assumes that the variable s, on which the energy depends, can have continuous values. But this is not really the case: The states described by s are discrete, so these integrals should be viewed as approximations of sums like $\sum_i \exp(-as_i^2/kT)$. If we have a dense energy staircase, i.e., shifting the index by one gives a small shift of the variable s, replacing the sum with an integral is justified, and the equipartition theorem will hold. But if the index shift gives a large jump in the variable s, at some lower temperature the term $\exp(-as_1^2/kT)$ can already be negligible compared to $\exp(-as_0^2/kT)$. The sum reduces to a single term referring to the ground state, so this degree of freedom can no longer receive heat. It is said to be "frozen." Obviously, the measure of the staircase density is just kT: By increasing the temperature, we will theoretically always reach a regime in which the difference between as_1^2/kT and as_0^2/kT is arbitrarily small. This is the classical limit of high temperatures. The only question is whether this temperature is easily accessible, which depends on the nature of the excitation, i.e. the size of a in the expression as^2.

Thus, for translational motion, we have levels

$$\varepsilon_{n_x,n_y,n_z} = \frac{\hbar^2}{2M}(k_{n_x}^2 + k_{n_y}^2 + k_{n_z}^2), \tag{10.7}$$

where $k_j = j\pi/\ell$. If we put the side of the box $\ell = 1$ cm and the mass 10^{-23} g, then the first energy step for translation is $\varepsilon_{1,0,0} - \varepsilon_{0,0,0} \sim 10^{-14}$ K. Obviously, the translation degrees of freedom will contribute to the heat capacity according to the equipartition theorem at all available temperatures.

The contribution of the rotational degrees of freedom, which are also kinetic, can be evaluated from the Schrödinger equation for a free particle in spherical coordinates:

$$-\frac{\hbar^2}{2M}\nabla^2\Psi = -\frac{\hbar^2}{2M}\left(\frac{1}{r^2}\nabla_{\theta,\phi}^2 + \nabla_r^2\right)\Psi = E\Psi, \tag{10.8}$$

We know that the eigenfunctions of the angular part are spherical harmonics $Y_{lm}(\theta,\phi)$, with eigenvalues $-l(l+1)$:

$$-\frac{\hbar^2}{2Mr^2}\nabla_{\theta,\phi}^2 Y_{lm} = E_{rot}Y_{lm} = \frac{\hbar^2 l(l+1)}{2\mathcal{I}}Y_{lm}, \tag{10.9}$$

where the characteristic quantity $Mr^2 = \mathcal{I}$, the effective moment of inertia, has appeared. We find

$$E_{rot} = \frac{\hbar^2 l(l+1)}{2\mathcal{I}}, \tag{10.10}$$

so the scale of the step is $\hbar^2/2\mathcal{I} \sim 10\,\mathrm{K}$, if we take the same mass as before, and the size of the hydrogen atom for r. This degree of freedom is also classical at room temperature, but it is possible to observe it "freezing out" at low temperatures. This is especially the case for the small and light molecule H_2, whose small moment of inertia raises the estimate above to $\sim$100 K.

Finally, we cannot evaluate the energy of the vibrational contributions directly, because we would need to know something about the forces that hold molecules together. Nevertheless, we know that they must be at least strong enough to overcome the increase in momentum uncertainty, which comes from forcing the atoms to stay close together. The kinetic energy corresponding to this uncertainty is

$$\frac{\Delta p^2}{2M} = 3\frac{h^2}{2M\Delta r^2} \approx 2500\ \text{K}, \tag{10.11}$$

if the mass and diameter of the hydrogen atom are inserted. Experimentally, the excitation energy for vibration of the hydrogen molecule is 6100 K, while for heavier molecules, such as carbon monoxide (3070 K), it is typically half as much, all corresponding to wavelengths in the microwave range. These states will obviously not be excited at room temperature, so we can understand why vibrational degrees of freedom usually do not contribute to the heat capacity.

10.2.1 Molecules of Identical Atoms

If a diatomic molecule consists of identical atoms, we have additional complications due to indistinguishability of identical particles. Hydrogen nuclei are fermions, so the overall wave function of the molecule H_2 must be antisymmetric under exchange of nuclei. Since the states of relative motion are symmetric, or antisymmetric, with the parity of the angular momentum l, i.e., their parity is $(-1)^l$, it would follow that a hydrogen molecule can only have states of odd angular momentum.

This analysis is incomplete, because it does not take into account that hydrogen nuclei are protons that themselves have spin $1/2$, and can couple to states of total spin $S = 0$ or $S = 1$. The spin state $S = 0$, called parahydrogen, is antisymmetric under exchange of protons, so that only states of even orbital angular momentum are allowed. On the other hand, states of total momentum $S = 1$, or orthohydrogen, are symmetric with respect to spin exchange, so only odd orbital momenta are allowed. Finally, the sums over states of these two types of hydrogen molecules are

$$\begin{aligned} Z_{para} &= \sum_{\text{even } l} (2l+1)\exp\left(-\frac{\hbar^2 l(l+1)}{2IkT}\right), \\ Z_{ortho} &= 3\sum_{\text{odd } l} (2l+1)\exp\left(-\frac{\hbar^2 l(l+1)}{2IkT}\right), \end{aligned} \tag{10.12}$$

where I is the moment of inertia, and the factor $2l + 1$ comes from the $2l + 1$ degenerate projection states, $m = -l, -l+1, \dots, l$, of each spherical harmonic. The factor $3 = 2S + 1$ comes from the degeneracy of the total spin of the orthohydrogen nuclei. The question is how to combine them into one sum over states of the hydrogen gas.

If ortho- and parahydrogen are in thermodynamic equilibrium with each other, i.e., transitions between ortho- and para- states are unhindered, the sum of states per molecule of H_2 is simply the sum of these two,

$$Z = Z_{para} + Z_{ortho}. \tag{10.13}$$

On the other hand, if the transition between the two species is suppressed, and we have prepared a mixture of N_p molecules of parahydrogen and N_o molecules of orthohydrogen, the total sum of states is, taking kinetic energy into account:

$$Z = \left(\frac{V}{\lambda_T^3}\right)^{N_p+N_o} \frac{Z_{para}^{N_p}}{N_p!} \frac{Z_{ortho}^{N_o}}{N_o!}, \tag{10.14}$$

the same as if we had a mixture of two different non-interacting gases. In this case, the free energies, i.e., logarithms of the sums of states, are added in the appropriate ratio.

In reality, ortho- and parahydrogen cannot pass into each other by mutual scattering, because this requires too large a change in the spatial wave function. Therefore, equilibration is very slow, and they behave like a mixture of two gases. However, if we put a platinized sponge in a container with them, the H_2 molecules can adsorb on it, with equal probability for adsorption and desorption of both isomers. They thus come into thermodynamic equilibrium, and all the H_2 behaves as a single gas. Platinum is said to catalyze their reaction.

Surfaces can be catalysts because the energy of adsorption changes the energy barrier of the reaction. Free electrons in metals are an infinite reservoir of angular momentum L, because a plane wave is a superposition of spherical waves (harmonics) with coefficients that slowly decrease with L. Both effects can facilitate the reconfiguration of molecules in chemical reactions. In particular, it has long been known that platinum catalyzes many reactions with hydrogen. The detailed mechanism of this phenomenon is under intense investigation because of its potential for producing hydrogen from water.

Blowing a mixture of hydrogen and nitrogen into hot osmium shavings under high pressure, Haber obtained ammonia.[2] This reaction changed the course of history, because ammonia is the raw material for both synthetic fertilizers and explosives, *nitro*cellulose (flash cotton), from which smokeless gunpowder is made, *nitro*glycerin, from which Nobel made dynamite, and tri*nitro*toluene, the artillery explosive of choice in World War I. Haber received the Nobel prize for it in 1918.

[2] In industry, iron is used instead of osmium, which is called the Haber-Bosch process.

10.3 Van der Waals Model of Gas Liquefaction

The phenomenon of gas liquefaction is familiar to anyone who has ever cooked with camping gas. While there is liquid in the bottle, the gas comes out at constant pressure, as evident from the fact that the flame is steady. As soon as the liquid disappears, the pressure at the outlet begins to drop sharply, so that the flame goes out quickly. This means that there is a range of concentrations and temperatures in which the gas and the liquid coexist at some equilibrium pressure. This coexistence is called *phase separation*. Phase separation is crucial to the water cycle, because the equilibrium state of water and vapor above it, 100% air humidity, is rarely achieved in nature. Water surfaces begin to evaporate as soon as winds disperse the vapor above them, which sets the cycle in motion.

In 1873, Van der Waals proposed an equation of state of a non-ideal gas that explained this phenomenon. It includes principal corrections to the ideal gas equation, which arise from the fact that gas particles attract over long distances and repel over short ones. The ideal-gas assumption is that the pressure,

$$p = \frac{N}{V}kT, \tag{10.15}$$

is due only to kinetic energy, without interactions. If there are attractive forces between molecules, the pressure will decrease by some amount, as molecules from the interior of the gas attract those approaching a wall. This amount is proportional to the number of pairs that attract, which is the square of the concentration for long-range forces, longer than the container. Thus, kTN/V becomes $kTN/V - a(N/V)^2$ with $a > 0$, which means that p in the equation of state should be replaced by $p + a(N/V)^2$. On the other hand, if the particles repel at small distances, each will have an effective volume *smaller* than V/N available, because part of the total volume is occupied by other particles. Thus, V/N should be replaced by $V/N - b$, which together gives the van der Waals equation of state,

$$p \cdot \frac{V}{N} \to \left[p + a\left(\frac{N}{V}\right)^2\right] \cdot \left(\frac{V}{N} - b\right) = kT. \tag{10.16}$$

If the parameters a and b are small in some sense, at high temperatures (kT large) the value of the right-hand side is dominated by the product pV/N, so we expect behavior similar to an ideal gas. Figure 10.1 shows the isotherms of the Eq. (10.16) in the (p, V) diagram. When the temperature drops below a certain critical value $T = T_c$, the volume is no longer uniquely determined for a given pressure and temperature, because the isotherm is not a monotonic function. In that regime, the model describes an unstable state, in which a decrease in volume causes a decrease in pressure, so the compressibility is negative.

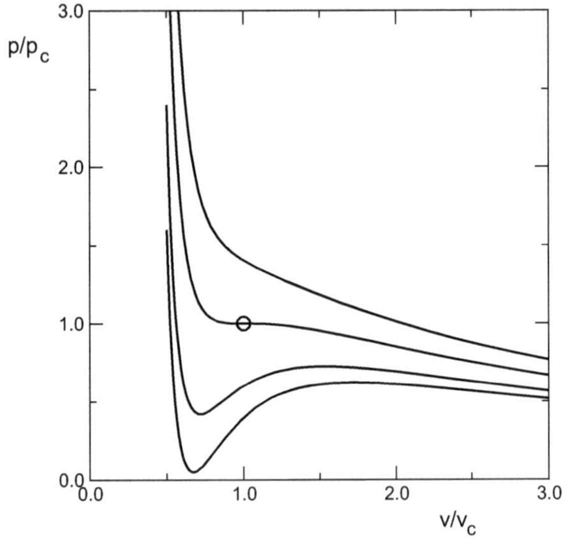

Fig. 10.1 Isotherms of the reduced van der Waals equations (10.20). The temperature values are, in order from bottom to top, $T/T_c = 0.85, 0.9, 1, 1.1$, with $v = V/N$. The inflection point (p_c, v_c) is marked with an open circle

10.3.1 *The Law of Corresponding States*

It is easy to understand the behavior of the isotherms in Fig. 10.1. At given temperature and pressure, the Eq. (10.16) is a cubic equation for the volume (we write $v := V/N$):

$$v^3 - (b + \frac{kT}{p})v^2 + \frac{a}{p}v - \frac{ab}{p} = 0. \tag{10.17}$$

For high temperatures, it has one real and two conjugate complex solutions; for low temperatures, pressures are possible at which three real solutions appear. The critical point is the one where all three solutions are equal. To wit, when $p = p_c$ and $T = T_c$ are inserted into the polynomial (10.17), it factorizes:

$$v^3 - (b + \frac{kT_c}{p_c})v^2 + \frac{a}{p_c}v - \frac{ab}{p_c} = (v - v_c)^3. \tag{10.18}$$

Comparing the coefficients of the same powers of v gives three equations in three unknowns, with a unique solution:

$$kT_c = \frac{8a}{27b}, \quad p_c = \frac{a}{27b^2}, \quad v_c = 3b. \tag{10.19}$$

Introducing reduced variables $\bar{p} = p/p_c$, etc., the van der Waals equation reads

$$\left(\bar{p} + \frac{3}{\bar{v}^2}\right)\left(\bar{v} - \frac{1}{3}\right) = \frac{8\bar{T}}{3}. \tag{10.20}$$

Here there are no more constants that would characterize the material. This is a very interesting property of the van der Waals model: There are natural units of the model, in which the equation of state is universal. All information about a specific substance is contained in the values of the state variables at the critical point. The statement that the equation of state is universal if expressed in suitable units is called the *law of corresponding states*. In contemporary terminology, the equation of state *scales* with the critical parameters.

The appearance of natural units is expected, because thermodynamics cannot depend on an arbitrary choice of units. Robust models have the additional property that their natural units can be fixed by measurements, so that there is no doubt about the physical meaning of the model parameters, nor about the agreement achieved.

10.3.2 Maxwell's Construction

The van der Waals model has a critical temperature, but the isotherms do not describe a stable state when $T < T_c$. The instability persists below the transition because we persist in modelling a single (homogeneous) phase, while separation into two phases occurs in reality. Can we somehow "save" the model below the critical point, so that it also predicts the separation pressure?

Of two possibilities, the one that gives the lower total free energy will be realized. It is easy to find the free energy if we know the equation of the isotherm, $p(V, T = const.)$, because $(dF)_T = -pdV$ at fixed temperature, so

$$F(V) - F(V_0) = -\int_{V_0}^{V} p(V, T = const.)dV, \tag{10.21}$$

where V_0 is a choice of origin of the free energy. If the phases are separated, the isotherm is also an isobar $p = p_{sep}$ (constant), so

$$F_{sep}(V) - F_{sep}(V_0) = -p_{sep}(V - V_0). \tag{10.22}$$

The curve $F_{sep}(V)$ for the separated phases is a straight line. This line should be compared with the curve given by the Eq. (10.21) with the van der Waals expression (10.16) for the pressure in a homogeneous phase.

The homogeneous phase separates when its free energy goes from being smaller to being larger than that of the separated phases. For some pressure to be the constant p_{sep}, it must intersect the isotherm in two particular points V_1 and V_2, such that the free energies of the homogeneous and separated phases are equal at these two points, as they cross each other:

$$\int_{V_0}^{V_1} p_{vdW}(V, T = const.)dV = p_{sep}(V_1 - V_0),$$

$$\int_{V_0}^{V_2} p_{vdW}(V, T = const.)dV = p_{sep}(V_2 - V_0),$$

where p_{vdW} is given by the model expression (10.16). The arbitrary origin is eliminated by subtraction:

$$\int_{V_1}^{V_2} p_{vdW}(V, T = const.)dV = p_{sep}(V_2 - V_1). \tag{10.23}$$

It is easy to interpret this condition geometrically: p_{sep} is the mean value of the integral of $p_{vdW}(V)$ from V_1 to V_2. It is the value at which the area of the rectangle of height p_{sep} and width $(V_2 - V_1)$ is equal to the area under the van der Waals isotherm in the range from V_1 to V_2. This geometric construction is called Maxwell's. It is shown in Fig. 10.2. Another way to express it is that the areas subtended by the arcs of the van der Waals isotherm above and below p_{sep} must be equal.

In the separation region, the substance does not resist a change in volume. Pushing a piston into a cylinder containing separated phases merely pushes molecules from the vapor into the liquid, at no cost because both are at the same pressure and chemical potential. Similarly, drawing it out has the reverse effect, with no work or heat transfer. One way to describe this situation is that the matrix of

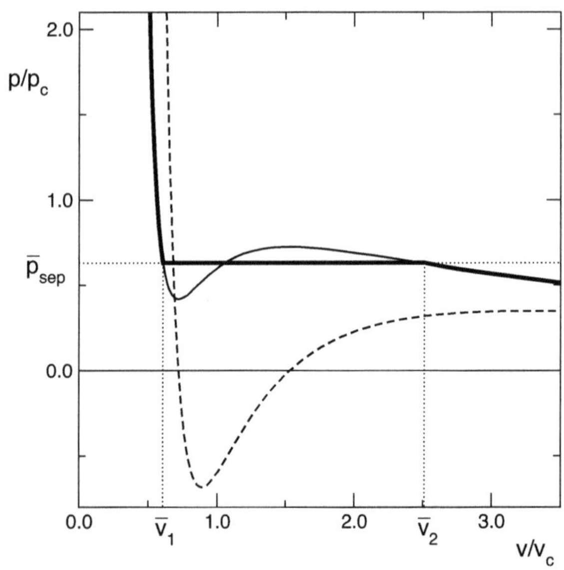

Fig. 10.2 Simultaneous depiction of the isotherm $T = 0.9T_c$ from Fig. 10.1 (*thin full line*) and the reciprocal of the corresponding compressibility, $1/\kappa_T$ (*dashed line*, arbitrary units). The thick full line is Maxwell's construction

susceptibilities (2.61) is doubly degenerate: The parabola that provided the restoring force to a change of volume has flattened out at fixed total number of particles. Equivalently, one may note that the free energy (10.22) is already proportional to the volume, rather than the grand potential (2.45), i.e., one step too soon in the textbook sequence of changes of variables.

10.4 Paramagnetism

If we place a sample in a given magnetic field, e.g., of a permanent magnet, the energies of individual macroscopic states will change, $E_i =: E_i(0) \to E_i(\boldsymbol{B})$. This changes the sum over the states, and thus the internal energy,

$$Z(\boldsymbol{B}) = \sum_i \exp(-\beta E_i(\boldsymbol{B})), \quad U(\boldsymbol{B}) = -\frac{\partial}{\partial\beta} \ln Z(\boldsymbol{B}). \tag{10.24}$$

This internal energy is sometimes called *spectroscopic*, because it is related to the shift of spectral lines in a magnetic field. If we keep the other variables fixed,

$$dU = \sum_{i=1}^{3} \frac{\partial U}{\partial B_i} dB_i \equiv -\boldsymbol{M} \cdot d\boldsymbol{B}, \tag{10.25}$$

where the conjugate variable, called *magnetization*, appears:

$$\boldsymbol{M} = -\left(\nabla_B U\right)_{N,V,S}. \tag{10.26}$$

The sign is conventional. If the energy of the system increases with increasing field, as if the sample "resisted" the field, we say that it is magnetized negatively, i.e., opposite to the direction of the field; such responses are called *diamagnetic*. If the energy decreases, the sample is attracted by the field, the magnetization is positive, and the response is called *paramagnetic*.

Magnetic induction, which is a natural variable of internal energy, is intensive: A sample twice the size is exposed to the same induction, but has twice the magnetization. Therefore, one should guard against the prejudice that the natural variables of internal energy are always extensive. Magnetic induction is the independent variable in a laboratory situation where the field sources are permanent magnets. If the fields are induced by electric coils connected to sources of fixed voltage,[3] one should employ a different internal energy,

$$U' = U + \boldsymbol{M} \cdot \boldsymbol{B}, \quad dU' = \boldsymbol{B} \cdot d\boldsymbol{M} + \ldots \tag{10.27}$$

[3] If the coil is connected to a source of fixed current, it also behaves like a permanent magnet.

Here, magnetization is the independent variable, and $\boldsymbol{M} \cdot \boldsymbol{B}$ is the interaction (potential) energy of the sample of magnetization $\boldsymbol{M}$, which did work on the currents in the coils through the induction $\boldsymbol{B}$ as it was brought in from infinity. Therefore, U' is the "internal" energy of a larger thermodynamic system, which includes the currents in the coils.

Classical physics cannot explain the magnetic response of a macroscopic system. Namely, due to this response, the magnetic field can do work on the sample. Thus, if we hang a light stick, e.g., a match, horizontally on a thin thread, and place one end in an inhomogeneous magnetic field, the thread will rotate in one direction or the other, depending on whether the response of the material is paramagnetic (so a stronger field attracts it) or diamagnetic, like match wood (so a stronger field repels it). The work done by the field is stored in the potential energy of the twisted thread.

Classically, however, a magnetic field cannot do work on a charge, because the Lorentz force due to the field is always perpendicular to its direction of motion. Because there is classically no other explanation of magnetic moments of matter, except as the moments of currents originating from charges in motion, the result follows. It is called the Bohr-van Leuven theorem.

Quantum-mechanically, the Bohr-van Leuven theorem does not hold, for two reasons. First, there are intrinsic magnetic moments of particles, associated with their spin, that do not originate from charges in motion. Second, the orbital motion of the electron, which corresponds to classical currents in matter, is quantized: The orbital angular momentum can only assume some discrete values. This means that the electron cannot perfectly adapt its motion to the external field, so as to make its direction of motion always perpendicular to it. Therefore, an electron in an atomic orbital behaves as if it had a *given* angular velocity, i.e., the same as an intrinsic magnetic moment.

10.4.1 Equations of State

If we have a system of non-interacting particles with magnetic moment $\boldsymbol{\mu}$, the energy of each such particle in an external field $\boldsymbol{B}$ will be $-\boldsymbol{\mu} \cdot \boldsymbol{B}$. If the other degrees of freedom do not depend on the orientation of the magnetic moment,[4] the sum over states of the system factorizes in the form $Z_{tot} = Z_{magn.} Z_{other}$, so we can consider the magnetic part only. Assume that the external field is homogeneous, $\boldsymbol{B} = B\hat{z}$. Then, classically speaking, the interaction energy with the field is

$$E_{int} = -\boldsymbol{\mu} \cdot \boldsymbol{B} = -\mu_0 B \cos\theta, \tag{10.28}$$

[4] If that were always the case, transformers would not hum.

where θ is the continuous angle between the magnetic moment and the direction of the field. The classical sum over states of one particle is then

$$Z_{C,class.}(1, \beta) = \int e^{\beta\mu_0 B\cos\theta} d\Omega = 2\pi \int_{-1}^{1} e^{\beta\mu_0 Bs} ds, \tag{10.29}$$

The integral is easy to calculate, so the classical sum over states of N magnetic moments is

$$Z_{C,class.}(N, \beta) = Z_{C,class.}(1, \beta)^N = \left(4\pi \frac{\sinh \beta\mu_0 B}{\beta\mu_0 B}\right)^N, \tag{10.30}$$

where we have not added any factors $N!$ for indistinguishability. This is because, for spin models, it is usually imagined that the corresponding particles are fixed at N given locations in space, so that the locations are distinguishable, even if the particles are not. Such modelling is in accord with the most common experimental realization, where metal ions in some crystals have magnetic moments due to electrons, localized in open d or f shells, that do not exchange place during observations.

The quantum derivation is quite different. There, the magnetic moment of a particle is related to its angular momentum,

$$\boldsymbol{\mu} = g\mu_B \boldsymbol{J}, \tag{10.31}$$

which can take on $2J + 1$ discrete values with respect to the direction of the field,

$$E_{int} = -\boldsymbol{\mu} \cdot \boldsymbol{B} = -g\mu_B B J_z, \quad J_z = -J, -J+1, \ldots, J. \tag{10.32}$$

The quantity μ_B is the standard *Bohr magneton*, which is for electrons

$$\mu_B = \frac{e\hbar}{2m_e} = 9.27 \times 10^{-24} JT^{-1}, \tag{10.33}$$

while the actual magnitude of the magnetic moment of the electronic state is introduced via the *gyromagnetic factor* g; for most simple situations, $g = 2$, so we will assume this value from now on.

Now the one-particle sum over states can be calculated as a partial sum of the geometric series,

$$Z_C(1, \beta) = \sum_{J_z=-J}^{J} e^{\beta 2\mu_B B J_z} = \frac{\sinh(2J+1)\beta\mu_B B}{\sinh \beta\mu_B B}. \tag{10.34}$$

Specifically, if $J = 1/2$, then

$$Z_C(N, \beta) = (2 \cosh \beta \mu_B B)^N , \tag{10.35}$$

which is our familiar result for N balls in two boxes, derived in Sect. 3.1.

The internal energy is $U = -\partial \Psi_C / \partial \beta$, giving the equation of state

$$U = -N\mu_B B \left[(2J+1) \coth(2J+1) \frac{\mu_B B}{kT} - \coth \frac{\mu_B B}{kT} \right], \tag{10.36}$$

which again simplifies for $J = 1/2$:

$$U = -N\mu_B B \tanh \frac{\mu_B B}{kT}. \tag{10.37}$$

At low temperature, $kT \ll \mu_B B$, the hyperbolic tangent tends to unity, and the energy becomes $-N\mu_B B$, which is easy to understand: All the spins are aligned along the field, because there is no thermal energy to overturn even one. At high temperature, $kT \gg \mu_B B$, the internal energy tends to zero. Namely, then the energy difference between spins along the field and against the field is negligible, so there are equally many positive and negative single-particle energies in both directions; the average of these is zero.

Magnetization is, by definition, conjugate to induction,

$$M = -\left(\frac{\partial U}{\partial B}\right)_{N,S} = -\left(\frac{\partial F}{\partial B}\right)_{N,T} = kT \left(\frac{\partial \Psi_C}{\partial B}\right)_{N,T} . \tag{10.38}$$

Instead of calculating the derivative, we notice from (10.34) that the derivatives with respect to β (internal energy) and with respect to B (magnetization) differ only by a factor. Therefore, the equation of state is

$$M = -\frac{U}{B} = N\mu_B \cdot \mathcal{B}_J(\frac{\mu_B B}{kT}), \tag{10.39}$$

where the square bracket in (10.36) is called the Brillouin function $\mathcal{B}_J(x)$. This equation of state can be verified by measurement. In particular, magnetization saturates as a function of the field,

$$\lim_{\mu_B B \gg kT} \frac{M}{N\mu_B} = 2J, \tag{10.40}$$

so the value of the local moment in paramagnetic crystals can be found simply by reading the asymptote of the magnetization curve, scaled by the total number of magnetic atoms.

10.4.2 Negative Temperatures

We may ask whether it is possible to make the expectation value of the energy (10.36) positive. This will happen when higher energy states are more likely to be occupied than lower energy states. Formally, we can model this by allowing *negative* temperatures in the Gibbs distribution. But since the temperature is always $1/T = \partial S/\partial U$, negative temperatures can only be realized if entropy decreases with increasing internal energy. This means that entropy will increase if the system releases this energy; thus, a negative temperature always implies a thermodynamic instability, because the lower energy state is statistically more likely, so the system will certainly find it and shed the excess energy.

An increase in internal energy can reduce entropy if the single-particle excitation spectrum is bounded from above. This is precisely the case for spin systems. The state in which all spins are oriented opposite to the field has zero entropy, just like the state in which all are along the field: It can be realized in only one way. The distribution in which there are more particles in higher energy states is called *inverse population*. The state of maximum entropy is the one where it does not matter which way the spin is oriented, which is achieved at infinite temperature.

Due to thermodynamic instability, states of formally negative temperature can be realized only as non-equilibrium, but if such a state is maintained by compensating the system for the energy it is spontaneously losing, these non-equilibrium states can be stationary. This is the case of inversely populated atomic states in laser tubes, when the laser operates in the so-called continuous mode. It should be borne in mind that the description of such systems by negative temperatures is purely formal; if we put a thermometer in the laser tube, it will register a positive temperature.

10.4.3 Heat Capacity and Classical Limit

The heat capacity is easy to find,

$$c_V = k\beta^2 \frac{\partial^2 \Psi_C}{\partial \beta^2} = Nk \left[\frac{(\mu_B B/kT)^2}{\sinh^2 \mu_B B/kT} - \frac{(2J+1)^2 (\mu_B B/kT)^2}{\sinh^2 (2J+1)\mu_B B/kT} \right], \tag{10.41}$$

or, for $J = 1/2$,

$$c_V = Nk \left[\frac{\mu_B B/kT}{\cosh \mu_B B/kT} \right]^2. \tag{10.42}$$

This curve is shown in Fig. 10.3. It has two characteristic regimes. At low temperatures, $kT \ll \mu_B B$, the heat capacity decreases exponentially. That is characteristic of any system that has a gap in the energy spectrum between the ground and first excited state. At high temperatures, it vanishes again, this time as a power of

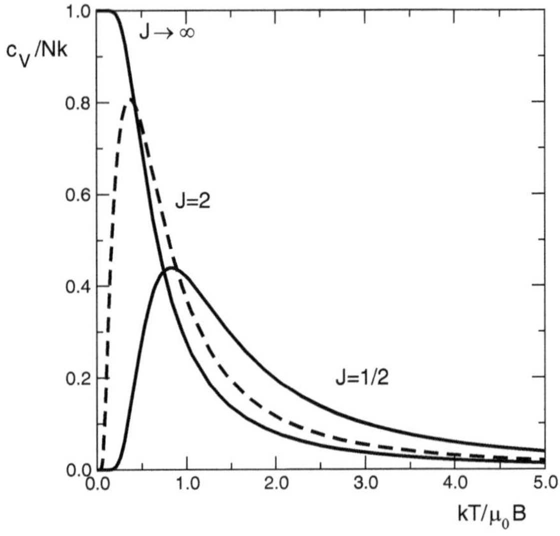

Fig. 10.3 Paramagnetic heat capacitance (10.41), for various spin values, with scaling (10.44)

temperature. The reason is that the excitation space is finite, so when all states are excited with equal probability, the system can no longer absorb heat. In between, a maximum appears, known as the *Schottky anomaly*. The temperature at which it appears is about 1 K for a field of 1 T, which is a good mnemonic for the very low energy scales associated with magnetic fields.

It is interesting to compare this behavior with the classical heat capacity,

$$c_{V,class.} = Nk\left[1 - \left(\frac{\mu_0 B/kT}{\sinh \mu_0 B/kT}\right)^2\right], \tag{10.43}$$

which is obtained from (10.30). In order to make a meaningful comparison, one must replace

$$\mu_B \to \mu_0/(gJ), \tag{10.44}$$

in the previous expressions, so that the same energy range, $\pm\mu_0 B$, is maintained in the classical and quantum cases. With this substitution, the gaps Δ in the spin spectrum become increasingly dense as the magnetic moment increases, so the classical limit $\Delta \ll kT$ can be reached. Then

$$\lim_{J\to\infty} c_V = c_{V,class.}, \tag{10.45}$$

and the classical heat capacity is denoted by $J \to \infty$ in Fig. 10.3. The curve for $J = 2$ is also depicted, which shows that the transition to the classical regime is quite fast.

The classical curve is not difficult to understand. The position of the classical spin is described by a point on the unit sphere, so it has two degrees of freedom. A classical ideal gas in two dimensions would have $c_V = Nk$, by the equipartition theorem. This is exactly what is observed for low temperatures, $kT \ll \mu_0 B$. The behavior is similar to an ideal gas only as long as the sphere (spin phase space) is well approximated by the tangential plane (phase space of an ideal gas), i.e., while the spin orientation has a small scatter in relation to the direction along the field, which corresponds to low temperatures. But as soon as the temperature rises, $kT \sim \mu_0 B$, it becomes evident that the phase space is *compact*, so that it cannot absorb an arbitrary amount of heat. (The occupation probabilities p_M of the projection states M saturate at high temperature, so $dp_M = 0$.) The *compactness* of the classical phase of space corresponds to the *finite excitation spectrum* of the quantum system. Conversely, quantum systems whose spectrum of single-particle excitations is unlimited, such as the harmonic oscillator or ideal gas, have an unbounded classical phase space.

10.4.4 Susceptibility and Entropy

The magnetic susceptibility is defined per unit volume, as the ratio of magnetization to field,

$$\chi = \frac{1}{V} \lim_{B \to 0} \frac{M}{B}, \tag{10.46}$$

for small (test) fields. According to this definition, the susceptibility diverges if the system remains magnetized when the field vanishes. Then we say that there is spontaneous magnetization. Formally, it is obvious that

$$\chi = -\frac{1}{V} \left(\frac{\partial^2 F}{\partial B^2} \right)_{N,V,T} \Bigg|_{B=0}, \tag{10.47}$$

all under the assumption that $M(0) = 0$, i.e., that there is no spontaneous magnetization. This equation characterizes the susceptibility as the curvature of the extremum of the free energy as a function of the field, at zero field. In our case (10.39), we obtain explicitly

$$\chi = \frac{N}{V} \frac{\mu_B^2}{kT} \frac{4J(J+1)}{3}. \tag{10.48}$$

This result is called *Curie's law*. In words, the product $\chi \cdot T$ is a constant for a given material, called the Curie constant.

The entropy is most easily read from the thermodynamic relation (2.48), $S/k = \Psi_C + \beta U$, or, in our case, according to (10.34) and (10.36),

$$S/Nk = \ln\left[\frac{\sinh(2J+1)\beta\mu_B B}{\sinh\beta\mu_B B}\right] - \beta\mu_B B \cdot \mathcal{B}_J(\frac{\mu_B B}{kT}). \tag{10.49}$$

For the case $J = 1/2$ we have again a somewhat more transparent expression,

$$S/Nk = \ln(2\cosh\beta\mu_B B) - \beta\mu_B B\tanh\beta\mu_B B. \tag{10.50}$$

In the high-temperature limit, the entropy tends to a maximum value of $k\ln 2$ per spin, $k\ln(2J+1)$ in general. This is simply the logarithm of the $(2J+1)^N$ ways in which we can assign one of the $2J+1$ projection states to each of the N spins.

We can interpret the entropy statistically using the expectation value of the spin projection,

$$\langle J_z\rangle = \frac{1}{Z(1,\beta)}\sum_{J_z} J_z e^{\beta 2\mu_B B J_z} = \frac{kT}{2N\mu_B}\frac{\partial\Psi_C}{\partial B} = \frac{M}{2N\mu_B} = \frac{1}{2}\mathcal{B}(\frac{\mu_B B}{kT}), \tag{10.51}$$

For $J = 1/2$, this is

$$\langle J_z\rangle = \frac{1}{2}\tanh\beta\mu_B B. \tag{10.52}$$

Inverting this expression and inserting into (10.50), we get

$$S/Nk = -\left(\frac{1}{2} - \langle J_z\rangle\right)\ln\left(\frac{1}{2} - \langle J_z\rangle\right) - \left(\frac{1}{2} + \langle J_z\rangle\right)\ln\left(\frac{1}{2} + \langle J_z\rangle\right), \tag{10.53}$$

all for the case $J = 1/2$. This is the expression (3.28); namely, the probability that the projection state $J_z = \pm 1/2$ will be occupied is exactly

$$p_\pm = \frac{1}{2} \pm \langle J_z\rangle. \tag{10.54}$$

10.5 Ferromagnetism

Samples of some metals, such as iron, nickel and cobalt, possess macroscopic magnetization even without an external field. Such magnetization is called spontaneous. In 1895, Pierre Curie presented the hypothesis that it originates from the interaction of the same magnetic moments that are responsible for paramagnetism in other circumstances. Curie's main idea is correct: Interactions between spins are the basis of all explanations of magnetic phases of matter today. However, in his time, no one could imagine what could make this interaction so strong as to explain the values of observed transition temperatures, which can reach 1000 K. The classical Ampère interaction of magnetic dipoles is a thousand times too weak for that, which was an insurmountable problem before the discovery of quantum mechanics.

10.5.1 Weiss's Phenomenological Model

Weiss' molecular field theory (1906) is the simplest phenomenological formulation of Curie's hypothesis. If there is a spin interaction, claims Weiss, then a single spin feels not only the external field B, but also some "molecular field" B_m, which originates from the other spins. Recalling the paramagnetic equation of state (10.39), we can write

$$M = M_0 \mathcal{B}(\mu_B B/kT) \tag{10.55}$$

where M_0 is the magnetization value at zero temperature. Weiss's idea is that instead of the field B, the spin sees the total field $B + B_m$, and that the magnetization itself is proportional to that field, $M \sim B_m$:

$$M = M_0 \mathcal{B}(\mu_B (B + B_m)/kT), \quad M(T)/M_0 = B_m(T)/B_0. \tag{10.56}$$

If this *mean field equation* has a solution $B_m \neq 0$ when the external field $B = 0$, the model will exhibit spontaneous magnetization. It can be solved graphically, as in Fig. 10.4. Obviously, there is a critical temperature T_c such that for $T < T_c$ there are two solutions, $B_m = 0$ and $B_m \neq 0$, while for $T > T_c$ we have only the solution $B_m = 0$. Thus T_c behaves as an experimental critical temperature, below which we have a phase transition to a magnetic state. It is easy to see that the free energy of the magnetized state will be lower; we show that for spin $1/2$. The sum over states is, according to (10.35),

$$Z_C(N, \beta) = (2 \cosh \beta \mu_B B_m)^N, \tag{10.57}$$

and since the hyperbolic cosine is always ≥ 1, the solution with $B_m \neq 0$ has a larger Massieu potential $\Psi_C = \ln Z_C$, i.e., a smaller free energy, $F = -kT \ln Z_C$.

The critical temperature of the model is easy to find: If the slope of the Brillouin function at the origin is less than $1/B_0$, we only have the solution $B_m = 0$, and if it is greater, we also have a solution $B_m \neq 0$. The critical temperature is the one at which the slope is exactly equal to $1/B_0$:

$$\frac{4J(J+1)}{3} \frac{\mu_B}{kT_c} = \frac{1}{B_0} \implies kT_c = \frac{4J(J+1)}{3} \mu_B B_0. \tag{10.58}$$

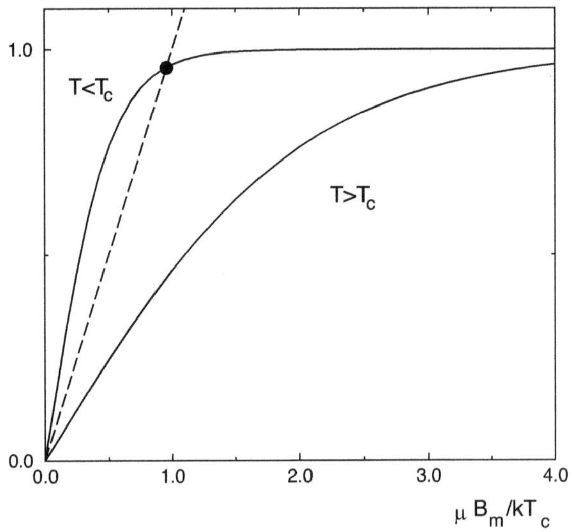

Fig. 10.4 Solving the mean field Eq. (10.56) for $J = 1/2$. *Solid lines* are Brillouin's functions for two temperatures, and the *dashed line* is B_m/B_0. The *filled circle* indicates a solution with spontaneous magnetization

For $J = 1/2$, this gives $kT_c = \mu_B B_0$, which is physically particularly transparent: The critical temperature is the one at which the thermal disorder manages to destroy the largest order that the molecular field can create, which is expressed by the energy gain in the ordered state at $T = 0$.

10.5.1.1 Energy of the Transition

In the previous section we saw how the transition temperature can be used to estimate the condensation energy associated with the ordered phase. If T_c is the transition temperature, the gain from ordering is $\Delta E \approx kT_c$ per particle.

This estimate appears quite general, so it is difficult to imagine how it could fail significantly. There is one famous case where it is wrong by four orders of magnitude. At the superconducting transition in metals, the energy gain is not kT_c, but kT_c^2/T_F, where T_F is the Fermi temperature. Why? Because only electrons near the Fermi energy participate in superconductivity. The estimate is actually correct, only when it is said "per particle," one should remember that the number of particles is not N but NT_c/T_F.

10.5.2 Susceptibility and Magnetization

The susceptibility above the critical temperature can be obtained by expanding the Brillouin function in a small field $B + B_m$ (namely, B_m vanishes for $T > T_c$, and the susceptibility is defined in the limit $B \to 0$):

$$B_m = B_0 \frac{4J(J+1)}{3} \frac{\mu_B(B+B_m)}{kT} = (B+B_m)\frac{T_c}{T}. \tag{10.59}$$

From here,

$$B_m = B\frac{T_c}{T-T_c}, \tag{10.60}$$

and the susceptibility is

$$\chi = \frac{1}{V}\lim_{B\to 0}\frac{M}{B} = \frac{C}{T-T_c}, \tag{10.61}$$

which is called the Curie-Weiss law. If we put $M_0 = N\mu_B$, the constant C is the same as in the paramagnetic Curie law,

$$C = \frac{N}{V}\frac{\mu_B^2}{k}\frac{4J(J+1)}{3}. \tag{10.62}$$

We see that for $T \to T_c$ the susceptibility diverges as $\chi \sim (T-T_c)^{-1}$. This power of -1 is called the *critical exponent* for the susceptibility. For iron, the measured $\chi \sim (T-T_c)^{-1.3}$, so the critical exponent of mean field theory deviates significantly from experiment.

When $T < T_c$, but close to T_c, the molecular field is small, so the mean field equation can be solved by expanding it to the next order,

$$B_m = B_0\left[\frac{4J(J+1)}{3}\frac{\mu_B B_m}{kT} - \frac{(2J+1)^4-1}{45}\left(\frac{\mu_B B_m}{kT}\right)^3\right], \tag{10.63}$$

which can be rearranged:

$$\frac{1+(2J+1)^2}{15}\left(\frac{\mu_B B_m}{kT}\right)^2 = 1-\frac{T}{T_c}. \tag{10.64}$$

Detailed numerical factors are unimportant here; the essential prediction is that the magnetization vanishes like a parabola near T_c:

$$M(T) \sim (1-T/T_c)^{1/2}, \quad T < T_c. \tag{10.65}$$

This power $1/2$ is another example of a critical exponent. In the majority of real materials, the critical exponent for magnetization is about 0.3, thus again different from the predictions of mean field theory.

10.6 The Origin of Magnetism

From the point of view of classical physics, one cannot understand the appearance of atomic magnetic moments, nor the strength of their interactions, responsible for the transition temperatures. Both are due to an interplay of the Coulomb force and the Pauli principle. Because the total wave function must be antisymmetric, different symmetries for the spin part imply different configurations for the space part, where the Coulomb potential acts. The Coulomb force is strong, so when it enters the energy-balance calculation between different spin configurations, one easily obtains large effects. Here, only the appearance of magnetic moments will be discussed, as expressed by so-called Hund's rule. The explanation for the magnetic interactions between magnetic moments, localized on ions in insulating crystals, is similar in principle, but becomes varied in detail, so that the elucidation of the mechanism at work in some specific material, among several plausible ones, may be the fruit of considerable debate. The explanation for the ferromagnetism of conducting electrons, such as in elemental iron, is quite different again. The physical difference between the origins of magnetism in metals and insulators is observable in transition temperatures, which are of the order of 1000 K for the former, and 100 K for the latter.

10.6.1 Hund's Rule

Hund's rule is a mnemonic describing how orbitals are filled in multi-electron atoms. It says that the lowest energy (ground) state is the one where the electrons in an open shell have the highest possible total spin. For neutral atoms, this rule is without exception throughout the periodic table.

Hund's rule means that electrons fill open shells in such a way that all their spins point in the same direction, until the maximal spin is achieved for a half-open shell. After that, they start filling them in the opposite direction, lowering the spin towards the closing of the shell. The closed shell of course has spin zero, because each suborbital is filled with a pair of electrons with opposing spins, after which the process starts again with the next shell. For example, manganese, with five electrons in the open $3d$ shell, has spin $5/2$ in the ground state, while iron, with one electron more, has spin $5/2 - 1/2 = 2$.

The open-shell electrons may be considered as orbiting a closed-shell ion, whose positive charge is equal to their number. The maximal-spin wave function is symmetric in spin space, because when all spins are pointing "up," and they are indistinguishable, exchanging them cannot change the wave function. However, electrons are fermions, hence the space part must antisymmetric for the total wave function to be antisymmetric.

When the space wave function is fully antisymmetric, the electrons avoid each other in space, i.e., no two electrons can be found near each other. In that case, each electron sees the total charge of the closed-shell ion, which attracts it by the Coulomb force. The consequence is that the whole atom contracts, lowering the Coulomb energy. This energy gain is the mechanism driving Hund's rule.

A further twist in the story is that "avoid each other in space" should not be understood literally, as if there were *angular* correlations between the electrons, like keeping them on opposite sides of the ion. The antisymmetric spatial wave function exhibits the same effect even if it consists of spherical (s) orbitals alone.

Conversely, when the space function is symmetric, electrons will pass close to each other, so that some will see the core charge reduced by the charge of others. This effect is called screening, and plays a significant role in metals, where the conducting electrons typically do not have a net magnetic moment, because "up" and "down" electrons fill the Fermi-liquid ground state to the same Fermi level. Thus, to explain the magnetism of metals is to understand how metallic screening is frustrated in some exceptional cases. The largest transition temperatures, such as in iron, are achieved when the conducting electrons carry part of the magnetic moment, so the Fermi levels for the two spin orientations are not the same. When that is not the case, such as in gadolinium, whose magnetism is due to localized $4f$ electrons, the transition temperature (293 K) is much lower, similar to that of magnetic insulators.

10.7 Landau's Description of Phase Transitions

Landau extended Weiss' approach to a general mean-field description of phase transitions, capable of handling transitions of any order and symmetry. His main idea is that, however complicated the free energy may be as a function of an input (microscopic) field, such as the magnetic induction $\boldsymbol{B}$, it must be a simple function of the output (macroscopic) field, called the order parameter, such as the magnetization $\boldsymbol{M}$, because it is extremal at the equilibrium value of the latter, so it is a parabola there. In other words, instead of writing the free energy in terms of the problem ($\boldsymbol{B}$), one writes it in terms of the solution ($\boldsymbol{M}$). The transformation from one to the other is a Legendre transform,

$$\mathcal{F}(\boldsymbol{M}) = F(\boldsymbol{B}) + \boldsymbol{B} \cdot \boldsymbol{M}. \tag{10.66}$$

The standard exposition, which we follow, is in terms of a magnetic transition as the most intuitive. The approach depends in general on experimental information about the order parameter (i.e., being told the solution). Its power is in encoding this information most efficiently, so that one readily derives equations of state which can be used to fix the parameters of the model by measurements.

For the ferromagnetic transition, it is sufficient to write

$$\mathcal{F}(\boldsymbol{M}) = \mathcal{F}(0) + a|\boldsymbol{M}|^2 + b|\boldsymbol{M}|^4, \tag{10.67}$$

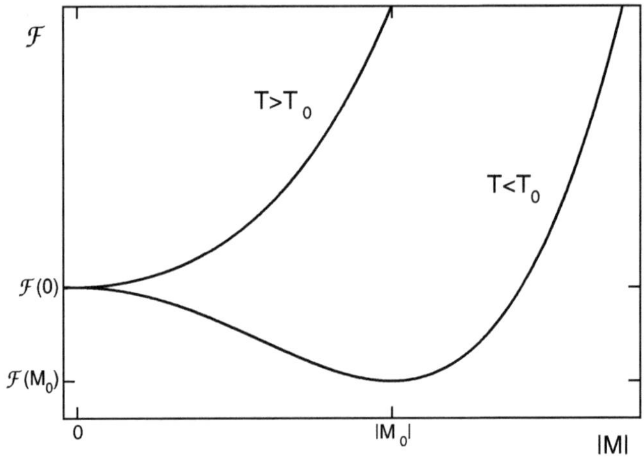

Fig. 10.5 Landau's description of phase transitions according to Eq. (10.67)

dropping all higher-order terms. Because the field is the conjugate variable to the magnetization, $\boldsymbol{B} = \nabla\mathcal{F}(\boldsymbol{M})$, the second derivative is the *reciprocal* of the magnetic susceptibility,

$$a \propto \chi^{-1} \implies a = a'(T - T_0), \tag{10.68}$$

according to the Curie-Weiss law (10.61). For $T > T_0$, the coefficient a is positive, and the stable solution of the equation of state is $\boldsymbol{M} = 0$. For $T < T_0$, this solution is unstable, $\mathcal{F}(0)$ is a maximum, but there appears a new solution at finite magnetization $\boldsymbol{M} = \boldsymbol{M}_0$, stabilized by the fourth-order term. The situation is depicted in Fig. 10.5.

The formal simplicity of the example above hides some sophisticated ideas. First, a phase transition is typically accompanied by a change in symmetry. In second-order transitions, this change is a spontaneous symmetry breaking associated with the appearance of an order parameter. In the case of magnetization, that is a vector breaking rotation symmetry by selecting a direction in space. The type of symmetry-breaking dictates the terms that can be postulated in Eq. (10.67). One can appeal to group theory to be certain that all invariants of the group of interest (e.g., $\boldsymbol{M} \cdot \boldsymbol{M}$) have been found, up to the required order. In structural phase transitions, the apparatus can get quite involved. Sometimes, like when the lattice couples to the magnetic field (the reason transformers hum), several order parameters and their combined invariants, corresponding to off-diagonal susceptibilities, need to be considered.

Second, Landau's approach neatly separates the macroscopic (thermodynamic) and microscopic (quantum-mechanical) aspects of a particular system. Quantum many-body theory provides a systematic way to calculate the coefficients in the expansion (by the *vertex functions*, of which χ^{-1} above is the simplest). However, the form of the expansion, and the concomitant equations of state, are fixed macroscopically, so that theoretical calculations are maximally constrained by the phenomenology, which provides a direct connection to experiment.

Finally, the system is stabilized by *nonlinear* responses at the transition itself, because the divergence of the susceptibility at the transition point corresponds to the vanishing of the quadratic term in $\mathcal{F}(\boldsymbol{M})$, i.e., of the harmonic restitutive force against changes in magnetization. However, as soon as it passes the critical point, a new quadratic minimum develops around $\boldsymbol{M}_0$, so the response is linear again, but with a different susceptibility.

For second-order transitions, these nonlinear responses are subtler than they appear, the vanishing of the harmonic term being only the first sign of theoretical difficulties. As indicated in the discussion of Bose-Einstein condensation, near the transition point nonlinear responses of arbitrary order all compete on equal footing, in the form of critical fluctuations. Such a situation is implied by the divergence of the susceptibility, which means physically that a vanishingly small field causes a macroscopic change in magnetization. The net effect is that the curve $\chi^{-1}(T)$ deviates from linearity in the critical region slightly above T_c, and approaches this point at an oblique angle, like a hockey stick, meaning that the critical exponent is a real number that cannot be found by truncating expansions of the free energy, as in Weiss' and Landau's approaches. A consistent description of second-order phase transitions in the critical regime that can obtain such irrational critical exponents was developed by Wilson, who was awarded a Nobel prize for this work.

10.7.1 Superconductivity

Landau and Ginzburg have applied the ideas above to develop a thermodynamic description of a superconductor. In the phenomenological spirit, all one needs to know is that the order parameter of a superconductor is a complex number. For an isotropic system, the corresponding expansion of the free energy is

$$\mathcal{F}[\Psi(\boldsymbol{r})] = \mathcal{F}_0 + \alpha(T - T_c)|\Psi(\boldsymbol{r})|^2 + \frac{1}{2}\beta|\Psi(\boldsymbol{r})|^4 + \gamma|\nabla\Psi(\boldsymbol{r})|^2, \tag{10.69}$$

where we have allowed the order parameter to vary in space, because we want to be able to describe the supercurrent, which is proportional to the gradient of its phase. In this way, we see immediately that the current in a superconductor is limited because it increases the free energy reversibly, by climbing in a potential landscape, not by resistance (Joule heating) as in a normal conductor. For the equilibrium state at $T < T_c$, the extremum at $\Psi = \Psi_0$ is determined by

$$\alpha(T - T_c) + \beta|\Psi_0|^2 = 0, \tag{10.70}$$

whence the order parameter vanishes parabolically, $|\Psi_0| \propto \sqrt{T_c - T}$, like the magnetization in Weiss's theory, which is the universal behavior of order parameters in the mean-field approach (why?). The condensation energy is

$$\mathcal{F}[\Psi_0] - \mathcal{F}_0 = -\frac{\alpha^2}{2\beta}(T - T_c)^2, \tag{10.71}$$

which vanishes quadratically with the temperature, predicting a step in the heat capacity at T_c.

Allowing a phase modulation now, $\Psi = \Psi_0 e^{i\phi(\boldsymbol{r})}$, the energy in the gradient term is nothing but the energy of the supercurrent:

$$E_s := \gamma|\Psi_0|^2|\nabla\phi(\boldsymbol{r})|^2 = \frac{\gamma\alpha}{\beta}(T_c - T)|\nabla\phi(\boldsymbol{r})|^2. \tag{10.72}$$

On the other hand, one may interpret this energy as the kinetic energy of the superconducting density:

$$E_s = \frac{1}{2}Mn_s v_s^2 = \frac{1}{2}Mn_s\left|\frac{\hbar}{2M}\nabla\phi\right|^2, \tag{10.73}$$

whence one obtains the concentration of the superconducting carriers:

$$n_s = \frac{8M}{\hbar^2}\frac{\gamma\alpha}{\beta}(T_c - T), \tag{10.74}$$

thus fixing the parameters of the model physically, because n_s can be inferred from the penetration depth (5.14):

$$\lambda_L^2 = \frac{Mc^2}{n_s e^2} = \frac{\hbar^2 c^2}{8e^2}\frac{\beta}{\gamma\alpha(T_c - T)}. \tag{10.75}$$

Finally, and most interestingly, comparing the condensation energy and the energy of the current gives a physical estimate of the critical current that destroys superconductivity:

$$-\frac{\alpha^2}{2\beta}(T - T_c)^2 + \frac{\gamma\alpha}{\beta}(T_c - T)|\nabla\phi(\boldsymbol{r})|_c^2 = 0. \tag{10.76}$$

A neat way to express this estimate is to write the phase as a wave, $\phi = \exp(ir/2\xi)$, whose wavelength plays the role of a *correlation length*, given by the above expression:

$$\xi(T) = \sqrt{\frac{\gamma}{\alpha(T_c - T)}}. \tag{10.77}$$

Macroscopically, the superconductor conducts electricity by a modulation of the phase of the order parameter, but the phase also needs to be correlated over a distance $\xi(T)$ for the superconducting state to be stable, so the current cannot "twist" the phase over shorter lengths, or it will destroy the coherence. When the required correlation length diverges, the maximal allowed current must go to zero.

Notably, the correlation length diverges as $T \to T_c$ with the same square root as the penetration depth, so the two scales are comparable in the vicinity of T_c:

$$\lim_{T \to T_c} \frac{\lambda_L(T)}{\xi(T)} = \frac{\hbar c}{2e}\sqrt{\frac{\beta}{2\gamma^2}} =: \kappa. \tag{10.78}$$

The competition between these two scales is quantitative, because their temperature dependence is qualitatively the same. Physically, they point to two ways a superconductor can transition to a normal state, one if the temperature is too high, the other if the magnetic field is too strong. Indeed, there are two kinds of superconductors in nature. Penetration by the field is sufficient to destroy superconductors of the *first kind*, while superconductivity survives even total penetration, from end to end, in superconductors of the *second kind*. In the latter, the correlation length is shorter than the penetration depth, so the field enters in the form of thin tubes of the normal state, surrounded by the superconductor shielding itself with internal Meissner currents in the form of vortices. It can be shown that the limiting value of the ratio (10.78) for this to occur is $\kappa > 1/\sqrt{2}$. Superconductors of the second kind are preferred for applications because they can be used in the presence of magnetic fields.

10.8 Particle-hole Excitations

Particle-hole excitations are excitations of a Fermi liquid near its ground state, which consist of a fermion with energy above the Fermi energy (the particle) that was excited from some other level below the Fermi energy (the hole). Small admixtures of particle-hole excitations to the Fermi liquid ground state correspond to density oscillations in real space, and their energy tends continuously to zero as the wave vectors of the particle and the hole tend to wave vectors at the Fermi surface. Both these properties are similar to sound, however, there is an important difference. Because the particle and the hole can occupy energy levels independently, subject only to the constraint that one is above and the other below the Fermi surface, the energy spectrum of particle-hole excitations is a continuum. For this reason, they are the principal source of dissipation for probes of Fermi liquids near the ground state.

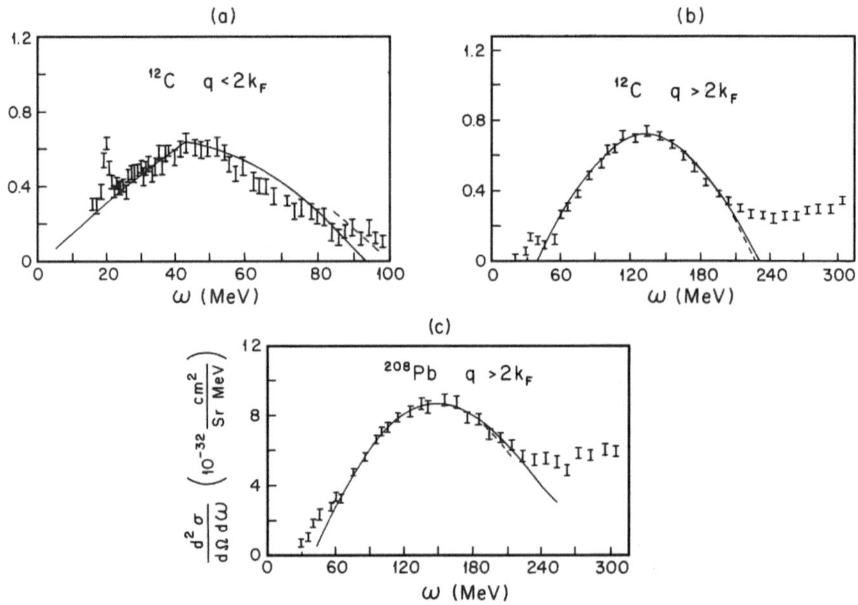

Fig. 10.6 Scattering of electrons on nuclei. *Full lines* are the calculated expressions (10.91), shifted by the binding energy of the nucleus. The discrepancy at high energy is due to pion production, which opens an additional dissipation channel [John W. Negele and Henri Orland, Quantum Many-Particle Systems. CRC Press Taylor and Francis Group, 2019. Fig. 5.19. Reprinted with permission.]

10.8.1 Scattering of Electrons on Nuclei

When electrons are scattered by atomic nuclei, two regimes for the cross section are possible. If the momentum transfer is less than some critical value, the cross section increases linearly with energy, before starting to fall off in a parabolic curve. If it is higher, scattering does not occur at lower energies, so there is a gap in the scattering spectrum, and when it does appear, it follows a parabola both ascending and descending (Fig. 10.6).

The observed curves can be obtained by a simple geometric calculation, as the density of states of particle-hole excitations. In order to obtain the exact shape of the curve, it is not even necessary to take into account the interaction between nucleons explicitly, which means that they are initially a Fermi liquid in the ground state. It is however necessary to assume that the space of quantum configurations, accessible to the excitations, is filled randomly, as if it were thermalized. This seems surprising, because the target nucleus is in the ground state, corresponding to zero temperature, and the energy and momentum of the incoming and scattered particles are controlled. However, as already said, the energy of particle-hole excitations is a continuum at given momentum transfer. The success of the calculation means that the excited nucleus is a thermodynamic system at fixed energy, which is randomly distributed over the available excitations.

10.8.2 Intuitive Calculation

The energy of an ideal Fermi gas is distributed over the fermion single-particle states of motion, which are of course kinetic, because there are no interactions, so their dispersion is

$$\varepsilon(\boldsymbol{k}) = \frac{\hbar^2 k^2}{2M}. \tag{10.79}$$

A particle-hole pair appears by moving a fermion from a filled single-particle state of wave vector $\boldsymbol{k}$, $|\boldsymbol{k}| < k_F$, to an empty state of wave vector $\boldsymbol{k}+\boldsymbol{q}$, $|\boldsymbol{k}+\boldsymbol{q}| > k_F$. The energy of such an excitation is clearly

$$\omega(\boldsymbol{k}, \boldsymbol{q}) := \varepsilon(\boldsymbol{k}+\boldsymbol{q}) - \varepsilon(\boldsymbol{k}), \tag{10.80}$$

while its momentum is equal to $\boldsymbol{q}$, because a particle of momentum $\boldsymbol{k}$ has been removed and replaced by one of momentum $\boldsymbol{k}+\boldsymbol{q}$. Explicitly,

$$|\boldsymbol{k}+\boldsymbol{q}|^2 - |\boldsymbol{k}|^2 = (\boldsymbol{k}+\boldsymbol{q}) \cdot (\boldsymbol{k}+\boldsymbol{q}) - \boldsymbol{k} \cdot \boldsymbol{k} = q^2 + 2\boldsymbol{k} \cdot \boldsymbol{q}, \tag{10.81}$$

which also depends on $\boldsymbol{k}$, meaning that $\omega(\boldsymbol{k}, \boldsymbol{q})$ is a continuum for a given $\boldsymbol{q}$.

The phase space accessible to an excitation of energy ω and momentum $\boldsymbol{q}$ consists of all points $\boldsymbol{k}$ which satisfy the constraints

$$\omega(\boldsymbol{k}, \boldsymbol{q}) = \omega, \quad |\boldsymbol{k}| < k_F < |\boldsymbol{k}+\boldsymbol{q}|. \tag{10.82}$$

Calculating the size of this space is a geometric task, which comes down to the visualization of a surface that satisfies the stated constraints.

The task is solved in two steps. First, we consider the case when $q > 2k_F$, so the condition $k_F < |\boldsymbol{k}+\boldsymbol{q}|$ is trivially satisfied. From now on, we measure ω in units of ε_F, and wave vectors in units of k_F. The energy condition defines a plane, $k \cos\theta = const.$:

$$k \cos\theta = \frac{\omega}{2q} - \frac{q}{2}, \quad k < 1. \tag{10.83}$$

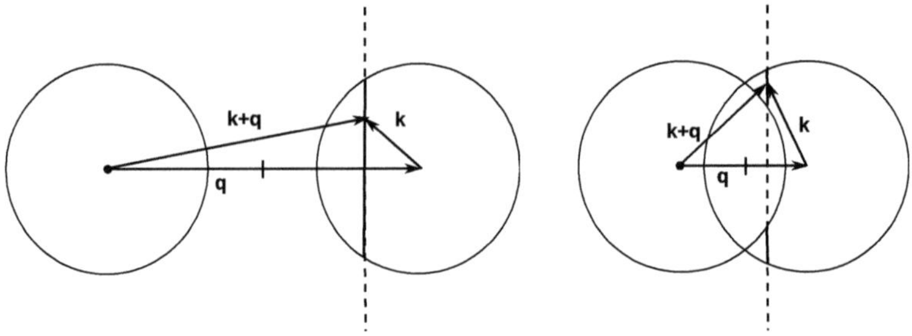

Fig. 10.7 The locus of allowed states for particle-hole excitations is depicted as *vertical solid lines*. *Dashed lines* are the locus of given energy. Allowed states are outside the first Fermi sphere and inside the second. Left: $q > 2k_F$, the locus is a disk. Right: $q < 2k_F$ and $\omega < 2q - q^2$, the locus is a circular annulus

The inequality is satisfied for all points of intersection of the unit (Fermi) sphere with this plane. The intersection is a disk whose area is

$$\left[1 - \left(\frac{\omega}{2q} - \frac{q}{2}\right)^2\right]\pi, \tag{10.84}$$

as seen in Fig. 10.7. If the expression in square brackets is negative, the plane has missed the Fermi sphere, so the phase space is zero. When $q > 2$, that can happen when ω is either too small or too large. The intersection is finite when

$$q^2 - 2q < \omega < q^2 + 2q, \quad q > 2. \tag{10.85}$$

The other case is $q < 2$. Then the two Fermi spheres in the figure begin to overlap, and we need the area of a part of the plane, which is *inside* one sphere ($k < 1$) but *outside* the other ($|\boldsymbol{k} + \boldsymbol{q}| > 1$). If the energy condition is such that the plane lies outside the second sphere, the calculation is the same as before. That will happen when

$$2q - q^2 < \omega < 2q + q^2, \quad q < 2. \tag{10.86}$$

But if the plane intersects both spheres, the phase space is a circular annulus, obtained when the intersection with one sphere is subtracted from the intersection with the other. Its surface is

$$\left[1 - \left(\frac{\omega}{2q} - \frac{q}{2}\right)^2\right]\pi - \left[1 - \left(\frac{\omega}{2q} + \frac{q}{2}\right)^2\right]\pi = \omega\pi, \tag{10.87}$$

when $\omega < 2q - q^2$, and $q < 2$.

We have obtained all the main trends of the experimental data. The critical value of the momentum transfer is exactly $q = 2k_F$. When $q < 2k_F$, the scattering cross section increases linearly, $\sigma \propto \omega$, as long as the energy is not too large, $\omega < 2q - q^2$, while it falls parabolically to zero for $2q - q^2 < \omega < 2q + q^2$. For higher values of the momentum transfer, $q > 2k_F$, we see the entire parabola in the range $q^2 - 2q < \omega < q^2 + 2q$.

10.8.3 Precise Calculation

The previous calculation appears quite convincing, but it has a pitfall. Namely, we cannot calculate the volume of a shell of a given energy in phase space as its surface: One dimension is missing! In fact, we need the change in the cumulative volume with energy, for a given change in energy, which is the density of states:

$$d\mathcal{N} = \frac{d\mathcal{N}}{d\omega} d\omega =: g(\omega) d\omega. \tag{10.88}$$

Thus, the third dimension is understood as the thickness of the shell in momentum space, which is obtained by a small change in energy, divided by that change. The measured cross section "at a given energy" discussed earlier was in fact the *differential* cross section $d\sigma/d\omega$; the integral of these measurements would count all events up to some energy.

Note that we may have already done the main part of the work when we naively calculated the surface. After all, we did obtain the observed trends. However, the factors that appear in the derivative may have a physical meaning. We cannot be sure that nothing will change in the above interpretation when the density of states is calculated properly.

The proper way to calculate the density of states is to apply the formula (6.11), adapted to the present situation:[5]

$$g(\boldsymbol{q}, \omega) = \frac{(2s+1)V}{(2\pi)^3} \int_{\omega(\boldsymbol{k},\boldsymbol{q})=\omega} \frac{dS}{|\nabla_k \omega(\boldsymbol{k}, \boldsymbol{q})|}. \tag{10.89}$$

Our previous calculation corresponded to $\int dS$, so the reason for concern is the velocity in the denominator, which we forgot to take into account. The velocity of a particle-hole excitation is

$$\nabla_k \omega(\boldsymbol{k}, \boldsymbol{q}) = \frac{\hbar^2 \boldsymbol{q}}{M} \to \frac{\hbar^2 k_F}{M} \boldsymbol{q}, \tag{10.90}$$

[5] If the particle-hole excitations include spin flips, one should put $(2s+1)^2$ in front.

where the arrow means the transition to dimensionless units for q. Similarly, $dS \to k_F^2 dS$. Since the velocity does not depend on the integration variable $\boldsymbol{k}$, we just copy the previous results to obtain the density of states of particle-hole excitations ($s = 1/2$):

$$g(\boldsymbol{q}, \omega) = V \frac{Mk_F}{h^2} \times \begin{cases} \frac{1}{q}\left[1 - \left(\frac{\omega}{2q} - \frac{q}{2}\right)^2\right], & q > 2,\ q^2 - 2q < \omega < q^2 + 2q, \\ \frac{1}{q}\left[1 - \left(\frac{\omega}{2q} - \frac{q}{2}\right)^2\right], & q < 2,\ 2q - q^2 < \omega < 2q + q^2, \\ \frac{\omega}{q}, & q < 2,\ 0 < \omega < 2q - q^2, \end{cases} \tag{10.91}$$

where q and ω are in units of k_F and ε_F, respectively.

The factor q in the denominator, which comes from the velocity, is not physically harmless, although it does not spoil the calculation of the area. It gives the total scale of the density of states, as the apex of the parabola at $\omega = q^2$ is at $1/q$, while $1/q$ is the slope in the linear part. This factor is obviously a constant in the energy dependence at given momentum transfer, unlike the dependence on the momentum transfer at given energy, where it changes the shape of the curve. That is why our previous calculation of the surface gave the correct shapes of the experimental curves as a function of energy. On the other hand, the $1/q$ is not as dangerous as it might seem. The readers can show that the level density vanishes in the limit of small momentum transfer at fixed energy.

10.8.4 Thermodynamics of Atomic Nuclei

The statement that the scattering cross section is proportional to the level density is Fermi's golden rule, already discussed in the context of blackbody radiation. Because adiabatic factorization assumes thermal relaxation of the final states, the question arises how this is possible under the given experimental circumstances.

When a projectile of known energy hits a target in the ground state, the composite system is an isolated system at fixed energy. If there are uncontrolled degrees of freedom, we must make the assumption of equal probability for the distribution of energy among them, which is the standard construction of thermodynamics. Thus, Planck's natural radiation hypothesis evidently applies to nuclear particle-hole excitations as well. Nuclear physicists may call the fits in Fig. 10.6 "just phase space," but they are a surprising testimony to the scope of thermodynamic principles.

Electrons interact only with protons in the nucleus, by the Coulomb interaction that is much weaker than the strong nuclear force, which keeps the nucleons together. In terms of time scales, strong translates to fast. The time the electron takes to transit the nucleus is sufficient for the nucleons to reach thermal equilibrium. To them, the $\sim 100\,\text{MeV}$ electron is an adiabatically slow probe.

The density of states of particle-hole excitations was calculated above for a free Fermi gas. We have already noted that the same formalism pertains to a Fermi liquid. The single-particle quantum numbers in our calculation refer to quasiparticles. Thus, our success in calculating the density of states of particle-hole excitations of an atomic nucleus is a success of the quasiparticle picture. Notably, it was achieved for a target carbon nucleus with only 12 nucleons.

Interactions did not explicitly enter into the comparison with measurements, because it was only important to count the degrees of freedom accurately. They enter through aggregate quantities, such as the absolute amount of scattering, or the precise value of the critical momentum, which corresponds to the Fermi wave vector of the interacting system. The statistical calculation that we carried out could determine the trends of the curves, but not these absolute scales. In particular, the fit was achieved after shifting the energy axis to compensate for the binding energy of the nucleus.

One worry may remain. Even if we had to shift the energy origin, the effective-mass concept evidently survived, given that it was sufficient to use the free-particle dispersion. How did that happen? One would expect the effective dispersion to contain some information about the structure of the nucleus. Instead of dwelling on its absence, note that the fits in Fig. 10.6 are within the error bars, up to systematic discrepancies at both ends, which are known to be extraneous. Any theoretical discussion can either justify the simplest version of Landau's approach that we used, or be wrong. There is nothing left to explain here, so whoever thinks that the issue is worth considering should think of some other measurement against which various proposals could be cross-checked.

In his commentary on Boetius' book about the Trinity, from which the quote at the beginning of this chapter is taken, Aquinas introduces a classification of the sciences. Mathematics is on top, because it is purely abstract. Music and astronomy are in the middle, because they study mathematical principles as they appear in natural phenomena. Physics and agriculture are at the bottom, because they are wholly practical.

Chapter 11
Fluctuations and Nonequilibrium Processes

Never did I dream that the Nobel Prize could be awarded for the reciprocal relations.

Lars Onsager

The study of fluctuations is usually relegated to the end of a study of thermodynamics. That creates the impression that it is an advanced subject, not required for a basic understanding of the equilibrium state. On the contrary, it is not possible even to understand temperature without invoking fluctuations, under the guise of Carnot engines. Nevertheless, the impression persisted long after Carnot that equilibrium and nonequilibrium phenomena were somehow fundamentally different, so that when Kamerlingh Onnes discovered superconductivity in 1911, his claim that it was a new phase of matter was contested on the grounds that he found the transition in resistivity, hence, not at equilibrium. Today, everyone knows that a transition in a transport coefficient implies a transition in some equilibrium susceptibility, because the two are proportional to each other, the factor of proportionality being the temperature. The reason for that connection is that static and dynamic friction have the same physical origin. This insight is known as the fluctuation-dissipation theorem, whose first version was discovered by Einstein in his analysis of Brownian motion in 1905. It took another half-century to understand the full implications of Einstein's relation, culminating in the work of Kubo in the 1950s, who developed the first well-founded quantum expressions for transport coefficients.

The history of nonequilibrium investigations instills modesty. Most of what can be said fundamentally about second-order, or two-particle, response functions was said long ago, and took a long time to be understood. The vast majority of observations are intrinsically limited to such responses, notwithstanding special cases, such as measurements of multiparticle coincidences in scattering. The foundational period of investigations into second-order responses ended with the work of Onsager.

Onsager showed that the number of independent transport coefficients, connecting n potential differences and their corresponding n currents, is not n^2, as one might naturally assume, but rather $n(n+1)/2$, because the coefficient matrix is symmetric.

D. Sunko, *Conceptual Thermodynamics*,
https://doi.org/10.1007/978-3-032-20877-4_11

The reason for that symmetry is the symmetry of microscopic equations of motion with respect to time reversal, despite, and coexisting with, the fact that macroscopic transport is a source of Joule heat, i.e., irreversible. This surprising resurrection of a fundamental symmetry to reduce the number of dissipation channels at the macroscopic level connects directly to applied research today.

Many contemporary investigations in electronics, and storage and transport of electricity, aim to reduce Joule heating. Onsager's result is a generic insight into this problem, noting that a conserved symmetry prevents noise by the corresponding kinematic constraint. Quantum computing, spintronics, topological insulators, single-atom switches, superconducting cables, solid-state electrolytes, to name but some applications, all handle heat by closing dissipation channels before it is created. Hence the excitement when it was realized that naturally evolved light-harvesting complexes in green plants conduct electrons coherently at room temperature, somehow evading the fluctuation-dissipation theorem. It is still not clear how they do it, but it is heartening to know that it is possible. The conversation between physics and chemistry, which began with Newton and Boscovich, is well on its way to become the wave of the future.

11.1 Brownian Motion

Einstein's study of Brownian motion in 1905 laid to rest all doubts one may have had about the statistical foundations of thermodynamics. By a careful layout of elementary arguments, not quite as elementary as those of Carnot, but well within the reach of a beginning student of physics, he showed that the random motion of a small particle in a liquid, such as the pollen observed by Robert Brown in 1827, was visible static friction, i.e., thermalization. From the thermodynamic point of view, the pollen particle is just another molecule, on equal footing with the molecules of water.

By the same token, the particle is a probe into the microscopic world of molecules. In particular, Einstein's analysis of its motion provided a way to measure Boltzmann's constant k directly, making it finally possible for physicists to accept Avogadro's understanding of a mole in good conscience. The modest pollen has evolved into all sorts of scattering measurements today, where the scattering cross section is a susceptibility, related to the dissipation of the beam in the target.

We shall introduce Brownian motion as a mathematical game first, called the random walk, extending the coin-flipping procedure by which we have previously introduced equilibrium statistical physics. The random walk will be connected to Einstein's derivation in the next step. The main idea is to analyze the thermostat for times shorter than the equilibration time, but not much shorter, so that second-order correlations are not thermalized, but all higher-order ones are.

Table 11.1 Random walk in one dimension

	Walker coordinate							
Step	−3	−2	−1	0	1	2	3	Norm
0	·	·	·	1	·	·	·	1
1	·	·	1	·	1	·	·	1/2
2	·	1	·	2	·	1	·	1/4
3	1	·	3	·	3	·	1	1/8

11.1.1 One-dimensional Random Walk

Imagine someone walking along a path, flipping a coin before each step to decide whether to take a step forward or backward. Obviously, one cannot predict precisely where such a walker will be after t steps; if we want to describe his evolution, we must use a statistical description. First, what is the probability that the walker will be at distance $x = x_1$ after t steps, if he was initially at $x = x_0$? If the steps are all the same length, we have the handy construction in Table 11.1. Initially, the walker is at the origin. After the first step, he can only be at one of two places, left and right of the origin, with equal probability $1/2$. In the next step, he will be at one of the even coordinates $x = -2, 0, 2$, but the probability of being at coordinate 0 is higher, because he could have reached it from both previous positions $x = -1, 1$. The probability to find the walker at a certain point is equal to the sum of the probabilities of finding him in the two adjacent points in the previous step, divided by two:

$$P(x, t+1|x_0, t_0) = \frac{1}{2}\left[P(x-1, t|x_0, t_0) + P(x+1, t|x_0, t_0)\right]. \tag{11.1}$$

We can easily guess the solution of the recursion, noticing that binomial coefficients appear in Table 11.1. Indeed, this recursive construction is the well-known Pascal's triangle. The probability to find the walker is given by the binomial distribution:

$$P(x_1, t_1|x_0, t_0) = \frac{1}{2^{t_1-t_0}}\binom{t_1-t_0}{(x_1-x_0+t_1-t_0)/2}. \tag{11.2}$$

After a given number of steps $t_1 - t_0 = t$, what is the expected value of the walker's position, $\overline{x_1 - x_0}$? Obviously zero, because the distribution is symmetric; it is more informative to look at

$$\overline{(x_1-x_0)^2} = \frac{1}{2^t}\sum_x x^2\binom{t}{(x+t)/2}. \tag{11.3}$$

We perform an explicit calculation for the even positions, $t = 2k$ and $x = 2m$. We have

$$\sum_m m^2\binom{2k}{m+k} = \left[\left(\frac{1}{s}+2+s\right)^k\right]''_{s=1} = 4^{k-1}\cdot 2k, \tag{11.4}$$

where we used the generating function, provable by induction if insight fails,

$$\sum_m s^m \binom{2k}{m+k} = \left(\frac{1}{s} + 2 + s\right)^k. \tag{11.5}$$

One obtains

$$\overline{(x_1 - x_0)^2} = t_1 - t_0. \tag{11.6}$$

The linear deviation of the walker is proportional to the square root of the elapsed time, not the time, as in deterministic motion.

11.1.2 Continuum Limit

As the time intervals and step size become smaller, we expect that the binomial distribution will converge to a Gaussian. We replace $t_1 - t_0 \to (t_1 - t_0)/\tau$ and $x_1 - x_0 \to (x_1 - x_0)/a$, and consider the limit $\tau, a \to 0$. It is convenient to rewrite the recursion (11.1) by subtracting $P(x, t|x_0, t_0)$ from both sides,

$$\begin{aligned} &P(x, t+1|x_0, t_0) - P(x, t|x_0, t_0) \\ &\quad = \frac{1}{2}\left[P(x-1, t|x_0, t_0) + P(x+1, t|x_0, t_0) - 2P(x, t|x_0, t_0)\right]. \end{aligned} \tag{11.7}$$

Transforming to dimensionful variables, we introduce the density

$$p(x, t) := P(x/a, t/\tau|x_0/a, t_0/\tau)/a. \tag{11.8}$$

The recursion becomes

$$p(x, t+\tau) - p(x, t) = \frac{1}{2}\left[p(x-a, t) + p(x+a, t) - 2p(x, t)\right]. \tag{11.9}$$

If a and τ tend to zero independently, both sides vanish. Notice, however, that the first difference in t appears on the left, and the second difference in x on the right. To turn them into derivatives, this expression can be rewritten as

$$\frac{p(x, t+\tau) - p(x, t)}{\tau} = \alpha^2 \frac{p(x-a, t) + p(x+a, t) - 2p(x, t)}{a^2}. \tag{11.10}$$

If a and τ tend to zero so that the ratio $a^2/2\tau =: \alpha^2$ stays constant, it tends to

$$\frac{\partial p(x, t)}{\partial t} = \alpha^2 \frac{\partial^2 p(x, t)}{\partial x^2}. \tag{11.11}$$

This is the diffusion equation, with diffusion constant α^2. It can be understood as the continuum limit of the binomial recursion. Its solution is the Gaussian

$$p(x, t|x_0, t_0) = \frac{1}{\sqrt{4\pi\alpha^2(t - t_0)}} \exp\left(-\frac{(x - x_0)^2}{4\alpha^2(t - t_0)}\right), \tag{11.12}$$

which is normalized and satisfies the condition that the walker is initially at x_0:

$$\int p(x, t|x_0, t_0)\, dx = 1, \tag{11.13}$$

$$\lim_{t_1 \to t_0} p(x_1, t_1|x_0, t_0) = \delta(x_1 - x_0). \tag{11.14}$$

It also satisfies a less trivial condition, consistent with its interpretation as a conditional probability, namely, at some time t' between t_0 and t_1, the walker must have been somewhere, and it does not matter where:

$$\int p(x_1, t_1|x', t')p(x', t'|x_0, t_0)\, dx' = p(x_1, t_1|x_0, t_0), \quad t_0 < t' < t_1. \tag{11.15}$$

This is called the *Smoluchowski equation*. It is actually not clear a priori that some integral kernel, depending on x' and t', will not appear in this reconstruction of the probability $p(x_1, t_1|x_0, t_0)$. Its absence in the Smoluchowski equation is a consequence of the independence of the walker's decision, which will be the next step, on previous decisions. The position of the walker at the moment $t + \tau$ depends only on the position at moment t, not on how he got there. Processes, in which the value of a random variable x_{n+1} depends only its value x_n, and not on n or x_k for $k < n$, are called *Markov chains*. For them, a Smoluchowski equation holds. It is said that such processes have no memory.

Finally, we note that the discrete random walker model does not have a continuum limit in general, but only for one specific, critical scaling of temporal and spatial steps, such that $a^2/2\tau$ is kept constant. Physically, we reasonably expect that a model will depend on the microscopic lattice on which it is defined; but sometimes there are special, critical trajectories in parameter space, along which the behavior of the model does not depend on microscopic scales.

11.1.3 Multi-dimensional Walker

Consider a cubic lattice in d dimension, on which the walker can choose a step forward or backward along any axis. The recursion (11.1) generalizes straightforwardly to

$$P(\boldsymbol{x}, t + 1|\boldsymbol{x}_0, t_0) - P(\boldsymbol{x}, t|\boldsymbol{x}_0, t_0) = \boldsymbol{\Delta} P(\boldsymbol{x}, t|\boldsymbol{x}_0, t_0), \tag{11.16}$$

where the discrete Laplacian,

$$\boldsymbol{\Delta} f(\boldsymbol{x}) = \frac{1}{2d}\sum_{i=1}^{d}\left[f(\boldsymbol{x}-\boldsymbol{e}_i) + f(\boldsymbol{x}+\boldsymbol{e}_i) - 2f(\boldsymbol{x})\right], \tag{11.17}$$

appears in place of the second difference, and $\boldsymbol{e}_i$ are unit vectors in the d directions. In the continuum limit, it gives rise to the diffusion equation

$$\frac{\partial p(\boldsymbol{x},t)}{\partial t} = \alpha^2\nabla^2 p(\boldsymbol{x},t), \tag{11.18}$$

where we have introduced the scaling $a^2/(2d\tau) = \alpha^2$ and $p = P/a^d$. Its solution is the same Gaussian as before, where one only needs to replace the space variables with vectors, and take the d-th power of the norm.

We can solve the discrete recursion (11.16). Because the probabilities P are defined on integer lattice nodes $\boldsymbol{x}$, we consider them as Fourier coefficients of some function $\tilde{P}(\boldsymbol{k},t)$, periodic in all d components of the variable $\boldsymbol{k}$, with period 2π:

$$P(\boldsymbol{x},t|\boldsymbol{x}_0,t_0) = \int_{-\pi}^{\pi}\frac{d^d\boldsymbol{k}}{(2\pi)^d}e^{i\boldsymbol{k}\cdot\boldsymbol{x}}\tilde{P}(\boldsymbol{k},t). \tag{11.19}$$

Inserting this *ansatz* into the recursion, it is easy to obtain

$$\tilde{P}(\boldsymbol{k},t+1) = \left(\frac{1}{d}\sum_{i=1}^{d}\cos k_i\right)\cdot\tilde{P}(\boldsymbol{k},t), \quad \tilde{P}(\boldsymbol{k},t_0) = e^{-i\boldsymbol{k}\cdot\boldsymbol{x}_0}, \tag{11.20}$$

where the initial condition for $\tilde{P}$ follows from the initial condition that P must be a Kronecker delta symbol. The recursion for $\tilde{P}$ is trivial, so we get

$$P(\boldsymbol{x}_1,t_1|\boldsymbol{x}_0,t_0) = \int_{-\pi}^{\pi}\frac{d^d\boldsymbol{k}}{(2\pi)^d}e^{i\boldsymbol{k}\cdot(\boldsymbol{x}_1-\boldsymbol{x}_0)}\left(\frac{1}{d}\sum_{i=1}^{d}\cos k_i\right)^{t_1-t_0}. \tag{11.21}$$

This expression can be used to obtain a fundamental property of the random walk: The effective dimension of its trajectory is two, not one! The walker covers surfaces rather than drawing lines. We can see that by considering the total number of visits to a fixed lattice node $\boldsymbol{x}_1$:

$$G(\boldsymbol{x}_1-\boldsymbol{x}_0) = \sum_{t_1=t_0}^{\infty}P(\boldsymbol{x}_1,t_1|\boldsymbol{x}_0,t_0). \tag{11.22}$$

If a node is visited with probability, say, 1/4, we count that as a quarter of a visit. Inserting the integral expression for P, we have

$$G(\boldsymbol{x}_1 - \boldsymbol{x}_0) = \int_{-\pi}^{\pi} \frac{d^d \boldsymbol{k}}{(2\pi)^d} \frac{e^{i\boldsymbol{k}\cdot(\boldsymbol{x}_1 - \boldsymbol{x}_0)}}{1 - \frac{1}{d}\sum_i \cos k_i}. \tag{11.23}$$

The integral *diverges* for $d = 1$ and $d = 2$, because the volume element is $\sim k^{d-1}dk$, and the denominator is $\sim k^2$. It means that every node in one and two dimensions is certain to be visited, many times: The walker fills surfaces.

This result can be understood intuitively. Imagine each step as a link in a chain placed on the lattice, and let the linear distance between the first and the last links be r. What is the expected length t of the chain, its number of links? The result (11.6) says $t \sim r^2$. The chain has enough links to cover a *surface area* of linear extension r. Like a child, the walker has scribbled over a whole surface spanned by the initial and final positions.

It is less intuitive that the natural two-dimensionality of the walker's trajectory implies two characteristic dimensions of space, not one. The first is $d = 2$; it is obviously important whether the walker fills all space, or not. The other is $d = 4$, for a more subtle reason. It answers the question, when it becomes improbable that two walkers will meet. Just as lines generically do not intersect in three and more dimensions, so planes do not intersect when $d \geq 4$.[1] The physical significance of this observation is that we can model thermal fluctuations in a sample as random walks, so it is pertinent to ask whether a fluctuation will always evolve in an averaged background, or it will encounter other fluctuations regularly. The former is the mean-field assumption, so we can infer that it fails in one and two dimensions, and is correct in four and more dimensions. That is a surprise, because one would imagine that the mean-field idea becomes better gradually as the number of dimensions increases, not that it is abruptly fully justified at $d = 4$.

11.1.4 The Policeman's Lecture on One-dimensional Fluctuations

After the war ended in 1995, tourists started returning to the Croatian coast in large numbers, before the network of modern highways was built. One summer day, there was a massive traffic collapse, more than a hundred kilometers long, on the main road, which took twelve hours to clear. Everyone noticed that nothing really happened, however. There was no accident, or any physical obstruction on the road. A television reporter asked a traffic policeman to explain that if he could. Seated on

[1] In $d = 4$, they intersect in isolated points, which does not affect the physical interpretation.

his motorcycle by the side of the road, the policeman delivered the following lecture without hesitation, which made the evening news (quoting from memory):

> This road can take 90 km/h at most, but driving in a column, it is more like 70. Now, when people start looking for a place to have coffee, they slow down to maybe 40. By the time they find a place and stop, there is a column of slowed-down cars behind, which is at least ten kilometers long. For the last car in that column to accelerate back to 70, it takes half an hour, but by that time, someone else is already slowing down for coffee.

The fluctuations collide, even if the cars do not. The collapse is caused by a statistically induced long-range correlation. The correlation length diverges at sufficiently high density, when every car on the road "sees" every single car slowing down. The only velocity on which they can all agree is zero. That would not happen with ships, because they can go around each other.

11.1.5 *Massive Walker*

The function $G(\boldsymbol{x}_1 - \boldsymbol{x}_0)$ satisfies the equation

$$\boldsymbol{\Delta} G(\boldsymbol{x}_1 - \boldsymbol{x}_0) = -\delta_{\boldsymbol{x}_1 \boldsymbol{x}_0}, \tag{11.24}$$

so it is the Green's function for the discrete Laplace operator $-\boldsymbol{\Delta}$. We can view it as the limiting case ($\lambda = 1$) of the generating function for the probabilities P,

$$G(\boldsymbol{x}_1 - \boldsymbol{x}_0, \lambda) = \lambda \sum_{t_1 = t_0}^{\infty} \lambda^{t_1 - t_0} P(\boldsymbol{x}_1, t_1 | \boldsymbol{x}_0, t_0). \tag{11.25}$$

When $|\lambda| < 1$, this function is always convergent, as can be seen from the integral representation (no more zero in the denominator):

$$G(\boldsymbol{x}_1 - \boldsymbol{x}_0, \lambda) = \int_{-\pi}^{\pi} \frac{d^d \boldsymbol{k}}{(2\pi)^d} \frac{e^{i\boldsymbol{k}\cdot(\boldsymbol{x}_1 - \boldsymbol{x}_0)}}{\frac{1}{\lambda} - \frac{1}{d}\sum_i \cos k_i}. \tag{11.26}$$

It is also a Green's function, but for a slightly different equation:

$$\left[-\boldsymbol{\Delta} + \left(\frac{1}{\lambda} - 1\right)\right] G(\boldsymbol{x}_1 - \boldsymbol{x}_0, \lambda) = \delta_{\boldsymbol{x}_1 \boldsymbol{x}_0}. \tag{11.27}$$

We say that a walker obeying this equation is *localized*. It is easy to understand why, when one realizes that the corresponding recursion has $\lambda/2$ instead of $1/2$ in the Eq. (11.1). It means that part of the walkers will never take a step. The probability that the walker disappears is $1-\lambda$. As soon as this probability exists, however small, a drastic change in the walker's properties occurs. He can no longer go very far, and is unable to cover the entire space, even in one dimension. The probability of finding him decreases by geometric progression. A localized walker can therefore be seen only in the vicinity of the starting point.

Such a walker is also called "massive." This name can be understood by considering the continuum limit. If we put $\boldsymbol{x} \to \boldsymbol{x}/a$, $\boldsymbol{k} \to \boldsymbol{k}a$, and implement the critical scaling $1/\lambda - 1 \to m^2a^2/2d$, the limit $a \to 0$ gives

$$g(\boldsymbol{x}, m^2) = \lim_{a\to 0} \frac{1}{2da^{d-2}} G(\boldsymbol{x}/a, \lambda) = \int_{-\infty}^{\infty} \frac{d^d\boldsymbol{k}}{(2\pi)^d} \frac{e^{i\boldsymbol{k}\cdot\boldsymbol{x}}}{\boldsymbol{k}^2 + m^2}. \tag{11.28}$$

This is the Green's function for the (static) Klein-Gordon equation $(-\nabla^2 + m^2)\Psi(\boldsymbol{x}) = 0$, which is analogous to Laplace's equation in electrostatics, but for photons that have mass; explicit calculation of the integral in (11.28) gives Yukawa's potential

$$g(\boldsymbol{x}, m^2) \sim \frac{1}{|\boldsymbol{x}|} e^{-m|\boldsymbol{x}|} \tag{11.29}$$

instead of the Coulomb $1/|\boldsymbol{x}|$, so the massive walker has finite range. This can be seen nicely when ink drips on paper: The sharp edge of the stain is characteristic of exponential localization.

It is said that Fermi once commented on a discussion about galactic civilizations by asking, where is everybody? His point was that with quite mild assumptions about the diffusion of such civilizations—say, 10,000 years per star-hop—the Milky Way was old enough for everybody to arrive everywhere (a million hops per civilization). However, if there is even a small probability that hopping stops, the galaxy is also big enough for the civilizations never to overlap, like drops of ink splattered on paper.

11.1.6 Motion of a Brownian Particle

Now we shall see how the results of the random-walk analysis appear in a physical context. Observe a small particle suspended in a liquid. The equation of motion is

$$M\dot{v} = K(t), \tag{11.30}$$

where $K(t)$ represents a random force, which originates from our particle being pushed by molecules of the liquid. When the particle moves, the average effect of this force is viscous resistance, so subtract it:

$$K(t) = -v/\mu + MA(t), \tag{11.31}$$

where $A(t)$ is that part of the random force which averages to zero. When the particle is at rest, it is all that remains of the force $K(t)$. The coefficient μ is called the *mobility*. The equation of motion is now

$$\dot{v} = -\frac{1}{M\mu}v + A(t), \tag{11.32}$$

and we need to find the time evolution of the coordinate $x(t)$. Recalling that the average of $x(t)$ vanishes, we rearrange the equation of motion to find the evolution of $x^2(t)$. First, multiply it by x:

$$x\ddot{x} = -\frac{1}{M\mu}x\dot{x} + xA(t). \tag{11.33}$$

Note that

$$x\dot{x} = \frac{1}{2}\frac{d}{dt}x^2, \quad x\ddot{x} = \frac{1}{2}\frac{d^2}{dt^2}x^2 - \dot{x}^2.$$

The equation of motion becomes

$$\frac{1}{2}\frac{d^2}{dt^2}x^2 - \dot{x}^2 = -\frac{1}{2M\mu}\frac{d}{dt}x^2 + xA(t). \tag{11.34}$$

The time average $\overline{x(t)A(t)}$ vanishes, because the random force is not correlated with the position of the particle. Einstein's key step in the derivation was to invoke the equipartition theorem for the *macroscopic* suspended particle:

$$\boxed{\frac{M\overline{\dot{x}^2}}{2} = \frac{kT}{2}}, \tag{11.35}$$

because time and ensemble averages are equal. Finally, denote the time average of the derivative x^2 by γ,

$$\gamma = \overline{\frac{d}{dt}x^2} = \frac{d}{dt}\overline{x^2}.$$

Then the equation of motion becomes

$$\dot{\gamma} = \frac{2kT}{M} - \frac{1}{M\mu}\gamma. \tag{11.36}$$

The asymptotic solution ($t \to \infty$) is the one for which $\dot{\gamma} = 0$:

$$\gamma = \frac{d}{dt}\overline{x^2} = 2\mu kT, \tag{11.37}$$

or, by integrating,

$$\overline{x^2} = 2\mu kT \cdot t. \tag{11.38}$$

We see, therefore, that the mean *square* excursion of a Brownian particle is proportional to time, just like for a random walker. Furthermore, for a spherical particle of radius R, one can include the relation between the mobility μ and the viscosity constant η, so-called Stokes' law:

$$\mu = \frac{1}{6\pi R\eta}, \tag{11.39}$$

so one gets

$$\overline{x^2} = \frac{kT}{3\pi R\eta} t. \tag{11.40}$$

This equation can be applied to determine Boltzmann's constant k, measuring the size of the Brownian particle under a microscope and inserting the known viscosity. Alternatively, the magnitude of μ can be determined experimentally, from the limiting velocity at a constant external force K_0, $\overline{\dot{x}} = \mu K_0$. In such a way, the ratio of the gas and Boltzmann constants, Avogadro's number, was determined for the first time.

11.1.7 Einstein's Relation and the Fluctuation-dissipation Theorem

From the solution of the diffusion Eq. (11.12), it is easy to obtain

$$\overline{x^2} = \int x^2 p(x,t) = 2\alpha^2 t, \tag{11.41}$$

so the diffusion constant turns out to be

$$\alpha^2 = kT \cdot \mu. \tag{11.42}$$

We have derived a direct relation between the diffusion constant and the mobility, i.e. the microscopic and macroscopic characteristics of the particle's motion. This result is called Einstein's relation. It is the prototype for a class of relations collectively known as the fluctuation-dissipation theorem. It means that the physical origin of the limiting velocity of a falling particle and of its Brownian motion is the same, random collisions with molecules. By the same token, it means that an electron in a metal cannot distinguish the electric field of another electron from an external field causing the electric current. All versions of the fluctuation-dissipation theorem, proved later for particular circumstances, are conceptually the same as Einstein's relation.

11.1.8 Comparison of Brownian and Thermodynamic Fluctuations

If we make the usual identification of measurements of some thermodynamic quantity at time intervals greater than the thermalization time with a random selection of representatives from the ensemble, we can estimate the fluctuation of that quantity over time from its fluctuation in the ensemble. The essential difference between these fluctuations and those where the time evolution is modelled by a process like Brownian motion is that there is no dependence of one measurement on the previous ones in thermodynamic fluctuations. Physically, the thermalization time separates mutually correlated measurements of microscopic quantities, such as the position of a Brownian particle, from uncorrelated measurements of macroscopic quantities, such as the internal energy of a system of particles. Therefore, one should distinguish the time correlations of a Markovian random variable from correlations within the ensemble.

The difference between these two types of fluctuations is evident in Fig. 11.1. It stems from the fact that the walker's position at any given time is determined by the previous position, so if it was, say, $x = -50$ in one step, it can only be $x = -51$ or -49 in the next; the walker will take a long time to get to $x = 50$, even though it is certain (in one dimension) that he will get there eventually. On the other hand, because ensemble members are chosen at random, they have an equal chance of being positive and negative at each step. One interpretation of this difference is that in a coin-toss game, in which two players give each other the same amount,

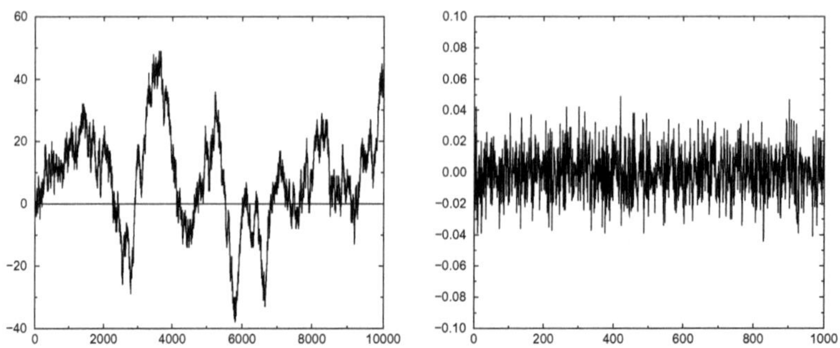

Fig. 11.1 Evolution of a single one-dimensional Brownian walker (left), and of samples from an ensemble (right). The figure at right shows the sum of spin orientations of a thousand randomly oriented spins $J = 1/2$, for a thousand randomly chosen members of the ensemble

depending on the outcome of the toss, one expects very long periods in which the same player has an advantage. Precise analysis shows that the number of crossings of the abscissa (change of advantage) in N tosses scales as $\sqrt{N}$, meaning that the walker in Fig. 11.1 should cross it about 100 times for the 10,000 tosses depicted, but only a thousand times for a million tosses.[2] The naive expectation that the advantage should change regularly is in fact fulfilled for many games, i.e., for the ensemble. Indeed, if there are many pairs of players, the players who win on heads will have the advantage in approximately half of the games, at any moment in time. Their total advantage at each moment behaves like the total spin in Fig. 11.1. As Feller put it, in a population of normal coins, each individual coin is abnormal.

11.2 Wiener-Khintchine Theorem

In the study of Brownian-type temporal fluctuations, two approaches are fundamentally distinguished. One starts with a concrete model of the walker, and tries to calculate all possible consequences from it, i.e. to describe the walker's behavior in as much detail as possible. The other attempts to relate the individual characteristics of the random process to each other, trying to remain as independent as possible from the choice of the concrete model in which these characteristics could be realized. While the former approach is usually chosen when modelling traffic and stock markets, the latter is more fruitful theoretically; it started with Einstein's relation, which connected the mobility and the diffusion constant. One of the important results of this approach is the Wiener-Khintchine theorem, which connects the spectrum of noise with its correlations in time.

Random processes can be divided into classes according to how many independent moments need to be determined in order to define the process. Thus, for uncorrelated process, the first moment is sufficient, the probability $p_1(x, t)$ that a particle will be found at a given place at a given time. For Markov processes, one also needs to specify $p_2(x_1, t_1; x_2, t_2)$, the probability that both events will occur, finding a walker at x_1 at time t_1, and at x_2 at time t_2. For uncorrelated processes, this would simply be the product of probabilities p_1, but not for Markov processes, as we saw in Fig. 11.1. For processes with memory, higher moments are required, while in the case of the Markov process, one has

$$P(x_3, t_3|x_2, t_2; x_1, t_1) = P(x_3, t_3|x_2, t_2), \quad t_3 > t_2 > t_1,$$

because it does not matter to the walker at t_3 what happened before t_2. Approaches that aim to be model-independent limit themselves to assumptions about low moments, spectral distributions, and the like. Wiener and Khintchine's result is important because it shows that some of these assumptions are not mutually

[2] This is the foundation of the common experience that the other queue always moves faster.

independent. Besides this theoretical significance, it also has a practical aspect: Listening to noise, we can learn something about the process that generates it.

11.2.1 Derivation of the Theorem

We represent the random path of a Brownian walker over a long time T by a Fourier integral in the continuum limit:

$$x(t) = \frac{1}{2\pi} \int_{-\infty}^{\infty} A(\omega) e^{i\omega t}\, d\omega. \tag{11.43}$$

Because $x(t)$ is a real number, $A^*(\omega) = A(-\omega)$. Note that, due to the known result $\overline{x^2(t)} \sim t$, the expected value of the square of the amplitude A diverges linearly with the interval T. We are interested in the temporal correlations of the walkers, i.e. the expected behavior of the *autocorrelation function* for a walker randomly selected from the ensemble:

$$C(\tau) := \langle x(t)x(t+\tau)\rangle = \lim_{\mathrm{T}\to\infty} \frac{1}{\mathrm{T}} \int_{-\mathrm{T}/2}^{\mathrm{T}/2} x(t)x(t+\tau)dt, \tag{11.44}$$

because we believe that the time and ensemble averages are the same if they are calculated over a time T, longer than the relaxation time. This means that we can also average both, $C(\tau) = \overline{\langle x(t)x(t+\tau)\rangle}$, without changing the result. Inserting the representation of $x(t)$ into the definition,

$$C(\tau) = \frac{1}{4\pi^2} \int_{-\infty}^{\infty} d\omega \int_{-\infty}^{\infty} d\omega' \lim_{\mathrm{T}\to\infty} \frac{1}{\mathrm{T}} A(\omega)A(\omega') \int_{-\mathrm{T}/2}^{\mathrm{T}/2} dt e^{i\omega t} e^{i\omega'(t+\tau)}. \tag{11.45}$$

The inversion formula for the Fourier transform implies

$$\int_{-\infty}^{\infty} dt e^{i\omega t} e^{i\omega'(t+\tau)} = e^{i\omega\tau} 2\pi \delta(\omega + \omega'),$$

so

$$C(\tau) = \frac{1}{\pi} \int_{0}^{\infty} d\omega \cos \omega\tau \lim_{\mathrm{T}\to\infty} \frac{1}{\mathrm{T}} |A(\omega)|^2, \tag{11.46}$$

because $|A(\omega)|^2$ is an even function of frequency if A is real. Introduce the power spectral function of the process,

$$G(\omega) \equiv \lim_{\mathrm{T}\to\infty} \frac{2}{\mathrm{T}} \left\langle |A(\omega)|^2 \right\rangle. \tag{11.47}$$

Then one can write

$$C(\tau) = \frac{1}{2\pi} \int_0^\infty G(\omega) \cos\omega\tau \, d\omega, \tag{11.48}$$

which is the Wiener-Khintchine theorem: The autocorrelation function of a process is the Fourier transform of its power spectrum. The inverse Fourier transform then gives, of course,

$$G(\omega) = 4 \int_0^\infty C(\tau) \cos\omega\tau \, d\tau. \tag{11.49}$$

Thus, if one knows the power radiated by the noise in a range of frequencies, one can determine the autocorrelation of the process that produced that noise.

11.2.2 Examples

Consider the most common case, when the autocorrelation function of the process decays exponentially:

$$C(\tau) = e^{-\tau/\tau_c} \implies G(\omega) = \frac{4\tau_c}{1 + (\omega\tau_c)^2}. \tag{11.50}$$

This means that there is no noise at high frequencies that correspond to times significantly shorter than the characteristic time τ_c, over which the process is correlated. This is clear: Strong autocorrelations mean deterministic behavior, hence the absence of noise. At low frequencies, which effectively average over times longer than τ_c, the forgetfulness of our walker manifests itself as noise in the power spectrum.

On the other hand, suppose that the process forgets immediately:

$$C(\tau) \sim \delta(\tau) \implies G(\omega) \sim \mathit{const.}, \tag{11.51}$$

which is called "white noise;" if there are no times so short that the walker is correlated over them, then high-frequency processes are just as susceptible to noise as slow ones. This approximation is very good in real situations. In metals, the relaxation time, i.e., the time over which the random electron scatterings are correlated, is about 10^{-13} s, so $G(\omega)$ in (11.50) is practically constant, i.e., white, up to microwave frequencies.

We can further imagine a low-pass filter, allowing white noise up to some frequency ω_m, above which it is cut off:

$$G(\omega) \sim \theta(\omega_m - \omega) \implies C(\tau) \sim \frac{\sin \omega_m \tau}{\tau}, \tag{11.52}$$

which corresponds to correlations that slowly decay on the scale $1/\omega_m$, while for $\omega_m \to \infty$ this scale disappears, so the process is reduced to the previous one. The oscillations in the correlation function are a consequence of the imposed sharp cut in the power spectrum; they are smoothed out if $G(\omega)$ decays across some range of frequencies, rather than as a step.

11.3 Nyquist Theorem

The Nyquist theorem is one of the most important results in the analysis of time-related fluctuations of thermodynamic systems. Historically, it is the original expression of the fluctuation-dissipation theorem in the context of electric circuits, which attracted much wider attention than Einstein's relation because of this practical context.

The theorem gives a precise expression for the magnitude of thermal voltage fluctuations that occur in a resistor. They are the minimum possible noise in a resistor at a given temperature. At the same time, one should keep in mind that this noise will not necessarily load the rest of the electric circuit. The transmission of noise to other elements depends on the tuning of the circuit. Thus, although the resistor produces white noise, the rest of the circuit may only "hear" noise in some frequency range. Namely, because the expectation value of the fluctuating voltage is zero, it does not give rise to a gradient of the chemical potential ("electromotive force"). The fact that the square of the voltage does not vanish on average makes it a power dimensionally, but it is important to understand that this power is incapable of macroscopic work, so the noise can be suppressed by a passive filter, much like a gas is kept inside a bottle.

11.3.1 Macroscopic Derivation of the Theorem

If a resistor is connected by two wires to a sensitive galvanometer, the needle will not be stationary; the resistor is a source of voltage fluctuations. Let us imagine, for the sake of simplicity, a second resistor, the same as the first, in place of the galvanometer. Then we have two resistors connected by wires, in thermal equilibrium with the surroundings. Electromagnetic waves, thermally excited in such a system, will travel either towards one resistor, or the other. In both cases, they will be absorbed without reflection, because the resistances are equal. The number of such waves is given by the Bose-Einstein distribution, so their energy at frequency ω is equal to

$$\frac{h\nu}{e^{h\nu/kT}-1} \approx kT,$$

because room temperature corresponds to frequencies $\sim 10^{12}$ Hz, so the relevant frequencies are in the classical regime. The total energy for the two modes is then $2kT$, but each of them is absorbed by a single resistor, which thus receives an amount of energy kT per each mode of oscillation of the circuit, i.e., for each frequency. In a range of frequencies $\Delta\nu$, it will absorb the energy $kT\Delta\nu$ per unit of time. In equilibrium, it must radiate the same amount of energy. If its nominal resistance is R, this radiation will appear as a fluctuating current, whose average power is equal to the absorbed energy per unit of time:

$$\left\langle I^2 \right\rangle R = kT\Delta\nu, \tag{11.53}$$

or alternatively

$$\left\langle V^2 \right\rangle = 4RkT\Delta\nu, \tag{11.54}$$

because the current appears in a circuit of total resistance $2R$, so the corresponding voltage drop is $I/(2R)$. The expression (11.54), or (11.53), is the Nyquist theorem.

An interesting contemporary application of Nyquist's theorem is to measure the temperature of a very small sample at very low temperatures. Such a measurement is difficult because one never knows how the thermometer itself perturbs the sample. However, if one can measure both current and voltage fluctuations, which is possible in a standard (four-wire) arrangement, one can eliminate the resistance from the two equations to find the temperature without a thermometer.

This application points to a conceptual aspect of Nyquist's theorem, namely that the resistance R is the actual physical resistance at that temperature, just like the temperature is the actual temperature. Because the fluctuation is at equilibrium, one can turn the formula around, to calculate the transport coefficient as an equilibrium fluctuation by Gibbs' approach. In this way, Kubo obtained a microscopic formula for the conductance.[3]

[3] Apparently, it took 50 years to realize that $\alpha^2 = \mu kT$ implied $\mu = \alpha^2/kT$.

11.3.2 Comparison with Einstein's Relation

Einstein's connection (11.42) between the diffusion constant and mobility is a prototype of all fluctuation-dissipation theorems. Namely, for a Gaussian walker, the diffusion constant is a measure of the expected value of his (fluctuating) amplitude; mobility, on the other hand, is directly related to the limiting speed at a constant force, i.e., average friction. Thus, we have

$$\frac{\overline{x^2}}{t} = (2kT) \cdot \mu. \tag{11.55}$$

In electric circuits, the charge Q is analogous to the displacement x. The Nyquist theorem for current fluctuations is

$$\frac{1}{\Delta\nu}\left\langle \left(\frac{dQ}{dt}\right)^2 \right\rangle = (kT) \cdot \frac{1}{R}. \tag{11.56}$$

The dimension of the left-hand side is square of the charge, divided by time; on the right, we have the temperature factor, multiplied by the conductance (reciprocal resistance). Conductance is analogous to mobility, so we have a direct analogy between Einstein's relation and Nyquist's theorem. As we shall see from the generalization of the Nyquist theorem in the next section, the only difference between them is that the Einstein relation refers to the situation where the molecular noise is probed by a Brownian particle, while in the Nyquist case it is probed by an oscillator tuned to some frequency. This difference in the measuring device, which in the second case has a harmonic restitutive force, means that the Nyquist theorem can be stated in the frequency domain, i.e., as a specific statement about the spectrum of thermal noise.

11.3.3 Output Noise and the Generalized Nyquist Theorem

The most general linear relation between a time-dependent excitation (input variable) I of a system, and its response, measured by a quantity V (output variable), is the convolution

$$V(t) = \int_{-\infty}^{\infty} Z(t-t')I(t')dt', \tag{11.57}$$

where $Z(t)$ is called the *response function*.[4] By the convolution theorem for Fourier transforms, the relation between the Fourier components is much simpler,

$$v(\omega) = z(\omega)i(\omega). \tag{11.58}$$

The prototype of such an expression is Ohm's law, or the relation $\boldsymbol{D}(\omega) = \varepsilon(\omega)\boldsymbol{E}(\omega)$ between the electric field and the dielectric displacement. The relation between the input and output power spectra follows immediately:

$$G_v(\omega) = |z(\omega)|^2 \, G_i(\omega). \tag{11.59}$$

If these are noise spectra, we have a relation between the noise at the input and output of our device.

To be specific, imagine a mirror suspended on a torsion thread, which is subjected to a random force from the impacts of air molecules ($\mathcal{I}$ is the moment of inertia):

$$\mathcal{I}\ddot{\theta} + \alpha\dot{\theta} + c\theta = A(t). \tag{11.60}$$

For any excitation $A(t)$, the Fourier transform gives $\theta_\omega = z(\omega)a(\omega)$, where the response function is

$$z(\omega) = \frac{1}{-\mathcal{I}\omega^2 + i\alpha\omega + c}. \tag{11.61}$$

From the Wiener-Kintchine theorem we have

$$\left\langle \theta^2 \right\rangle = C(0) = \frac{1}{2\pi}\int_0^\infty G_\theta(\omega)d\omega = \frac{G_A}{2\pi}\int_0^\infty |z(\omega)|^2 d\omega = \frac{G_A}{4\alpha c}, \tag{11.62}$$

where we assumed that the power spectrum of excitation fluctuations $A(t)$ is white noise, i.e., independent of frequency. That assumption is redundant, as we show next.

For an arbitrary power spectrum $G_A(\omega)$, we can obtain a simple expression for position fluctuations, if the damping is small ($\alpha \ll \sqrt{c\mathcal{I}}$), so the oscillator is tuned to the resonant frequency $\omega_0 = \sqrt{c/\mathcal{I}}$. Then

$$\left\langle \theta^2 \right\rangle = \frac{G_A(\omega_0)}{4\alpha c}. \tag{11.63}$$

Assuming that the ability to tune the system to an arbitrary frequency is independent of the friction force responsible for the constant α, and that the system thermalizes

[4] For reasons of causality we expect $Z(t) = 0$ for $t < 0$, which leads to Kramers-Kronig relations for $z(\omega)$, but we will not dwell on that.

because of the same random impacts that gave rise to the position fluctuations, we can use the equipartition theorem in thermodynamic equilibrium to conclude

$$\left\langle c\theta^2 \right\rangle = kT, \tag{11.64}$$

and comparing it with the previous expression we get

$$G_A(\omega_0) = (4kT) \cdot \alpha, \tag{11.65}$$

which is the fluctuation-dissipation theorem again, in a more general context. The spectrum of fluctuations is now derived to be white noise, assuming that ω_0 can be set arbitrarily, and that equipartition is valid, assumptions that fail at high frequencies. It is equal to kT multiplied by some measure of macroscopic friction, i.e., dissipation. Had we considered an electric LRC circuit as the oscillator instead, we would have obtained the original expression of the Nyquist theorem. If we were to drop the restitutive force from the equation of motion, we would return to the analysis of Brownian particles.

The important physical assumption in the preceding derivation is that the fluctuation $\langle \theta^2 \rangle$, due to random impacts in time, is the same as the one which occurs due to fluctuations within the ensemble, as given by the equipartition theorem. (If we prod the mirror randomly with a finger, that is no longer the case!) Again, we applied the equipartition theorem to the macroscopic mirror. As already mentioned, the first application of this understanding of thermodynamics was by Einstein himself, to the distribution of colloidal particles in a gravitational field, which he predicted to obey the barometric formula, even though the particles were large.

11.3.4 Galvanometer Fluctuations

If several different mechanisms produce thermal noise, and they all act on the same device, that device will have neither greater nor lesser thermal fluctuations than if all the noise comes from a single source. We show this using the example of a galvanometer needle, which is simultaneously affected by noise from a resistor, via a coil, and by random impacts of air molecules. Formally, it will be seen that each new source of noise introduces additional friction, so that the total noise with the total friction gives the same fluctuations of the needle as if it were subject to only one noise, with its own friction. The physical reason for this behavior is that thermal contact cannot depend on the microscopic mechanism by which it is realized. If it did, the difference would contain information, i.e., the most probable state a system found in contact with one thermal reservoir would not be the same as found in contact with another. In brief, all thermostats must be equivalent.

A galvanometer under the influence of a current i from a resistor R is described by the coupled system of equations:

$$\begin{aligned} I\ddot{\theta} + \alpha\dot{\theta} + c\theta &= A(t) + gi, \\ Ri &= V(t) - g\dot{\theta}, \end{aligned} \tag{11.66}$$

where $A(t)$ is the fluctuating force from the air molecules, and $V(t)$ is the fluctuating voltage; g is the effective current-deflection constant, which originates from the coil. When the current is eliminated, the equation for the deflection reads

$$I\ddot{\theta} + \left(\alpha + \frac{g^2}{R}\right)\dot{\theta} + c\theta = A(t) + \frac{g}{R}V(t). \tag{11.67}$$

To obtain the power spectrum of this composite random force, note that its two components are uncorrelated:

$$\overline{[A + (g/R)V]^2} = \overline{A^2} + (g^2/R^2)\overline{V^2}. \tag{11.68}$$

Thus, inserting in the generalized Nyquist theorem (11.65),

$$G_A + \frac{g^2}{R^2}G_V = 4kT\left(\alpha + \frac{g^2}{R}\right), \tag{11.69}$$

for all frequencies, so we can find out the fluctuation of the needle from (11.62):

$$\left\langle \theta^2 \right\rangle = \frac{4kT(\alpha + g^2/R)}{4(\alpha + g^2/R)c} = \frac{kT}{c}, \tag{11.70}$$

thereby recovering the equipartition theorem. It is clear that the increased noise is compensated precisely by the corresponding increased friction. Thus, our simple model conforms to the reality that all thermostats are equal.

11.3.5 Limits of Measurement Accuracy

If we measure a force F by the deflection θ of a needle, which is resisted by a harmonic force (e.g., mounted on a spring), the change in the force will be proportional to the change in deflection:

$$\Delta F = c\Delta\theta. \tag{11.71}$$

Due to thermal oscillations of the needle, the force appears to fluctuate:

$$\overline{\Delta F^2} = c^2 \overline{\Delta \theta^2} = ckT, \tag{11.72}$$

according to the equipartition theorem. This error can theoretically be reduced by reducing the spring constant, but then it will take a long time for the needle to come to rest. Instead of the force F, we will be measuring the average value of the needle position during some given time S, as long as we can wait:

$$F_S(t) = \frac{1}{S} \int_{t-S}^{t} F(t')dt'. \tag{11.73}$$

This relation between the input force $F(t)$ and the output (measured) force $F_S(t)$ is linear, so the response to a harmonic input force is given by the response function $z(\omega)$:

$$F(t) = Ae^{i\omega t} \implies F_S(t) = Ae^{i\omega t} \frac{1 - e^{-i\omega S}}{i\omega S} \implies z(\omega) = \frac{1 - e^{-i\omega S}}{i\omega S}. \tag{11.74}$$

The mean value of the square of the measured force is, therefore,

$$\left\langle F_S^2 \right\rangle = \frac{G_F}{2\pi} \int_0^\infty |z(\omega)|^2 d\omega = \frac{G_F}{2S} = \frac{2\alpha kT}{S}, \tag{11.75}$$

where we used the Nyquist theorem for the input force. This is the relevant measure of fluctuations, if we have a given time at our disposal, and we see that it is determined by the damping in the measuring device.

11.3.5.1 Error Reduction by Cooling

One way to get around the thermodynamic limit of measurement accuracy is to cool the dominant source of noise, when cooling the whole device would not be practical. Let the needle and the resistor of the galvanometer be at different temperatures. Their total power spectrum is then

$$4kT_1\alpha + 4kT_2\frac{g^2}{R} =: 4kT_{\mathit{eff}} \left(\alpha + \frac{g^2}{R}\right), \tag{11.76}$$

where we have defined some effective temperature of the circuit. In practice, the entire circuit behaves as if it were at that temperature. It is obviously between T_1 and T_2, with each element entering the average weighted with its own noise. This trick is used in heat-seeking missiles, when the infrared sensor is cooled with liquid nitrogen to obtain a better contrast.

11.3.5.2 Error Reduction by Feedback

We can try to reduce the fluctuations of the system by introducing a feedback loop, which introduces an additional damping α', but with no associated noise. The noise spectrum is still the "old" $G = 4\alpha kT$, which defines an effective temperature with the additional damping:

$$4\alpha kT =: 4(\alpha + \alpha')kT_{\mathit{eff}}. \tag{11.77}$$

It is, of course, lower than the one without feedback. This does not affect the limit of measurement accuracy, because that depends on the product of damping and temperature, i.e., the power spectrum, which remains the same. However, since a weakly damped instrument allows more accurate measurements, but requires more time, it can be convenient to increase the damping with a feedback to reduce the measurement time without losing accuracy.

11.3.6 Evolution of Clocks

The development of clocks delightfully encapsulates the relation between precision and damping. In antiquity, time was measured exclusively by diffusive processes, such as the time needed for water to drip out of a bucket, or for a glowing stick to burn out. In the middle ages, a variant of this idea was developed, in which a weight unwound a rope from an axis, on which a pointer was mounted. Evidently, such a clock depends critically on friction, which is a random process, so the clock's accuracy cannot be increased without it falling apart. It is fatally in contradiction with itself, so to say. In order to slow down the falling of the weight, the axis was connected to a setup of cylinders on the ends of a horizontal balance (*foliot*), which swung one way and the other, changing direction at each turn of the axis. Importantly, these oscillations were forced by the weight, so their period depended on the mechanical construction.

The next step was the true inflection point in clock evolution. Instead of the foliot forced by the weight, Huygens patented a pendulum that was allowed to oscillate freely with its natural frequency.[5] A feedback was introduced, noiseless by intention, between a spring powering the pendulum and the pendulum controlling the unwinding of the spring. Thus, for the first time, the signal and power trains in a clock were separated, opening the way to increase precision by reducing friction to this day, all thanks to noiseless feedbacks.

[5] He credited this idea to Galileo.

11.4 Return to Equilibrium as an Irreversible Process

We have so far considered the relation for the probability of a fluctuation statically, for example, we obtained single-particle distributions as probability fluctuations in which a certain single-particle state was certainly occupied. One can use it similarly to show that the spontaneous appearance of a non-equilibrium state is unlikely, such as all the air gathering in a corner of the room.

It is interesting, however, that it can also be understood dynamically. Namely, if a system is perturbed out of equilibrium, bringing it back is like sliding along the paraboloid of entropy, whose extremum is the equilibrium state. Forces that push the system into equilibrium are derivatives of entropy by macroscopic mechanical variables. The macroscopic dynamics of the return to equilibrium is described as the time change in entropy under the influence of these forces. The change is irreversible because entropy increases (returns to a maximum). A simple example of this scenario is heating the air by speaking.

11.4.1 Dissipation

Imagine a ball at the bottom of a paraboloid $z = a(x^2 + y^2)$, and choose $y = 0$ (axial symmetry). Move the ball to the position $x = x_0$ and hold it. In doing so, we have done work ax_0^2 and now we have to balance the force $2ax_0$ if we want the ball to remain at rest. What will happen when we release the ball? Obviously, the coordinate z, which is the potential energy, will change with the speed

$$\frac{dz}{dt} = 2ax\frac{dx}{dt} = F \cdot v. \tag{11.78}$$

If z is energy, then dz/dt is power. The statement that work is equal to force multiplied by distance becomes, in the dynamic case, that power is force multiplied by velocity.

If there are many particles, a current appears in place of velocity. This leads us to look for the time-dependent changes in entropy in the form

$$\frac{d\sigma}{dt} = X \cdot J, \tag{11.79}$$

where X is a generalized (thermodynamic) force, and J is a macroscopic current, driven by that force, which returns the system to equilibrium. Recalling that the force is the change in entropy along a generalized coordinate, $X = d\sigma/dx$, we have an interpretation of the current by dimensional analysis:

$$\frac{d\sigma}{dt} = \frac{d\sigma}{dx} \cdot \frac{dx}{dt}, \tag{11.80}$$

so the current is like the velocity of a generalized coordinate, an independent variable of the entropy. The time rate of change of entropy is called *dissipation*. It is the stochastic counterpart of deterministic power.

11.4.2 Susceptibilities and Correlations Out of Equilibrium

The change in entropy due to fluctuations in macroscopic quantities, its independent variables in an isolated system, can be described by the paraboloid

$$\Delta\sigma = -\sum_{i,k} g_{ik} x_i x_k, \tag{11.81}$$

where x_i are changes in the generalized coordinates, independent variables of the entropy. The origin is chosen so that all $x_i = 0$ in equilibrium. The generalized forces induced by these changes are

$$X_i = \frac{\partial\sigma}{\partial x_i} = -\sum_k g_{ik} x_k, \tag{11.82}$$

and these should also be understood as differences of the corresponding intensive variables from their equilibrium values. Since g_{ik} is a positive definite form, otherwise the equilibrium state would not be a maximum, it can be inverted, so we can also obtain the displacements from the forces:

$$x_i = -\sum_k g_{ik}^{-1} X_k. \tag{11.83}$$

We can interpret the matrix g_{ik} thermodynamically, as a matrix of susceptibilities, or second derivatives of thermodynamic potentials, e.g. $c_V^{-1} \sim \partial^2 S/\partial U^2$.

Statistically interpreted, as second derivatives of generating functions, the susceptibilities appeared as variances, or correlations of the corresponding thermodynamic quantities in equilibrium, e.g., $c_V \sim \langle E^2\rangle - \langle E\rangle^2$. Is the same true out of equilibrium? We can at least show that it is not an additional assumption, but follows from the assumption that the statistical interpretation of the entropy itself can be used out of equilibrium. In that case, the probability of fluctuation can be written again as

$$P(x_1, \ldots)d\boldsymbol{x} = \frac{e^{\Delta\sigma} d\boldsymbol{x}}{\int e^{\Delta\sigma} d\boldsymbol{x}}, \tag{11.84}$$

where $dx = dx_1 \cdots$ is now a volume element in the *macroscopic* state variables, from which we can calculate the correlation of intensive and extensive quantities in the ensemble:

$$\langle x_i X_j \rangle = \int x_i X_j P(x_1, \ldots) d\boldsymbol{x} = \int x_i \frac{\partial P}{\partial x_j} d\boldsymbol{x} = -\int P \frac{\partial x_i}{\partial x_j} d\boldsymbol{x} = -\delta_{ij}. \tag{11.85}$$

The integrated part vanishes because the probability of large fluctuations is small. From here we immediately obtain the correlations of forces

$$\langle X_i X_j \rangle = -\sum_k g_{ik} \langle x_k X_j \rangle = g_{ij}, \tag{11.86}$$

and of coordinates,

$$\langle x_i x_j \rangle = -\sum_k g_{ik}^{-1} \langle X_k x_j \rangle = g_{ij}^{-1}. \tag{11.87}$$

Indeed, they are given by the susceptibility matrix. We emphasize that these are correlations within the ensemble, where the time variable does not appear explicitly, although we imagine that the system evolves towards equilibrium. There is a hidden assumption that each such ensemble lasts on scales not only much longer than the thermalization scale, but also much shorter than the characteristic rates of change of the macroscopic variables. Thus, there appears a range of *intermediate* time scales for which the above expressions make physical sense even out of equilibrium. In particular, the velocities of macroscopic variables should be understood as averages over such intermediate ensembles, rather than mathematical derivatives:

$$\frac{dx}{dt} \leftrightarrow \left\langle \frac{x(t+\tau) - x(t)}{\tau} \right\rangle, \tag{11.88}$$

where τ is some time in the appropriate range.

In the formula for the fluctuation probability, we wrote the entropy as a function of macroscopic variables, the coordinates x_i. This was in accord with the understanding of entropy as a macroscopic thermodynamic function. Recall, however, that we calculated the entropy in equilibrium directly as the volume of the microscopic phase space. The latter approach makes no sense out of equilibrium; the microscopic phase space is always the same, and the entropy calculated in this way cannot change. Physically, the system is not in equilibrium precisely because we have not waited long enough to see all microscopic states occupied with the same probability. Thus, the formula for the entropy of the non-equilibrium states also depends on the hidden assumption about intermediate time scales, which are still shorter than the scale on which the macroscopic variables, including entropy itself, evolve. Such an entropy, defined for neither-too-short-nor-too-long times, is called the coarse-grained or relevant entropy, and it is the only one we have available for describing non-equilibrium processes. The use of the same term as for the equilibrium relevant entropy (3.28) is a natural extension of the coarse-graining to time intervals.

11.5 Time Evolution of Fluctuations and Onsager Relations

To close the description of the return of a system to equilibrium, we need one additional assumption, which cannot be obtained from the mechanical analogy. Namely, unlike mechanical systems, thermodynamical ones relax by diffusive processes, which have the characteristic time behavior of a decreasing exponential. This behavior means that the rate of change of quantities that measure the deviation from equilibrium is linearly proportional to themselves. When several variables deviate from equilibrium simultaneously, mutual influences also appear, so that the most general linear assumption is

$$\frac{dx_i}{dt} = \sum_j C_{ij} x_j . \tag{11.89}$$

The left-hand side is a current, and the right can be expressed by forces:

$$J_i = \sum_j L_{ij} X_j . \tag{11.90}$$

This general connection of forces and currents now enables a physical representation of the change entropy over time, a step further than we guessed by dimensional analysis:

$$\frac{d\sigma}{dt} = \sum_i \frac{\partial \sigma}{\partial x_i} \frac{dx_i}{dt} = \sum_i X_i J_i = \sum_{i,j} L_{ij} X_i X_j . \tag{11.91}$$

We see that the rate of change of entropy in time, or dissipation, is a quadratic form in thermodynamic forces. As already said, it is analogous to power in mechanical systems. Evidently, the dissipation will be large if the thermodynamic forces are large. But since the forces are the differences of the corresponding intensive variables (pressure, chemical potential, inverse temperature), it follows that for reversible processes (where dissipation vanishes) the changes in the intensive variables between different parts of the system must also be small. In brief, large gradients mean large dissipation, hence the efficiency of the process is diminished.[6]

[6] This notion of "efficiently" is the opposite of "quickly."

Because the derivatives in the previous expressions are not formal operations, but rather macroscopic averages of microscopic processes, one can obtain an interesting restriction on the coefficients L_{ij}. Because of the reversibility of microscopic processes with respect to time, every correlation function must satisfy

$$\left\langle x_i(t)x_j(t+\tau)\right\rangle = \left\langle x_i(t)x_j(t-\tau)\right\rangle = \left\langle x_i(t+\tau)x_j(t)\right\rangle, \tag{11.92}$$

where the second equality follows from the arbitrary choice of the origin of time with respect to the intermediate scale τ. Subtracting $\left\langle x_i(t)x_j(t)\right\rangle$ on both sides and using the definition of a current as a physical (average difference) derivative, we get

$$\left\langle x_i(t)J_j(t)\right\rangle = \left\langle J_i(t)x_j(t)\right\rangle, \tag{11.93}$$

which is a phenomenologically very strong limitation: The correlation of the jth current with the i-th coordinate is equal to the correlation of the i-th current with the j-th coordinate. The result is even more striking if the currents are expressed in terms of forces:

$$\sum_k L_{jk}\left\langle x_i(t)X_k(t)\right\rangle = \sum_k L_{ik}\left\langle X_k(t)x_j(t)\right\rangle, \tag{11.94}$$

and since the correlators of extensive and intensive quantities are Kronecker symbols, it boils down to

$$L_{ji} = L_{ij}. \tag{11.95}$$

These are called *Onsager reciprocal relations*. Because of the microscopic reversibility of time, dissipation cannot be just any quadratic form in the macroscopic forces, but must be symmetric!

Notably, we are dealing with two different interpretations of time here. In the expression (11.89), the time variable appears on a coarse scale, on which the diffusive process takes place, which is not reversible in time. On the other hand, the formula (11.92) refers to the microscopic Hamiltonian time that appears in the classical equations of motion, which are of course reversible in time. The expression (11.95) is a perennial monument to one who was not confused by this difference.

11.5.1 Thermoelectric Effect

If we heat one end of a metal rod, the heat current of electrons will cause a voltage difference between the two ends. Similarly, if we impose a voltage gradient, a temperature gradient will appear in the wake of the electric current. Phenomenologically, the situation can be described by a system of coupled equations:

$$
\begin{aligned}
I &= l_{11}\Delta\varphi + l_{12}\Delta T \\
W &= l_{21}\Delta\varphi + l_{22}\Delta T,
\end{aligned}
\tag{11.96}
$$

where I is the electric current, W is the heat flow, while $\Delta\varphi$ and ΔT are the corresponding voltage and temperature differences. The coefficients l_{ij} can be measured and are not symmetric, because these gradients are not the natural generalized forces related to the non-equilibrium entropy. When we write the same relation in these forces, the symmetry of the Onsager coefficients will result in a certain relation between the coefficients l_{ij}, which can in principle be verified.

While in equilibrium it is a matter of habit whether we choose the energy or the entropy as the initial thermodynamic potential, here we have no such choice. We search for generalized forces via the entropy, the only thermodynamic potential sensibly defined out of equilibrium. Consider the two ends of the rod as two parts of an isolated system, between which there is a transfer of energy $\Delta U = -\Delta U_1 = \Delta U_2$, and of particles $\Delta N = -\Delta N_1 = \Delta N_2$. During the transfer, the entropy of each part will change according to

$$
\begin{aligned}
\Delta\sigma_1 &= -\frac{\Delta U}{T} + \frac{\mu(T)}{T}\Delta N, \\
\Delta\sigma_2 &= \frac{\Delta U}{T+\Delta T} - \frac{\mu(T+\Delta T) + q\Delta\varphi}{T+\Delta T}\Delta N.
\end{aligned}
\tag{11.97}
$$

Because the system is not in equilibrium, the total entropy increases:

$$
\Delta\sigma = \Delta\sigma_1 + \Delta\sigma_2 \approx \Delta U\left[-\frac{\Delta T}{T^2}\right] - q\Delta N\left[\frac{\Delta\varphi}{T} + \frac{\Delta T}{q}\frac{\partial}{\partial T}\left(\frac{\mu}{T}\right)\right], \tag{11.98}
$$

where we cut off higher than linear terms, in accordance with thermodynamic phenomenology. The dissipation is then simply

$$
\frac{d}{dt}\Delta\sigma = W\left[-\frac{\Delta T}{T^2}\right] - I\left[\frac{\Delta\varphi}{T} + \frac{\Delta T}{q}\frac{\partial}{\partial T}\left(\frac{\mu}{T}\right)\right] \equiv I X_1 + W X_2, \tag{11.99}
$$

from where one can read off the generalized forces:

$$
\begin{aligned}
X_1 &= -\left[\frac{\Delta\varphi}{T} + \frac{\Delta T}{q}\frac{\partial}{\partial T}\left(\frac{\mu}{T}\right)\right], \\
X_2 &= -\frac{\Delta T}{T^2},
\end{aligned}
\tag{11.100}
$$

which go along with the electric current $I = J_1$ and thermal current $W = J_2$. The coefficients in Onsager's relation are by definition those that appear in the expression

$$\begin{aligned} I &= L_{11}X_1 + L_{12}X_2, \\ W &= L_{21}X_1 + L_{22}X_2. \end{aligned} \tag{11.101}$$

Comparing with the laboratory-bench formulation (11.96), we find

$$\begin{aligned} l_{11} &= -\frac{L_{11}}{T}, \\ l_{12} &= -\frac{L_{12}}{T^2} - \frac{L_{11}}{q}\frac{\partial}{\partial T}\left(\frac{\mu}{T}\right), \\ l_{21} &= -\frac{L_{21}}{T}, \\ l_{22} &= -\frac{L_{22}}{T^2} - \frac{L_{21}}{q}\frac{\partial}{\partial T}\left(\frac{\mu}{T}\right). \end{aligned} \tag{11.102}$$

The relation $L_{21} = L_{12}$ implies that the measured coefficients have to satisfy

$$l_{12} = \frac{l_{21}}{T} + l_{11}\frac{T}{q}\frac{\partial}{\partial T}\left(\frac{\mu}{T}\right). \tag{11.103}$$

A fundamental result of quantum field theory is the so-called CPT theorem, which states that a product of charge conjugation, mirror reflection, and time reversal is an exact symmetry of nature. Because the weak nuclear force violates the product CP, it can be inferred that it also violates time reversal, although such a phenomenon has not been observed directly. The Onsager relations prove that this weak-force violation is not responsible for macroscopic irreversibility, i.e., an arrow of time, because they would be violated experimentally if that were the case.[7] Macroscopic irreversibility is a statistical phenomenon, not related to any fundamental symmetry-breaking.

11.6 Relevant Time Scales in Society

Processes in nature can proceed independently if the relevant time scales are different. The Moon goes around the Earth, the Earth goes around the Sun, and the Sun goes around the center of the Milky Way. An animal feeds on the scale of

[7] It is not excluded that they could be violated in some cleverly designed weak-force version of the thermoelectric effect, but only there.

a day, breeds on the scale of a year, lives on the scale of a decade, and evolves on the scale of a million years. The processes with shorter time scales are bound to the longer ones, which they perceive as static and unchangeable.

The two examples above are qualitatively different, because they refer to different notions of time. The first concerns the reversible time of Hamiltonian dynamics, the second, irreversible time of diffusion. Diffusive processes naturally relax exponentially, but life depends on their becoming periodic. It is sustained by engines, dissipative cycles with hierarchically organized relevant time scales. The engines work to prevent decay into equilibrium. Human society has naturally evolved into a hierarchical structure, so it is a pertinent question, which are the relevant time scales.

11.6.1 The Playing Cards

In Italian playing cards, the top suit are swords (*spade*, spades), symbolizing the hereditary nobility, or second estate; second are chalices (*coppe*, hearts), the clergy, first estate; third are money (*denari*, diamonds), the merchants, third estate; fourth are clubs (*bastoni*), the peasants (Fig. 11.2). We assign them time scales by inspection: Nobility, $\sim$ 50–100 years, the scale of inheritance; clergy, $\sim$ 1000 years, the scale of religion; merchants, $\sim$ 5–15 years, return on investment; peasants, $\sim$ 1 year, the seasons.

The first estate is not socially on top, evidently because its scale is too long for effective governance. Similarly, the peasants' scale is too short. The physically relevant time scales are neither too long, nor too short. The political and commercial classes fall in that range. Crucially, they are well separated, so the nobility averages out the merchants, like the Sun averages out the Earth. This separation of time scales is the source of stability of feudal society, which has survived as a system of governance over more than 1000 years of considerable material change.

In representative democracy, the electoral cycle is shorter than the business cycle, so the merchants average out the politicians. On the other hand, the latter are still supposed to maintain the long-term material and immaterial infrastructure

Fig. 11.2 Time scales in society

that commerce simply takes for granted. We conclude that the political class in a democracy is unstable, because of this self-contradictory requirement on its characteristic time scale. This instability was noticed by Jefferson, who proposed to mitigate it by forbidding re-election. His idea was for the person of the president to be ephemeral, so the characteristic time scale would be that of the office, which outlasts all commercial and foreign interests that could try to capture the person.

11.6.2 The Jubilee

In Leviticus 25:8–10, we read:

> 8 You shall count off seven Sabbaths of years, seven times seven years; and there shall be to you the days of seven Sabbaths of years, even forty-nine years. 9 Then you shall sound the loud trumpet on the tenth day of the seventh month. On the Day of Atonement you shall sound the trumpet throughout all your land. 10 You shall make the fiftieth year holy, and proclaim liberty throughout the land to all its inhabitants. It shall be a jubilee to you; and each of you shall return to his own property, and each of you shall return to his family.

The time scale of the jubilee is the same as that of inheritance. A complete restart of the economy is mandated at least once in an expected lifetime (declared elsewhere to be 70 years). In metallurgy, such a restart is called annealing. When defects accumulate in a metal under cyclic loading, they begin to interlock, making it hard and brittle. This is material fatigue, a considerable latent danger in critical applications, such as airplane wings and bridge suspensions. Annealing recycles the metal to its original ductile state, relaxing the defects and making it workable again.

The economy accumulates injustices across business cycles as a metal accumulates defects, for the same reason: Random events are real. When everyone who has suffered without fault knows someone like that, society is fatigued. The remedy of the jubilee has been partially applied many times in history, in the form of selective debt relief. The closest to a complete jubilee in recent times was the restart of the German economy after World War II, when the old currency was abolished and everyone was given 1000 units of the new one. One could argue that treating everyone equally in fact targeted ill-gotten riches, making such a radical move dependent on a unique situation. However, the jubilee also treats everyone equally, because everyone knows when it will come. Physically, it amounts to an oscillatory driving force, which should be compared with the random force of economic crises, not with some theoretical equilibrium.

11.6.3 IKEA

An editorial written soon after the founder of IKEA died ended with the conclusion: The ability to present business as a deal from which both sides gain made him one of the greatest capitalists of history. But how to know whether he was only presenting it so, or both sides were really gaining?

The difference between presenting and gaining is material. The IKEA business model is to sell disassembled furniture, and pass some of the savings, in the form of lower prices, to the customers who assemble it themselves. Are the customers working for free, or not? The difference between the price of assembled and disassembled furniture is a potential. The conjugate current are people going through the stores. The question, whether the *same* customers keep returning, points to the difference between an open-circuit (relaxation) and closed-circuit (working engine) current. If both sides are gaining, the engine is turning over, which can be observed macroscopically. In this way, an enterprise can become a clock measuring reversible Hamiltonian time, theoretically extending its own lifetime to infinity under the same circumstances. Notably, the oscillations should be free like a pendulum, not forced like a foliot. Inducing customers to spend according to the needs of the enterprise is attempting to construct a *perpetuum mobile*, in which they pay to assemble furniture.

It takes a purposeful attitude to maintain that potential at a non-zero value, in face of rational temptation to usefully skim the customer by presenting. Natural evolution, a rational robot trained on the past, cannot restrain itself from usefully depleting every potential difference it encounters. The resulting equilibrium is called the Nash equilibrium, or equilibrium of selfish agents. The average payoff to all agents in a Nash equilibrium is smaller than it would be if they could maintain some agreement. The reason they cannot do so is that the first to break the agreement has a much larger payoff than the one on which they agreed. It is easy to say that they break it because they cannot imagine the future. However, all imaginable future can be included in the same calculation. (It is imaginable because something similar has happened before.) True hopeful behavior is restraint in face of the unimaginable future, which cannot be valuated.

11.6.4 The State as an Engine

In his classic work *Die Cultur der Renaissance in Italien*, Burckhardt calls the modern state a work of art. That work of art has moving parts, so it is an engine. The examples above point to two physical principles which underlie the design of such an engine.

First, order with purpose is achieved by closed-circuit currents which turn over many times. For that, one needs batteries with two poles. These poles are classes of the population whose motivations are inverted, like anode and cathode potentials, for example, the political and commercial classes, or universities and industry. One pole provides purpose and maintenance to the common enterprise, the other validation and irreversibility. Maintenance means keeping some potential differences fixed, neither too large, to avoid dissipation, nor too small, to preserve motivation. Mixing two such classes is literally shorting the battery.

Second, stability is achieved in a hierarchy of time scales. The dominant class is one that can outwait all others. For a state to be stable, that must be the political class, by definition. How that is achieved in a democracy is a problem in all democracies since ancient Athens. It is interesting that the earliest attempts to solve it involved effectively acknowledging that some temperature in society was finite. To wit, officials were chosen by lot from a pool of eligible citizens. This solution is physically similar to Jefferson's ephemeral president. The common thread is that one delimits the denominator of the fluctuation formula, presumably by some virtue, but the numerator, i.e., what actually happens in the course of time, is left to chance. In other words, virtue is used as a threshold, but not for ranking. The time scale is of the pool, not of individuals. If it does not matter who is chosen, the temperature is low, and the state is stable.

Chapter 12
Literature

I am most indebted to Kittel's textbook [131]. Kubo's problem-based repetitory of statistical physics [134] can supplement any prescribed textbook or problem set. As a next step for beginners, I can recommend Balian's graduate textbook [16] and the classic by Landau and Lifshitz [147]. I have also used the monographs by Gibbs [100] and Huang [120] to clarify particular points.

Below are specific references arranged by chapters, as a guide and invitation to further reading. Their ordering follows the main text. All have been selected either as gateways, or as elaboration of issues treated cursorily in the text. Any claims of superiority or priority relative to those not included are left to the readers. Original references to standard textbook material are generally not given.

Chapter 1

The quote is from [94].

The molecular hypothesis appeared in [15], Einstein's analysis of Brownian motion in [76].

The distinction among versions of ontology is taken from [60]. The fundamental exact result on the mediation of interactions is Eq. (12.14) in [91].

For cargo cults, see [122]. Quote: "As was pointed out earlier, the differences between the movements could be expressed in terms of naivety or sophistication, depending on how much the participants understood of the real relation between ends and means."

The need for both philosophy and revelation is stated in [9], Part 1, Question 1. Augustine's distinction between social constructs and objective reality can be found in [14], Book 6, Chapter 5, with a commentary in [203], pp. 271–276.

Naive viewers' perceptions of visual media are described in [97]. The relation between vision and cognition is discussed generally in [11].

The technical definitions of magic, mysticism, and heresy are the author's.

The Facebook susceptibility study was reported in [133].

D. Sunko, *Conceptual Thermodynamics*,
https://doi.org/10.1007/978-3-032-20877-4_12

Chapter 2

The quote is from [38], p. 68.

Boyle's results were published in [38], Mariotte's in [165], Avogadro's in [15].

Carathéodory gave the most influential axiomatic formulation of thermodynamics [47]. The microscopic interpretation of pulling a sled is from the textbook [73].

In 1890, Carnot's 1824 treatise [48, 49] was translated into English with a commentary by Kelvin [50]. The name entropy was introduced in [55], although Clausius' theorem was proven much earlier [54].

A summary of Massieu's contributions to thermodynamics is found in [17].

Useful engines are discussed in [4]. Maxwell's analysis of the centrifugal regulator appeared in [169]. The relation between cows, whales, and pigs is described in [65]. The early history of the Belousov-Zhabotinsky reaction is reviewed in [226]. One proposal for the spontaneous appearance of vesicles with lipid membranes is [52]. Earth's heat sources are described in [150], their consequences for plate tectonics in [58, 207]. The illustration of spontaneous symmetry breaking is from [174]. Hominins carrying tools to open grassland were described in [200]. Investigations of human prospective thinking are reviewed in [101].

Modern money creation is described in [118]. The evolution of crocodiles' circulation is revisited in [216].

Several modern authors have interpreted Leibniz's theodicy as mathematical optimization, for example [149]. A direct influence on Bernoulli is claimed in [106]. Voltaire's parody [220] has become part of the school canon in many languages.

Chapter 3

The quote is from [32].

An introduction to mathematical probability theory is found in [87]. Stirling's formula is derived in [46].

Newton's alchemical interests are described in [70]. The quote is taken from [172], where it is attributed to his *Opticks*. Boscovich's theory of matter is expounded in [34]. Boltzmann defended Boscovich by name from Lord Salisbury's simplifications as late as 1895 [32].

Gibbs explained the importance of Liouville's theorem [155] for statistical physics in [100]. Chaos in the gravitational three-body Sitnikov model, Fig. 3.2, is described in [117]. Chaos in phase space is extensively treated in [61]. A modern view of thermalization is found in [205].

Scattering of atoms on surfaces in the classical limit is investigated in [163].

Measurement of high-frequency noise in electronic devices is described in [204, 229]. The 1 THz quoted in the text is from [204].

The question, when a miracle becomes a discovery, is analyzed statistically in [23, 213].

Boltzmann's transport equation and his expression for the relevant entropy appeared in his proof of the H-theorem [29]. Boltzmann himself summarized the main idea in [32]. An extensive discussion of the H-theorem is found in [219]. Modern applications to the resistivity of metals are reviewed in [5].

A contemporary mathematical exposition of the hydrodynamic limit and transport equations is [104]. Poincaré invariants and their restriction of Hamiltonian flows are described in [61], Section 8.6. Boltzmann's equation has been systematically extended to higher-order correlations in the so-called Bogoliubov–Born–Green–Kirkwood–Yvon (BBGKY) hierarchy, by these authors working independently, of which [33, 129, 130] are easily accessible.

Boltzmann introduced the microscopic expression for the entropy in the classical work [30], up to an additive constant, and multiplied by the gas constant instead of Boltzmann's. A true microscopic expression, without the additive constant and multiplied by Boltzmann's constant, was first written by Planck for light, when he fitted the value of h to the measured blackbody spectrum [197], see also his comment in §120 of the book [198].

The expression (3.70) and many like it are derived in [158].

The story from the Netherlands was reported in [22]. The idea that the ideal job advertisment attracts exactly one candidate is from [190].

Chapter 4

The quote is from [113], p. 390.

Gibbs' explanation [99] is analyzed in [125], which also includes a discussion of non-extensivity.

The effect of the enfranchisement of women on government policy in the United States is studied in [175, 176].

Hawking's equation of state [114] was checked directly in an observation of merging black holes [1]. A laboratory analogue of thermal Hawking radiation was observed in [184].

The mathematics of the Laplace transform and saddle-point evaluation is explained in [46].

Molecular dynamics simulations of liquids are described in [6]. The liquid flow itself, a nonequilibrium phenomenon, is enabled by a subset of particles with non-Maxwellian distribution [56]. The Hellmann-Feynman theorem is proven in [92]. The quantum fluctuations of protons in ice are studied in [24].

For references to quasiparticles, see Chap. 6 below. The effective mass in liquid gadolinium is taken from [163].

Chapter 5

The quote is from [116], p. 81.

Neutron scattering in a magnetic field is described in [224].

Magellan's crew's date confusion is reported in [193], p. 185.

The Meissner effect was observed in [173]. London's explanation appeared in [156].

Electrons can cross over from indistinguishable to distinguishable within a single molecule [177, 178], instead of going to the Moon.

The time-dependent Schrödinger equation is derived from the time-independent one in [39]. The idea that there was no time in the early universe, because there were no clocks, is the author's.

Khalfin's work appeared in translation in [128]. It is still cited in the context of the so-called quantum Zeno effect [88].

The Bohm-Aharonov (spin) formulation of the EPR paradox appeared in [28]. The original paradox was in terms of continuous observables [82].

The version (5.71) of Bell's inequality is called the CHSH form [53]. The derivation here is due to Bell [21]. Bell's original theorem appeared in [20]. The simplified version of Bell's proof in the text is the author's.

Measurements of the violation of Bell's inequalities are reported in [13] and [12], the latter involving rapidly varying polarizers.

Quantum computation is described in [210].

The implication operation in partition logic is defined in [83]. An introduction to partition logic is found in [84]. The exposition here is adapted from [85]. Various aspects of logical entropy are presented in [162]. Von Neumann entropy was introduced in [183].

Decoherence of quantum systems is described in [232].

The rug-seller's story was recounted to the author by a rug-seller in Turkey. Models of the prisoner's dilemma are reviewed in [45]. Gleason's theorem appeared in [102]. The beneficial effect of one case of ostracism is described in [201], Chapters 14–15.

Two different mechanisms by which the brain could be a quantum computer are described in [95, 96, 111]. A more general discussion of free will is found in [107].

Chapter 6

The quote is from [139].

The condition below Eq. (6.20) is from Jordan's theorem [223].

Landau's theory of the Fermi liquid is found in [139], translated in [142]. An early review of applications to ^{3}He is [3]. Its implementation in many-body perturbation theory appeared in [159, 185].

The description of the BCS theory of superconductivity [19] is based on [59] and [161], pp. 627–675. Gor'kov's derivation [105] of the Landau-Ginzburg theory [146] from it is presented pedagogically in [62, 98]. The observed transition of ^{3}He from a Fermi liquid to a superfluid is described in [69]. The theory is given in [151, 152].

Chapter 7

The quote is from Einstein's letter to Schrödinger, 28 February 1925, as given in [192], p. 600.

Einstein's evolving views on boson statistics are studied in [192]. The Bose-Einstein condensate was observed for the first time 70 years after Einstein's prediction [8].

The early development of the two-fluid theory is described in [71]. A detailed history of the subject is laid out in [18]. London's proposal for a BEC fraction appeared in [157]. Landau's theory of superfluidity appeared in [138, 144]. The role of BEC was described via density matrices in [191]. A contemporary view of the role of BEC in the two-fluid behavior is [171].

Measurements of the BEC fraction in superfluid ^{4}He by neutron scattering are presented in [103, 214], and by surface evaporation in [40, 227]. A method to infer it from the sound velocity is described in [170]. A numerical calculation of the BEC fraction in the limit of strong interactions is presented in [206].

Chapter 8

The quote is from [195], p. 73.

Boltzmann's derivation of Stefan's law appeared in [31].

Buckingham has given elementary rederivations of Stefan's law [41] and Wien's displacement law [42]. His dimensional analysis of thermodynamics [43] is the foundation of all laws of corresponding states, or scaling. The first measurement of the Stefan-Boltzmann constant conforming with the theoretical value is [27].

Some of the early works of Planck [195–197] and Einstein's photon paper [77] are translated in the collection by Ter Haar [217]. The hypothesis of natural radiation was formulated in [195].

A pedagogical account of Einstein's analysis of Planck's distribution [80] is found in [67]. Inhibition of spontaneous emission by reducing the cavity size is described in [121]. The microscopic meaning of Einstein's B-coefficient is given in [68].

Bose's photon paper [35] is available in English translation [36]. In a brief letter to Bose, Einstein singled out his derivation of the photon density of states ("den Faktor") as a nice step forward [81]. The first observations of gravitating photons [202] were much earlier than of neutrons [57].

Planck's Nobel lecture appears on the web site of the Nobel foundation [199]. Bethe's seminal calculation of Lamb's shift [135] was published in [25]. His approach was extended systematically under the name "renormalization." Some criticisms of renormalization as "subtracting infinities" lack appreciation for his physical reasoning.

Chapter 9

The quote was made in closing remarks at a conference in 1985, which the author attended.

Einstein's model of crystal heat capacity appeared in [79], Debye's in [66]. The meaning of fits to the Debye model was analyzed in a series of publications beginning with [26]. Debye temperatures and elastic constants are tabulated for a number of cubic crystals in [211].

Zero sound is described in [140, 141]. The transition from first to zero sound in ^{3}He was observed both in speed and attenuation in [2], from where the numbers quoted are taken.

The historical references on superfluidity are cited in the literature for Chap. 7. The presentation of second sound here mostly follows [120, 153], see also [148]. A qualitative discussion of the suppression of short wavelengths by indistinguishability is found in [93].

Chapter 10

The quote is from [10], Pars 3 q. 5 a. 3 ad 6, translated in [168].

Einstein's second paper on Brownian motion is [78]. Reaction barriers in solution are discussed in [164].

Contemporary studies of platinum catalysis of hydrogen reactions are described in [230, 231]. Haber found it surprising that his process was much more productive with osmium [110] than with platinum [109]. An electrochemical alternative to the Haber-Bosch process is investigated in [222], with the idea to avoid high pressures by the trick mentioned below Eq. (4.16).

An authoritative review of Weiss' work appeared in [182]. A critical analysis of the concept of negative temperature is found in [74].

The mechanism behind Hund's rule was discovered in [63], discussed further in [37, 208, 228]. Unfortunately, an incorrect explanation based on electron repulsion [212] is habitually repeated in much more recent physics textbooks than the last to appear [167] before [63].

The standard reference for dynamic magnetic interactions between localized moments is [7]. A kinematic mechanism is described in [86].

A compact overview of the multiple issues involved in understanding metallic magnetism is found in [215], by an author from Ford Motor Company. Complications in alloys are considered in [72], while [115] also includes coupling to the lattice.

Numerous applications of the Landau theory of phase transitions [136, 137, 143] are described in [218]. The application to superconductivity [145, 146] here mostly follows [98], see also [62]. Wilson's approach is reviewed in [225]. A pedagogical presentation of its main ideas appears in [166]. A detailed operative exposition is given in [160].

Figure 10.6 is reprinted from [181] with permission. The thermodynamics of small systems at fixed energy is studied in [108]. Aquinas' division of the sciences is described in [10, 168].

Chapter 11

The quote is from [186].

The presentation of the random walk is taken from [124]. The classic reference on the subject is [89], Chapters 3 and 14, and [90], Chapter 1, from which Footnote 2 is taken.

Measurements of Boltzmann's constant suitable for a student laboratory are described in [119, 179]. A review of the state of the art is given in [194].

Light-harvesting complexes are described in [123, 209].

A detailed experimental verification of the Nyquist theorem appears in [127], by an author from Bell Telephone Laboratories. An application to low-temperature thermometry is described in [154].

A compact overview of the relation between Kubo's formula and Boltzmann's transport equation is found in [51], building on earlier similar work [112].

A contemporary analysis of coarse-graining is found in [221].

Onsager relations are described in [187, 188], and reviewed in [189].

Interlocking defects are discussed in [132].

Jefferson's opinion on re-election is expressed in [126]. Quote: "No foreign power, nor domestic party, will waste their blood and money to elect a person who must go out at the end of a short period."

The editorial on IKEA was published in [75]. The Nash equilibrium originally appeared in [180]. The equilibrium of hawks and doves, a generic example in evolutionary biology, is described in [64].

The original edition of Burckhardt's book is in the public domain [44].

References

1. Abac, A.G., et al.: GW250114: Testing Hawking's area law and the Kerr nature of black holes. Phys. Rev. Lett. **135**, 111403 (2025). https://doi.org/10.1103/kw5g-d732
2. Abel, W.R., Anderson, A.C., Wheatley, J.C.: Propagation of zero sound in liquid He^3 at Low Temperatures. Phys. Rev. Lett. **17**, 74–78 (1966). https://doi.org/10.1103/PhysRevLett.17.74
3. Abrikosov, A.A., Khalatnikov, I.M.: The theory of a fermi liquid (the properties of liquid ^3He at low temperatures). Rep. Progress Phys. **22**(1), 329 (1959). https://doi.org/10.1088/0034-4885/22/1/310
4. Alicki, R., Gelbwaser-Klimovsky, D., Jenkins, A.: The problem of engines in statistical physics. Entropy **23**(8) (2021). https://doi.org/10.3390/e23081095
5. Allen, P.B.: Boltzmann theory and resistivity in metals. In: Chelikowsky, J.R., Louie, S.G. (eds.) Quantum Theory of Real Materials, chap. 17, pp. 219–250. Kluwer, Boston (1996)
6. Allen, M.P., Tildesley, D.J.: Computer Simulation of Liquids, 2nd edn. Oxford University Press, Oxford (2017)
7. Anderson, P.W.: Antiferromagnetism. theory of superexchange interaction. Phys. Rev. **79**, 350–356 (1950). https://doi.org/10.1103/PhysRev.79.350
8. Anderson, M.H., Ensher, J.R., Matthews, M.R., Wieman, C.E., Cornell, E.A.: Observation of Bose-Einstein condensation in a dilute atomic vapor. Science **269**(5221), 198–201 (1995). https://doi.org/10.1126/science.269.5221.198
9. Aquinas, T.: Summa Theologica. Christian Classics, Burton (1981)
10. de Aquino (Santo), T.: Super Boetium de trinitate; Expositio libri Boetii de ebdomadibus. In: Opera omnia. Commissio Leonina (1992)
11. Arnheim, R.: Visual Thinking. University of California Press, Berkeley and Los Angeles (2004)
12. Aspect, A., Dalibard, J., Roger, G.: Experimental test of Bell's inequalities using time-varying analyzers. Phys. Rev. Lett. **49**, 1804–1807 (1982). https://doi.org/10.1103/PhysRevLett.49.1804
13. Aspect, A., Grangier, P., Roger, G.: Experimental realization of Einstein-Podolsky-Rosen-Bohm gedankenexperiment: a new violation of Bell's inequalities. Phys. Rev. Lett. **49**, 91–94 (1982). https://doi.org/10.1103/PhysRevLett.49.91
14. Augustine, A.: The city of god. T. & T. Clark, Edinburgh (1871). https://www.gutenberg.org/files/45304/45304-h/45304-h.htm
15. Avogadro, A.: Essai d'une manière de déterminer les masses relatives des molécules élémentaires des corps, et les proportions selon lesquelles elles entrent dans ces combinaisons. J. Phys. Chimie et d'Histoire Nat. **73**, 58–76 (1811)
16. Balian, R.: From Microphysics to Macrophysics, vols. I–II. Springer, Berlin (1982)
17. Balian, R.: François Massieu and the thermodynamic potentials. Comptes Rendus. Phys. **18**(9–10), 526–530 (2017). https://doi.org/10.1016/j.crhy.2017.09.011

18. Balibar, S.: The Discovery of Superfluidity. J. Low Temp. Phys. **146**(5), 441–470 (2007). https://doi.org/10.1007/s10909-006-9276-7
19. Bardeen, J., Cooper, L.N., Schrieffer, J.R.: Theory of Superconductivity. Phys. Rev. **108**, 1175–1204 (1957). https://doi.org/10.1103/PhysRev.108.1175
20. Bell, J.S.: On the Einstein Podolsky Rosen paradox. Phys. Phys. Fizika **1**, 195–200 (1964). https://doi.org/10.1103/PhysicsPhysiqueFizika.1.195
21. Bell, J.S.: The theory of local beables. In: CERN Preprint, pp. CERN–TH–2053. CERN, Geneva (1975). https://cds.cern.ch/record/980036
22. de Bellaigue, C.: Death on demand: has euthanasia gone too far? The Guardian (2019-01-18). https://www.theguardian.com/news/2019/jan/18/death-on-demand-has-euthanasia-gone-too-far-netherlands-assisted-dying
23. Benjamini, Y., Hochberg, Y.: Controlling the false discovery rate: a practical and powerful approach to multiple testing. J. R. Stat. Soc. Series B (Methodological) **57**(1), 289–300 (1995). https://doi.org/10.1111/j.2517-6161.1995.tb02031.x. http://www.jstor.org/stable/2346101
24. Berrens, M.L., Kundu, A., Calegari Andrade, M.F., Pham, T.A., Galli, G., Donadio, D.: Nuclear quantum effects on the electronic structure of water and ice. J. Phys. Chem. Lett. **15**(26), 6818–6825 (2024). https://doi.org/10.1021/acs.jpclett.4c01315
25. Bethe, H.A.: The electromagnetic shift of energy levels. Phys. Rev. **72**, 339–341 (1947). https://doi.org/10.1103/PhysRev.72.339
26. Blackman, M.: Contributions to the theory of the specific heat of crystals I—Lattice theory and continuum theory. Proc. R. Soc. Lond. Series A Math. Phys. Sci. **148**(864), 365–383 (1935). https://doi.org/10.1098/rspa.1935.0024
27. Blevin, W.R., Brown, W.J.: A precise measurement of the Stefan-Boltzmann constant. Metrologia **7**(1), 15 (1971). https://doi.org/10.1088/0026-1394/7/1/003
28. Bohm, D., Aharonov, Y.: Discussion of experimental proof for the paradox of Einstein, Rosen, and Podolsky. Phys. Rev. **108**, 1070–1076 (1957). https://doi.org/10.1103/PhysRev.108.1070
29. Boltzmann, L.: Weitere Studien Über das Wärmegleichgewicht unter Gasmolekülen. Sitzungsberichte der mathematisch-naturwissenschaftlichen Classe der Kaiserlichen Akademie der Wissenschaften, Wien, 2Abt. **66**(8), 275–370 (1872)
30. Boltzmann, L.: Über die Beziehung zwischen dem zweiten Hauptsatze des mechanischen Wärmetheorie und der Wahrscheinlichkeitsrechnung, respective den Sätzen über das Wärmegleichgewicht. Sitzungsberichte der mathematisch-naturwissenschaftlichen Classe der Kaiserlichen Akademie der Wissenschaften, Wien, 2Abt. **76**(8), 373–435 (1877)
31. Boltzmann, L.: Ableitung des Stefan'schen Gesetzes, betreffend die Abhängigkeit der Wärmestrahlung von der Temperatur aus der electromagnetischen Lichttheorie. Ann. Phys. **258**(6), 291–294 (1884). https://doi.org/10.1002/andp.18842580616
32. Boltzmann, L.: On certain questions of the theory of gases. Nature **51**(1322), 413–415 (1895). https://doi.org/10.1038/051413b0
33. Born, M., Green, H.S.: A general kinetic theory of liquids I. The molecular distribution functions. Proc. R. Soc. Lond. Series A Math. Phys. Sci. **188**(1012), 10–18 (1946). https://doi.org/10.1098/rspa.1946.0093
34. Boscovich, R.J.: Theoria philosophiae naturalis redacta ad unicam legem virium in natura existentium. Ex Typographia Remondiniana, Venetiis (1763)
35. Bose: Plancks Gesetz und Lichtquantenhypothese. Z. Phys. **26**(1), 178–181 (1924). https://doi.org/10.1007/BF01327326
36. Bose, S., Theimer, O., Ram, B.: The beginning of quantum statistics: a translation of "Planck's law and the light quantum hypothesis". Am. J. Phys. **44**(11), 1056–1057 (1976). https://doi.org/10.1119/1.10584
37. Boyd, R.J.: A quantum mechanical explanation for Hund's multiplicity rule. Nature **310**(5977), 480–481 (1984). https://doi.org/10.1038/310480a0
38. Boyle, R.: A defence of the doctrine touching the spring and weight of the AIR, propos'd by Mr. R. Boyle in his new physico-mechanical experiments against the objections of Franciscus Linus. Printed by F.G. for Thomas Robinson Bookseller in Oxon, London (1662)

39. Briggs, J.S., Rost, J.M.: On the derivation of the time-dependent equation of Schrödinger. Found. Phys. **31**(4), 693–712 (2001). https://doi.org/10.1023/A:1017525227832
40. Brown, M., Wyatt, A.F.G.: The surface boundary conditions for quantum evaporation in ^{4}He. J. Phys. Condensed Matter **2**(22), 5025 (1990). https://doi.org/10.1088/0953-8984/2/22/022
41. Buckingham, E.: Note on the deduction of Stefan's law. Phys. Rev. (Series I) **17**, 277–280 (1903). https://doi.org/10.1103/PhysRevSeriesI.17.277
42. Buckingham, E.: On the deduction of Wien's displacement law. Bull. Bureau Standards **8**(3), 545–557 (1912)
43. Buckingham, E.: On physically similar systems; illustrations of the use of dimensional equations. Phys. Rev. **4**, 345–376 (1914). https://doi.org/10.1103/PhysRev.4.345
44. Burckhardt, J.: Die Cultur der Renaissance in Italien. Schweighauser, Basel (1860). https://doi.org/10.11588/diglit.3478
45. Busemeyer, J., Asano, M., Lu, M.: Explaining interference effects in prisoner dilemma games. Exper. Econ. **27**(4), 743–765 (2024). https://doi.org/10.1007/s10683-024-09838-w
46. Butkov, E.: Mathematical Physics. Addison-Wesley, Reading (1968)
47. Carathéodory, C.: Untersuchungen über die Grundlagen der Thermodynamik. Math. Ann. **67**(3), 355–386 (1909). https://doi.org/10.1007/BF01450409
48. Carnot, S.: Réflexions sur la Puissance Motrice du feu et sur les Machines Propres à Développer Cette Puissance. Bachelier, Paris, Quai des Augustins 55 (1824)
49. Carnot, S.: Réflexions sur la puissance motrice du feu et sur les machines propres à développer cette puissance. Ann. Sci. l'É.N.S. 2e série **1**, 393–457 (1872)
50. Carnot, S., Thurston, R., Carnot, H., Kelvin, W.: Reflections on the Motive Power of Heat and on Machines Fitted to Develop that Power. John Wiley, Hoboken (1890)
51. Chen, L.Y., Su, Z.B.: Quantum Boltzmann equation and Kubo formula for electronic transport in solids. Phys. Rev. B **40**, 9309–9311 (1989). https://doi.org/10.1103/PhysRevB.40.9309
52. Cho, C.J., An, T., Lai, Y.C., Vázquez-Salazar, A., Fracassi, A., Brea, R.J., Chen, I.A., Devaraj, N.K.: Protocells by spontaneous reaction of cysteine with short-chain thioesters. Nature Chem. **17**(1), 148–155 (2025). https://doi.org/10.1038/s41557-024-01666-y
53. Clauser, J.F., Horne, M.A., Shimony, A., Holt, R.A.: Proposed experiment to test local hidden-variable theories. Phys. Rev. Lett. **23**, 880–884 (1969). https://doi.org/10.1103/PhysRevLett.23.880
54. Clausius, R.: Ueber eine veränderte Form des zweiten Hauptsatzes der mechanischen Wärmetheorie. Ann. Phys. **169**(12), 481–506 (1854). https://doi.org/10.1002/andp.18541691202
55. Clausius, R.: Ueber verschiedene für die Anwendung bequeme Formen der Hauptgleichungen der mechanischen Wärmetheorie. Ann. Phys. **201**(7), 353–400 (1865). https://doi.org/10.1002/andp.18652010702
56. Cockrell, C., Dicks, O., Todorov, I.T., Elena, A.M., Trachenko, K.: Fast dynamics and high effective dimensionality of liquid fluidity. Sci. Rep. **13**(1), 15664 (2023). https://doi.org/10.1038/s41598-023-41931-7
57. Colella, R., Overhauser, A.W., Werner, S.A.: Observation of gravitationally induced quantum interference. Phys. Rev. Lett. **34**, 1472–1474 (1975). https://doi.org/10.1103/PhysRevLett.34.1472
58. Coltice, N., Husson, L., Faccenna, C., Arnould, M.: What drives tectonic plates? Sci. Adv. **5**(10), eaax4295 (2019). https://doi.org/10.1126/sciadv.aax4295
59. Cooper, L.N.: Remembrance of superconductivity past. In: BCS: 50 Years, pp. 3–19. World Scientific, Singapore (2010). https://doi.org/10.1142/9789814304665_0001
60. Ćurić, J.: Filozofski Spisi. Filozofsko-teološki institut Družbe Isusove, Zagreb (2022)
61. Cvitanović, P., Artuso, R., Mainieri, R., Tanner, G., Vattay, G.: Chaos: Classical and Quantum. Niels Bohr Inst., Copenhagen (2016). http://ChaosBook.org/
62. Cyrot, M.: Ginzburg-Landau theory for superconductors. Rep. Progress Phys. **36**(2), 103–158 (1973). https://doi.org/10.1088/0034-4885/36/2/001
63. Davidson, E.R.: Single-configuration calculations on excited states of Helium. II. J. Chem. Phys. **42**(12), 4199–4200 (1965). https://doi.org/10.1063/1.1695919

64. Dawkins, R.: The Selfish Gene. Oxford University Press, New York (1976)
65. Dawkins, R., Wong, Y.: The Ancestor's Tale: A Pilgrimage to the Dawn of Life, updated edn. Weidenfeld & Nicolson, London (2017)
66. Debye, P.: Zur Theorie der spezifischen Wärmen. Ann. Phys. **344**(14), 789–839 (1912). https://doi.org/10.1002/andp.19123441404
67. Di Mauro, M., Esposito, S., Naddeo, A.: Introducing quantum and statistical physics in the footsteps of einstein: a proposal. Universe **7**(6) (2021). https://doi.org/10.3390/universe7060184
68. Dirac, P.A.M.: The quantum theory of the emission and absorption of radiation. Proc. R. Soc. Lond. Series A Containing Papers Math. Phys. Character **114**(767), 243–265 (1927). https://doi.org/10.1098/rspa.1927.0039
69. Dobbs, E.R.: What happens to Liquid Helium 3 at very low Temperatures? Phys. Blätter **32**(12), 591–600 (1976). https://doi.org/10.1002/phbl.19760321214
70. Dobbs, B.J.T.: The Janus Faces of Genius: The Role of Alchemy in Newton's Thought. Cambridge University Press, Cambridge (1991)
71. Donnelly, R.J.: The two-fluid theory and second sound in liquid helium. Phys. Today **62**(10), 34–39 (2009). https://doi.org/10.1063/1.3248499
72. Drittler, B., Stefanou, N., Blügel, S., Zeller, R., Dederichs, P.H.: Electronic structure and magnetic properties of dilute Fe alloys with transition-metal impurities. Phys. Rev. B **40**, 8203–8212 (1989). https://doi.org/10.1103/PhysRevB.40.8203
73. Dulčić, A., Poljak, N., Požek, M.: Mehanika. Školska knjiga, Zagreb (2023)
74. Dunkel, J., Hilbert, S.: Consistent thermostatistics forbids negative absolute temperatures. Nature Phys. **10**(1), 67–72 (2014). https://doi.org/10.1038/nphys2815
75. Editorial: The Guardian view on Ikea: furnishing the modern world. The Guardian (2018-01-29). https://www.theguardian.com/commentisfree/2018/jan/29/the-guardian-view-on-ikea-furnishing-the-modern-world
76. Einstein, A.: Über die von der molekularkinetischen Theorie der Wärme geforderte Bewegung von in ruhenden Flüssigkeiten suspendierten Teilchen. Ann. Phys. **17**, 549–560 (1905). http://echo.mpiwg-berlin.mpg.de/MPIWG:YRZG2WCY
77. Einstein, A.: Über einen die Erzeugung und Verwandlung des Lichtes betreffenden heuristischen Gesichtspunkt. Ann. Phys. **17**, 132–148 (1905).http://echo.mpiwg-berlin.mpg.de/MPIWG:9Y16XU4Q
78. Einstein, A.: Zur Theorie der Brownschen Bewegung. Ann. Phys. **19**, 371–381 (1906). http://echo.mpiwg-berlin.mpg.de/MPIWG:Y7RD0UGY
79. Einstein, A.: Die Plancksche Theorie der Strahlung und die Theorie der spezifischen Wärme. Ann. Phys. **22**, 180–190 (1907). http://echo.mpiwg-berlin.mpg.de/MPIWG:87XK0UF8
80. Einstein, A.: Quantentheorie der Strahlung. Mitteilungen Phys. Ges. Zürich **16**, 47–62 (1916)
81. Einstein, A.: Letter to Satyendra Nath Bose (1924-07-02). https://www.bose.res.in/Prof.S.N.Bose-Archive/objects/0038.jpg
82. Einstein, A., Podolsky, B., Rosen, N.: Can quantum-mechanical description of physical reality be considered complete? Phys. Rev. **47**, 777–780 (1935). https://doi.org/10.1103/PhysRev.47.777
83. Ellerman, D.: The logic of partitions: Introduction to the dual of the logic of subsets. Rev. Symbolic Logic **3**(2), 287–350 (2010). https://doi.org/10.1017/S1755020310000018
84. Ellerman, D.: An introduction to partition logic. Logic J. IGPL **22**(1), 94–125 (2013). https://doi.org/10.1093/jigpal/jzt036
85. Ellerman, D.: Partitions, Objective Indefiniteness, and Quantum Reality. Springer, Cham (2024). https://doi.org/10.1007/978-3-031-61786-7
86. Eskes, H., Jefferson, J.H.: Superexchange in the cuprates. Phys. Rev. B **48**, 9788–9798 (1993). https://doi.org/10.1103/PhysRevB.48.9788
87. Evans, M., Rosenthal, J.S.: Probability and Statistics: The Science of Uncertainty. W.H. Freeman and Co., New York (2004). https://utstat.utoronto.ca/mikevans/jeffrosenthal/. Textbook freely available at the URL

88. Facchi, P., Nakazato, H., Pascazio, S.: From the quantum Zeno to the inverse quantum Zeno effect. Phys. Rev. Lett. **86**, 2699–2703 (2001). https://doi.org/10.1103/PhysRevLett.86.2699
89. Feller, W.: An Introduction to Probability Theory and Its Applications, vol. I, 3rd edn, Revised Printing. Wiley Series in Probability and Mathematical Statistics. John Wiley & Sons, New York (1971)
90. Feller, W.: An Introduction to Probability Theory and Its Applications, vol. II, 2nd edn. Wiley Series in Probability and Mathematical Statistics. John Wiley & Sons, New York (1971)
91. Fetter, A.L., Walecka, J.D.: Quantum Theory of Many-Particle Systems. Dover Publications, New York (2003)
92. Feynman, R.P.: Forces in molecules. Phys. Rev. **56**, 340–343 (1939). https://doi.org/10.1103/PhysRev.56.340
93. Feynman, R.P.: Atomic theory of liquid helium near absolute zero. Phys. Rev. **91**, 1301–1308 (1953). https://doi.org/10.1103/PhysRev.91.1301
94. Feynman, R.P., Leighton, R.: Part 2: a map of the cat? Surely You're Joking, Mr. Feynman! W. W. Norton, Hoboken (1985)
95. Fisher, M.P.: Quantum cognition: The possibility of processing with nuclear spins in the brain. Ann. Phys. **362**, 593–602 (2015). https://doi.org/10.1016/j.aop.2015.08.020
96. Fisher, M.P.A.: Are we quantum computers, or merely clever robots? Int. J. Modern Phys. B **31**(07), 1743001 (2017). https://doi.org/10.1142/S0217979217430019
97. Forsdale, J.R., Forsdale, L.: Film literacy. Teachers College Record **67**(8), 1–10 (1966). https://doi.org/10.1177/016146816606700810
98. de Gennes, P.G.: Superconductivity of metals and alloys. W. A. Benjamin, New York (1966)
99. Gibbs, J.W.: On the equilibrium of heterogeneous substances. First part. Trans. Connecticut Acad. Arts Sci. **3**, 108–248 (1875). https://doi.org/10.5479/sil.421748.39088007099781
100. Gibbs, J.W.: Elementary Principles in Statistical Mechanics. Dover, New York (1960). Unabridged republication of the Yale University Press edition of 1902
101. Gilbert, D.T., Wilson, T.D.: Prospection: experiencing the future. Science **317**(5843), 1351–1354 (2007). https://doi.org/10.1126/science.1144161
102. Gleason, A.M.: Measures on the closed subspaces of a Hilbert space. J. Math. Mech. **6**(6), 885–893 (1957). http://www.jstor.org/stable/24900629
103. Glyde, H.R., Azuah, R.T., Stirling, W.G.: Condensate, momentum distribution, and final-state effects in liquid ^{4}He. Phys. Rev. B **62**, 14337–14349 (2000). https://doi.org/10.1103/PhysRevB.62.14337
104. Golse, F.: Hydrodynamic limits. In: A. Laptev (ed.) European Congress of Mathematics Stockholm, June 27-July 2, 2004, pp. 699–718. European Mathematical Society, Providence (2005). https://doi.org/10.4171/009-1/44
105. Gor'kov, L.P.: Microscopic derivation of the Ginzburg-Landau equations in the theory of superconductivity. Sov. Phys. JETP **9**(6), 1364–1367 (1959)
106. Grabiner, J.V.: "It's all for the best": Optimization in the history of science. J. Humanistic Math. **11**, 54–80 (2021). https://doi.org/10.5642/jhummath.202101.05
107. Grgić, F., Pećnjak, D. (eds.): Free Will & Action: Historical and Contemporary Perspectives. Springer International Publishing, Cham (2018). https://doi.org/10.1007/978-3-319-99295-2
108. Gross, D.H.E., Votyakov, E.V.: Phase transitions in "small" systems. Eur. Phys. J. B Condensed Matter Complex Syst. **15**(1), 115–126 (2000). https://doi.org/10.1007/PL00011030
109. Haber, F., Le Rossignol, R.: Bestimmung des Ammoniakgleichgewichtes unter Druck. Zeitschrift für Elektrochemie und Angewandte Physikalische Chemie **14**(15), 181–196 (1908). https://doi.org/10.1002/bbpc.19080141502
110. Haber, F., Rossignol, R.L.: Production of Ammonia (1910). US Patent 512,679. Application filed Aug. 13, 1909, patented Sept. 27, 1910
111. Hameroff, S.: How quantum brain biology can rescue conscious free will. Front. Integrative Neurosci. **6**, 93 (2012)
112. Hänsch, W., Mahan, G.D.: Transport equations for many-particle systems. Phys. Rev. B **28**, 1902–1922 (1983). https://doi.org/10.1103/PhysRevB.28.1902

113. Hastings, C.S.: Biographical memoir of Josiah Willard Gibbs, 1839–1903. In: Biographical Memoirs, vol. VI, p. 375. National Academy of Sciences, Washington (1909). https://www.nasonline.org/publications/biographical-memoirs
114. Hawking, S.W.: Black hole explosions? Nature **248**(5443), 30–31 (1974). https://doi.org/10.1038/248030a0
115. Hawkins, R.J., Sanchez, J.M.: Ferromagnetism and chemical ordering in cobalt-iron. J. Phys. F Metal Phys. **18**(4), 767 (1988). https://doi.org/10.1088/0305-4608/18/4/012
116. Heisenberg, W.: Physics and Philosophy. Harper & Brothers Publishers, New York (1958)
117. Hevia, D.F., Rañada, A.F.: Chaos in the three-body problem: the Sitnikov case. Eur. J. Phys. **17**, 295–302 (1996). https://doi.org/10.1088/0143-0807/17/5/009
118. Hook, A.: Examining modern money creation: An institution-centered explanation and visualization of the "credit theory" of money and some reflections on its significance. J. Econ. Educ. **53**(3), 210–231 (2022). https://doi.org/10.1080/00220485.2022.2075510
119. Horne, M., Farago, P., Oliver, J.: An experiment to measure Boltzmann's constant. Am. J. Phys. **41**(3), 344–348 (1973). https://doi.org/10.1119/1.1987221
120. Huang, K.: Statistical Mechanics. Wiley, New York (1963)
121. Hulet, R.G., Hilfer, E.S., Kleppner, D.: Inhibited spontaneous emission by a Rydberg atom. Phys. Rev. Lett. **55**, 2137–2140 (1985). https://doi.org/10.1103/PhysRevLett.55.2137
122. Inglis, J.: Cargo cults: the problem of explanation. Oceania **27**(4), 249–263 (1957). http://www.jstor.org/stable/40329058
123. Ishizaki, A., Fleming, G.R.: Quantum superpositions in photosynthetic light harvesting: delocalization and entanglement. New J. Phys. **12**(5), 055004 (2010). https://doi.org/10.1088/1367-2630/12/5/055004
124. Itzykson, C., Drouffe, J.M.: Statistical Field Theory, vol. I. Cambridge Monographs on Mathematical Physics. Cambridge University Press, Cambridge (1989)
125. Jaynes, E.T.: The Gibbs paradox. In: Smith, C.R., Erickson, G.J., Neudorfer, P.O. (eds.) Maximum Entropy and Bayesian Methods: Seattle, 1991, pp. 1–21. Springer Netherlands, Dordrecht (1992). https://doi.org/10.1007/978-94-017-2219-3_1
126. Jefferson, T.: Letter of Thomas Jefferson to Uriah Forest (1787-12-31). URL https://founders.archives.gov/documents/Jefferson/01-12-02-0490
127. Johnson, J.B.: Thermal agitation of electricity in conductors. Phys. Rev. **32**, 97–109 (1928). https://doi.org/10.1103/PhysRev.32.97
128. Khalfin, L.A.: Contribution to the decay theory of a quasi-stationary state. Soviet Phys. JETP **6**(33), 1053–1063 (1958)
129. Kirkwood, J.G.: The statistical mechanical theory of transport processes I. General theory. J. Chem. Phys. **14**(3), 180–201 (1946). https://doi.org/10.1063/1.1724117
130. Kirkwood, J.G.: The statistical mechanical theory of transport processes II. Transport in gases. J. Chem. Phys. **15**(1), 72–76 (1947). https://doi.org/10.1063/1.1746292
131. Kittel, C.: Elementary Statistical Physics. Wiley, New York (1958)
132. Kleman, M., Friedel, J.: Disclinations, dislocations, and continuous defects: a reappraisal. Rev. Mod. Phys. **80**, 61–115 (2008). https://doi.org/10.1103/RevModPhys.80.61
133. Kramer, A.D.I., Guillory, J.E., Hancock, J.T.: Experimental evidence of massive-scale emotional contagion through social networks. Proc. Natl. Acad. Sci. **111**(24), 8788–8790 (2014). https://doi.org/10.1073/pnas.1320040111
134. Kubo, R.: Statistical Mechanics, an Advanced Course with Problems and Solutions. North-Holland, Amsterdam (1965)
135. Lamb, W.E., Retherford, R.C.: Fine structure of the hydrogen atom by a microwave method. Phys. Rev. **72**, 241–243 (1947). https://doi.org/10.1103/PhysRev.72.241
136. Landau, L.: К теории фазовых переходов I J. Exptl. Theoret. Phys. **7**, 19 (1937)
137. Landau, L.: К теории фазовых переходов II. J. Exptl. Theoret. Phys. **7**, 627 (1937)
138. Landau, L.: Теория сверхтекучести гелия II. J. Exptl. Theoret. Phys. **11**, 592 (1941)
139. Landau, L.D.: Теория ферми-жидкости. J. Exptl. Theoret. Phys. **30**(6), 1058–1064 (1956)
140. Landau, L.D.: Колебания ферми-жидкости. J. Exptl. Theoret. Phys. **32**, 59 (1957)
141. Landau, L.D.: Oscillations in a Fermi liquid. Sov. Phys. JETP **5**(1), 101–108 (1957)

142. Landau, L.D.: The theory of a Fermi liquid. Sov. Phys. JETP **3**(6), 920–925 (1957)
143. Landau, L.D.: On the theory of phase transitions. In: Haar, D.T. (ed.) Collected Papers of L. D. Landau, pp. 193–216. Gordon and Breach, Science Publishers Inc. and Pergamon Press Ltd. New York, London (1965)
144. Landau, L.D.: The theory of superfluidity of Helium II. In: Haar, D.T. (ed.) Collected Papers of L. D. Landau, pp. 301–330. Gordon and Breach, Science Publishers Inc. and Pergamon Press Ltd. New York, London (1965)
145. Landau, L.D., Ginzburg, V.L.: К теории сверхпроводимости. J. Exptl. Theoret. Phys. **20**, 1064 (1950)
146. Landau, L.D., Ginzburg, V.L.: On the theory of superconductivity. In: Haar, D.T. (ed.) Collected Papers of L. D. Landau, pp. 546–225. Gordon and Breach, Science Publishers Inc. and Pergamon Press Ltd. New York, London (1965)
147. Landau, L.D., Lifshitz, E.M.: Statistical Physics. Butterworth-Heinemann, Oxford (1980)
148. Landau, L.D., Lifshitz, E.M.: Fluid Mechanics, 2nd edn. Pergamon Press, Oxford (1987)
149. Lane, W.C.: Leibniz's best world claim restructured. Am. Philos. Q. **47**, 57–84 (2010)
150. Lay, T., Hernlund, J., Buffett, B.A.: Core–mantle boundary heat flow. Nature Geosci. **1**(1), 25–32 (2008). https://doi.org/10.1038/ngeo.2007.44
151. Leggett, A.J.: A theoretical description of the new phases of liquid ^{3}He. Rev. Mod. Phys. **47**, 331–414 (1975). https://doi.org/10.1103/RevModPhys.47.331
152. Leggett, A.J.: Nobel Lecture: Superfluid ^{3}He: the early days as seen by a theorist. Rev. Mod. Phys. **76**, 999–1011 (2004). https://doi.org/10.1103/RevModPhys.76.999. https://www.nobelprize.org/uploads/2018/06/leggett-lecture.pdf. Nobel Prize lecture.
153. Leggett, A.J.: Quantum Liquids: Bose Condensation and Cooper Pairing in Condensed-matter Systems. Oxford University Press, Oxford (2006). https://doi.org/10.1093/acprof:oso/9780198526438.001.0001
154. Levitin, L.V., van der Vliet, H., Theisen, T., Dimitriadis, S., Lucas, M., Corcoles, A.D., Nyéki, J., Casey, A.J., Creeth, G., Farrer, I., Ritchie, D.A., Nicholls, J.T., Saunders, J.: Cooling low-dimensional electron systems into the microkelvin regime. Nature Commun. **13**(1), 667 (2022). https://doi.org/10.1038/s41467-022-28222-x
155. Liouville, J.: Note sur l'intégration des équations différentielles de la Dynamique, présentée au Bureau des Longitudes le 29 juin 1853. J. Math. Pures Appl. 137–138 (1855). http://eudml.org/doc/233980
156. London, F.: A New Conception of Supraconductivity. Nature **140**(3549), 793–796 (1937). https://doi.org/10.1038/140793a0
157. London, F.: The λ-Phenomenon of Liquid Helium and the Bose-Einstein Degeneracy. Nature **141**(3571), 643–644 (1938). https://doi.org/10.1038/141643a0
158. Lubarda, V.A.: Thermodynamic analysis based on the second-order variations of thermodynamic potentials. Theoret. Appl. Mech. **35**, 215–234 (2008). https://doi.org/10.2298/tam08032151
159. Luttinger, J.M., Nozières, P.: Derivation of the Landau theory of Fermi liquids. II. Equilibrium properties and transport equation. Phys. Rev. **127**, 1431–1440 (1962). https://doi.org/10.1103/PhysRev.127.1431
160. Ma, S.K.: Introduction to the renormalization group. Rev. Mod. Phys. **45**, 589–614 (1973). https://doi.org/10.1103/RevModPhys.45.589
161. Mahan, G.D.: Many-Particle Physics. Springer US, Boston (2000). https://doi.org/10.1007/978-1-4757-5714-9
162. Manfredi, G.: Logical entropy – special issue. 4open **5**, E1 (2022). https://doi.org/10.1051/fopen/2022005
163. Manson, J.R., Miret-Artés, S.: Atom–surface scattering in the classical multiphonon regime. Phys. Chem. Chem. Phys. **24**, 16942–16972 (2022). https://doi.org/10.1039/D2CP01144A

164. Marcus, R.A.: Electron transfer reactions in chemistry. Theory and experiment. Rev. Mod. Phys. **65**, 599–610 (1993). https://doi.org/10.1103/RevModPhys.65.599. https://www.nobelprize.org/uploads/2018/06/marcus-lecture.pdf. Nobel Prize lecture.
165. Mariotte, E.: Discours de la nature de l'air (1676). In: Oeuvres de Mr. Mariotte de l'Académie Royale des Sciences, Tome Premier, p. 149. Pierre van der Aa, Leide (1717)
166. Maris, H.J., Kadanoff, L.P.: Teaching the renormalization group. Am. J. Phys, **46**(6), 652–657 (1978). https://doi.org/10.1119/1.11224
167. Mattis, D.C.: The Theory of Magnetism: An Introduction to the Study of Cooperative Phenomena. Harper's Physics Series. Harper & Row, New York (1965)
168. Maurer, A.: St. Thomas Aquinas: The Division and Methods of the Sciences. The Pontifical Institute of Mediaeval Studies, Toronto (1963)
169. Maxwell, J.C.: I. On governors. Proc. R. Soc. Lond. **16**, 270–283 (1868). https://doi.org/10.1098/rspl.1867.0055
170. Mayers, J.: A new method of determining the Bose condensate fraction in superfluid ^{4}He. In: Technical Report, pp. RAL–TR–96–031. Council for the Central Laboratory of the Research Councils, Chilton Didcot Oxfordshire (1996)
171. Mayers, J.: Bose-Einstein condensation and two fluid behavior in ^{4}He. Phys. Rev. B **74**, 014516 (2006). https://doi.org/10.1103/PhysRevB.74.014516
172. McKie, D.: Newton and chemistry. J. RASC **37**, 17–23 (1943). https://ui.adsabs.harvard.edu/link_gateway/1943JRASC..37...17M/ADS_PDF
173. Meissner, W., Ochsenfeld, R.: Ein neuer Effekt bei Eintritt der Supraleitfähigkeit. Naturwissenschaften **21**(44), 787–788 (1933). https://doi.org/10.1007/BF01504252
174. Michel, L.: Symmetry defects and broken symmetry. Configurations Hidden Symmetry. Rev. Mod. Phys. **52**, 617–651 (1980). https://doi.org/10.1103/RevModPhys.52.617
175. Moehling, C.M., Thomasson, M.A.: Votes for women: an economic perspective on women's enfranchisement. J. Econ. Perspectives **34**(2), 3–23 (2020). https://doi.org/10.1257/jep.34.2.3
176. Morgan-Collins, M.: The electoral impact of newly enfranchised groups: the case of women's suffrage in the United States. J. Politics **83**(1), 150–165 (2021). https://doi.org/10.1086/708955
177. Nakatsuji, H.: Discovery of a general method of solving the Schrödinger and Dirac equations that opens a way to accurately predictive quantum chemistry. Acc. Chem. Res. **45**(9), 1480–1490 (2012). https://doi.org/10.1021/ar200340j
178. Nakatsuji, H., Nakashima, H.: Free-complement local-Schrödinger-equation method for solving the Schrödinger equation of atoms and molecules: Basic theories and features. J. Chem. Phys. **142**(8), 084117 (2015). https://doi.org/10.1063/1.4909520
179. Nakroshis, P., Amoroso, M., Legere, J., Smith, C.: Measuring Boltzmann's constant using video microscopy of Brownian motion. Am. J. Phys. **71**(6), 568–573 (2003). https://doi.org/10.1119/1.1542619
180. Nash, J.F.: Equilibrium points in n-person games. Proc. Natl. Acad. Sci. **36**(1), 48–49 (1950). https://doi.org/10.1073/pnas.36.1.48
181. Negele, J.W., Orland, H.: Quantum Many-particle Systems. Advanced Book Classics. CRC Press Taylor and Francis Group, Boca Raton (2019)
182. Néel, L.: Le champ moléculaire de Weiss et le champ moléculaire local. In: Colloque National de Strasbourg, commémoratif de l'oeuvre de Pierre Weiss, pp. 1–24. CNRS (1957). HAL-02888372
183. Neumann, J.V.: Thermodynamik quantenmechanischer Gesamtheiten. Nachrichten von der Gesellschaft der Wissenschaften zu Göttingen, Mathematisch-Physikalische Klasse **1927**, 273–291 (1927). http://eudml.org/doc/59231
184. Muñoz de Nova, J.R., Golubkov, K., Kolobov, V.I., Steinhauer, J.: Observation of thermal Hawking radiation and its temperature in an analogue black hole. Nature **569**(7758), 688–691 (2019). https://doi.org/10.1038/s41586-019-1241-0
185. Nozières, P., Luttinger, J.M.: Derivation of the Landau theory of Fermi liquids. I. Formal preliminaries. Phys. Rev. **127**, 1423–1431 (1962). https://doi.org/10.1103/PhysRev.127.1423

186. Onsager, L.: Lars Onsager's Speech at the Nobel Banquet in Stockholm, December 10, 1968. https://www.nobelprize.org/prizes/chemistry/1968/onsager/speech
187. Onsager, L.: Reciprocal relations in irreversible processes. I. Phys. Rev. **37**, 405–426 (1931). https://doi.org/10.1103/PhysRev.37.405
188. Onsager, L.: Reciprocal relations in irreversible processes. II. Phys. Rev. **38**, 2265–2279 (1931). https://doi.org/10.1103/PhysRev.38.2265
189. Onsager, L.: The motion of ions: principles and concepts. Science **166**(3911), 1359–1364 (1969). https://doi.org/10.1126/science.166.3911.1359. https://www.nobelprize.org/uploads/2018/06/onsager-lecture.pdf. Nobel Prize lecture
190. Parkinson, C.N.: Parkinson's Law or the Pursuit of Progress. Penguin Books, Harmondsworth (1965)
191. Penrose, O., Onsager, L.: Bose-Einstein condensation and liquid helium. Phys. Rev. **104**, 576–584 (1956). https://doi.org/10.1103/PhysRev.104.576
192. Pérez, E., Sauer, T.: Einstein's quantum theory of the monatomic ideal gas: non-statistical arguments for a new statistics. Arch. History Exact Sci. **64**(5), 561–612 (2010). https://doi.org/10.1007/s00407-010-0066-x
193. Pigafetta, A.: Magellan's Voyage Around the World, vol. II. The Arthur H. Clark Company, Cleveland (1906). The original text of the Ambrosian MS., with English translation, notes, bibliography and index by James Alexander Robertson
194. Pitre, L., Plimmer, M.D., Sparasci, F., Himbert, M.E.: Determinations of the Boltzmann constant. Comptes Rendus. Phys. **20**(1–2), 129–139 (2019). https://doi.org/10.1016/j.crhy.2018.11.007
195. Planck, M.: Ueber irreversible Strahlungsvorgänge. Ann. Phys. **306**(1), 69–122 (1900). https://doi.org/10.1002/andp.19003060105
196. Planck, M.: Zur Theorie des Gesetzes der Energieverteilung im Normalspectrum. Verhandlungen der Deutschen Physikalischen Gesellschaft **2**, 237–245 (1900)
197. Planck, M.: Ueber das Gesetz der Energieverteilung im Normalspectrum. Ann. Physik **309**(3), 553–563 (1901). https://doi.org/10.1002/andp.19013090310
198. Planck, M.: The Theory of Heat Radiation. P. Blakinston's Son & Co., Philadelphia (1914). Authorised Translation by Morton Masius
199. Planck, M.: The Genesis and Present State of Development of the Quantum Theory (1920). https://www.nobelprize.org/prizes/physics/1918/planck/lecture/. Nobel Prize lecture
200. Plummer, T.W., Ditchfield, P.W., Bishop, L.C., Kingston, J.D., Ferraro, J.V., Braun, D.R., Hertel, F., Potts, R.: Oldest evidence of toolmaking hominins in a Grassland-dominated ecosystem. PLOS ONE **4**(9), 1–8 (2009). https://doi.org/10.1371/journal.pone.0007199
201. Plutarch: Lives, Volume III, Pericles and Fabius Maximus. Nicias and Crassus (Loeb Classical Library). Harvard University Press (1916). Translated by Bernadotte Perrin
202. Pound, R.V., Rebka, G.A.: Apparent Weight of Photons. Phys. Rev. Lett. **4**, 337–341 (1960). https://doi.org/10.1103/PhysRevLett.4.337
203. Ratzinger, J.: Volk und Haus Gottes in Augustins Lehre von der Kirche. EOS Verlag, Erzabtei St. Ottilien (1992)
204. Regensburger, S., Mukherjee, A.k., Schönhuber, S., Kainz, M.A., Winnerl, S., Klopf, J.M., Lu, H., Gossard, A.C., Unterrainer, K., Preu, S.: Broadband terahertz detection with zero-bias field-effect transistors between 100 GHz and 11.8 THz with a noise equivalent power of 250 $\mathrm{pW}/\sqrt{\mathrm{Hz}}$ at 0.6 THz. IEEE Trans. Terahertz Sci. Technol. **8**(4), 465–471 (2018). https://doi.org/10.1109/TTHZ.2018.2843535
205. Roduner, E., Krüger, T.P.: The origin of irreversibility and thermalization in thermodynamic processes. Phys. Rep. **944**, 1–43 (2022). https://doi.org/10.1016/j.physrep.2021.11.002
206. Rovenchak, A.A., Vakarchuk, I.O.: Calculation of the condensate fraction in liquid helium-4. J. Phys. Stud. **11**(4), 404–407 (2007). https://doi.org/10.30970/jps.11.404
207. Rowley, D.B., Forte, A.M., Rowan, C.J., Glišović, P., Moucha, R., Grand, S.P., Simmons, N.A.: Kinematics and dynamics of the East Pacific Rise linked to a stable, deep-mantle upwelling. Sci. Adv. **2**(12), e1601107 (2016). https://doi.org/10.1126/sciadv.1601107

208. Sajeev, Y., Sindelka, M., Moiseyev, N.: Hund's multiplicity rule: from atoms to quantum dots. J. Chem. Phys. **128**(6), 061101 (2008). https://doi.org/10.1063/1.2837456
209. Sarovar, M., Ishizaki, A., Fleming, G.R., Whaley, K.B.: Quantum entanglement in photosynthetic light-harvesting complexes. Nature Phys. **6**(6), 462–467 (2010). https://doi.org/10.1038/nphys1652
210. Shor, P.W.: Polynomial-time algorithms for prime factorization and discrete logarithms on a quantum computer. SIAM J. Comput. **26**(5), 1484–1509 (1997). https://doi.org/10.1137/S0097539795293172
211. Shukla, M.M., Padial, N.T.: A calculation of the Debye characteristic temperature of cubic crystals. Revista Brasileira de Física **3**(1), 39–46 (1973)
212. Slater, J.C.: The theory of complex spectra. Phys. Rev. **34**, 1293–1322 (1929). https://doi.org/10.1103/PhysRev.34.1293
213. Sorić, B.: Statistical "discoveries" and effect-size estimation. J. Am. Stat. Assoc. **84**(406), 608–610 (1989). URL http://www.jstor.org/stable/2289950
214. Sosnick, T.R., Snow, W.M., Sokol, P.E.: Deep-inelastic neutron scattering from liquid ^{4}He. Phys. Rev. B **41**, 11185–11202 (1990). https://doi.org/10.1103/PhysRevB.41.11185
215. Stearns, M.B.: Why is iron magnetic? Phys. Today **31**(4), 34–39 (1978). https://doi.org/10.1063/1.2994993
216. Summers, A.P.: Warm-hearted crocs. Nature **434**(7035), 833–834 (2005). https://doi.org/10.1038/434833a
217. Ter Haar, D.: The Old Quantum Theory. Pergamon Press, Oxford (1967)
218. Tolédano, J.C., Tolédano, P.: The Landau Theory of Phase Transitions. World Scientific Lecture Notes in Physics, vol. 3. World Scientific, Singapore (1987). https://doi.org/10.1142/0215
219. Tolman, R.: The Principles of Statistical Mechanics. Dover Books on Physics. Dover Publications, New York (1979)
220. Voltaire: Candide Ou L'Optimisme. Gallimard - Educa, Paris (2015)
221. Šafránek, D., Aguirre, A., Deutsch, J.M.: Classical dynamical coarse-grained entropy and comparison with the quantum version. Phys. Rev. E **102**, 032106 (2020). https://doi.org/10.1103/PhysRevE.102.032106
222. Wang, J., Yu, L., Hu, L., Chen, G., Xin, H., Feng, X.: Ambient ammonia synthesis via palladium-catalyzed electrohydrogenation of dinitrogen at low overpotential. Nature Commun. **9**(1), 1795 (2018). https://doi.org/10.1038/s41467-018-04213-9
223. Weir, A.J.: Lebesgue Integration and Measure. Cambridge University Press, Cambridge (1973). https://doi.org/10.1017/9781139171458
224. Werner, S.A., Colella, R., Overhauser, A.W., Eagen, C.F.: Observation of the phase shift of a neutron due to precession in a magnetic field. Phys. Rev. Lett. **35**, 1053–1055 (1975). https://doi.org/10.1103/PhysRevLett.35.1053
225. Wilson, K.G.: The renormalization group and critical phenomena. Rev. Mod. Phys. **55**, 583–600 (1983). https://doi.org/10.1103/RevModPhys.55.583. https://www.nobelprize.org/uploads/2018/06/wilson-lecture-2.pdf. Nobel Prize lecture
226. Winfree, A.T.: The prehistory of the Belousov-Zhabotinsky oscillator. J. Chem. Educ. **61**(8), 661 (1984). https://doi.org/10.1021/ed061p661
227. Wyatt, A.F.G.: Evidence for a Bose–Einstein condensate in liquid ^{4}He from quantum evaporation. Nature **391**(6662), 56–59 (1998). https://doi.org/10.1038/34134
228. Yamanaka, S., Koizumi, K., Kitagawa, Y., Kawakami, T., Okumura, M., Yamaguchi, K.: Chemical bonding, less screening, and Hund's rule revisited. Int. J. Quantum Chem. **105**(6), 687–700 (2005). https://doi.org/10.1002/qua.20784
229. Yang, X., Vorobiev, A., Jeppson, K., Stake, J.: Describing broadband terahertz response of graphene FET detectors by a classical model. IEEE Trans. Terahertz Sci. Technol. **10**(2), 158–166 (2020). https://doi.org/10.1109/TTHZ.2019.2960678

230. Zamljen, A., Lavrič, Ž., Prašnikar, A., Teržan, J., Grilc, M., Meden, A., Likozar, B.: Understanding platinum-based H_2 adsorption/ desorption kinetics during catalytic hydrogenation or hydrogen storage-related reactions. Renew. Energy **226**, 120467 (2024). https://doi.org/10.1016/j.renene.2024.120467
231. Zhang, K.X., Liu, Z.P.: Electrochemical hydrogen evolution on Pt-based catalysts from a theoretical perspective. J. Chem. Phys. **158**(14), 141002 (2023). https://doi.org/10.1063/5.0142540
232. Zurek, W.H.: Decoherence, einselection, and the quantum origins of the classical. Rev. Mod. Phys. **75**, 715–775 (2003). https://doi.org/10.1103/RevModPhys.75.715

Index

A
Absolute zero
  BEC fraction in superfluid, 166
  Fermi distribution, 149
  heat capacity
    Bose gas, 163
    Fermi gas, 147
    sound, 184
    vanishing, 21, 75
  number fluctuation, 146
  observation, 12
  Pauli blocking, 142
Adiabatic
  compressibility, 65
  decomposition
    emergence of clocks, 114
    feedbacks, 34
    golden rule, 179, 228
  volume change, 30

C
Chemical potential
  adsorption, 90
  barometric formula, 198
  Bose-Einstein condensate, 161
  change of pressure, 74
  classical ideal gas, 75
  classical limit, 112
  Cooper pairs, 157
  definition, 61
  equilibrium, 62
  Fermi energy, 149
  fixed number of particles, 90
  fixing, 27
  light, 170
  number fluctuation, 147
  origin, 82
  phenomenology, 15
  phonon, 185, 188
  position in gap, 141
  positive, 142, 152
  superfluid screening, 192

D
Definition of
  BEC transition temperature, 160
  effective temperature, 252
  Fermi temperature, 149
  generalized force, 61
  heresy
    first kind, 6, 135
    second kind, 135
  humanity, 38
  incidence matrix, 130
  level density, 143
    thermodynamic, 145
  magic, 4
  momentum fluctuations, 75
  mysticism, 3
  order, 38
  politics, 37
  profit, 36
    human, 38
    physical, 38
  purpose, 37
  sum over states, 84, 89
  superfluid transition, 167
  usefulness, 34

D. Sunko, *Conceptual Thermodynamics*,
https://doi.org/10.1007/978-3-032-20877-4

Dissipation
  coherent current, 168
  coherent transport, 192
  definition, 255
  Einstein's B-process, 178
  exponential relaxation, 117
  fuel consumption, 33
  governance, 70
  kinematic restrictions, 184, 191, 232
  Meissner effect, 104
  particle-hole excitations, 223
  quadratic form, 257
    Onsager relations, 258
  quantum coherence, 115
  scattering experiments, 95
  stabilizing sound waves, 184
  thermoelectric effect, 259

E

Effective mass
  difference from bare mass, 155
  gadolinium, 95
  reversible effects, 155
Engines
  Carnot, 48, 51
    black hole, 82
    definition, 17
    equilibrium fluctuations, 22
    macroscopic, 33
  economic, 39, 263
  efficiency, 18
    reversible, 19
  living cell, 135
  political, 79
    design, 263
  purposeful, 39
  useful, 261
    feedbacks, 35
    synchronization, 34
Entropy
  black hole, 82
  Boltzmann, 56, 93
    maximum, 93
  Bose gas, 162
  Clausius' theorem, 21
  coarse-grained, 257
  degenerate fermions or bosons, 107
  dissipation, 255
  Einstein's photons, 176
  emergent, 80
  equations of state, 59
  extensivity, 76
  Fermi gas, 147
  Fermi liquid, 153
  first law of thermodynamics, 62
  Gibbs, 56, 87
  Gibbs' paradox, 77
  heat differential, 87
  logical, 125
    increase by measurement, 129
    as probability, 126
    quantum, 129
    vs. relevant, 125
  macroscopic fluctuations, 255
  Massieu potentials, 28
  maximum at equilibrium, 23
  microscopic, 54, 73
  minimal internal energy, 24
  mixing, 77, 79
  natural generalized forces, 259
  negative temperature, 211
  non-extensive, 81
  paramagnet, 214
  physical profit, 38
  Planck's resonators, 175
  probability of fluctuation, 64
  production, 184
  quadratic form, 257
  quantum hydrodynamics, 193
  relevant, 55
  second derivative by volume, 65
  size of a reservoir, 21
  subadditive, 52, 78
  thermodynamic, 16
  time-dependent changes, 254
  von Neumann, 129

F

Feedbacks, 25, 34–36, 135, 253
Fermi liquid
  adiabatic interaction, 154
  effective mass, 155
  equation of state, 153
  microscopic justification, 154
  particle-hole excitations, 229
  perturbation theory, 154
  quasiparticles, 153
Fluctuation-dissipation theorem, 231, 246
  Einstein's relation, 241
  Kubo formula, 247
  temperature measurement, 247
  white noise, 250
Fluctuations
  adiabatic decomposition, 34
  canonical ensemble, 91
  Carnot engines, 48

composite noise, 251
counting, 44
critical, 164, 221
ensemble, 242
entropy, 123
equilibrium, 31, 68
heat, 22
kinematic restriction, 70
logical entropy, 125
mechanical stability, 48
momentum, 75
non-equilibrium, 255
number, 28, 90, 146, 165, 166
occupation probability, 111, 112
particle-antiparticle pair, 82
probability formula, 65
quantum, 149, 165, 166
random walk, 237, 242
reversible, 22
sound, 51, 183
transverse waves, 69
susceptibilities, 66
symmetry breaking, 37
thermal, 148
traffic, 238
volume, 65
Friction
black holes, 82
diagonalizing density matrices, 128
effective mass, 155
ensemble averaging, 80
gauge invariance, 103
irreversibility, 116
spontaneous emission, 179
irreversible, 58
macroscopic coherence, 103
measuring wave functions, 124
permutations of particles, 78
Planck's resonators, 174
quasiparticle collective, 94
reversible, 48, 58
static *vs.* dynamic, 17, 68, 231
superfluidity, 165
Fugacity, 110, 161
Functions of several variables
Carnot engines, 17
Clausius' theorem, 21
mathematical apparatus, 8
second-law constraint, 13

G
Generating function
canonical, 85
grand canonical, 89, 110
random walk, 234
Green's function, 238

H
Heat capacity
BEC, 163
Carnot's analysis, 21
classical ideal gas, 75
Debye model, 186
definition
physical, 8
statistical, 86
thermodynamical, 31
diatomic molecules, 199
Dulong-Petit, 95
Einstein model, 188
Fermi gas, 147
gapped, 189
light, 181
paramagnetic, 211
superconductor, 222

I
Indistinguishability
classical limit, 112
corrections due to, 109
critical fluctuations, 165
crowding in phase space, 102
diatomic molecules, 201
extensivity, 86
ground state
Bose gas, 159
Fermi gas, 140
kinematic correlation, 99
light
Bose's derivation, 180
photons, 180
Planck's derivation, 175
marking a particle, 98
paramagnets, 209
in principle, 76, 99
second quantization, 98
sound at low temperatures, 184
sound attenuation, 191
spin and statistics, 100
superfluid, 168
vs. entanglement, 127
Instability
administration, 67
anharmonicity, 66
BEC, 160

boson chemical potential, 110
gravitational, 113
negative temperatures, 211
political, 262
superconducting, 157
susceptibility, 31
tripartition of power, 67
van der Waals model, 205

M

Molecular hypothesis, 41
barometric formula, 197
Boltzmann's constant, 63
connection with thermodynamics, 56
integrating physics and chemistry, 45
photons, 181
stability, 1
time scales, 9
vanishing heat capacities, 97

O

Order parameter, 157, 161, 164, 166, 168, 192, 219, 221

P

Partition function, 84
Phase transitions, 219

Q

Quasiparticles
classical limit, 94
definition, 153
Einstein's photons, 180
Fermi liquid, 153
sound waves, 94, 180

S

Stability
adiabatic decomposition, 34
chance *vs*. necessity, 2
ensemble theory, 51, 101
hierarchy of time scales, 264
Le Chatelier's principle, 32
mechanism, 1
perturbation approach, 154
quantum coherence, 184
self-consistency, 40
society, 261, 264
sound fluctuations, 69
Sum over states
canonical, 84
Bose gas, 163
classical ideal gas, 85
classical limit, 112
Einstein model, 188
generating function, 89
grand canonical, 89
ideal Fermi or Bose gas, 110
light, 170
microcanonical, 84
ortho- and parahydrogen, 201
paramagnetic, 208
Planck's *vs*. Bose's derivations, 180
Planck's discrete-packet model, 173
simple quantum systems, 106
spin 1/2, 215
Superconductivity, 156
Landau-Ginzburg theory, 157, 221
Superfluid, 192
Superfluidity, 166
chemical potential, 193
quantum hydrodynamics, 194
Susceptibility
BEC, 165
collision integral, 191
criterion for understanding, 68
critical fluctuations, 221
definition, 30
doubly degenerate matrix, 207
Fermi-liquid spectrum, 155
heat capacity, 87
interpretation, 95
Jacobian matrix, 33
Le Chatelier's principle, 31
magnetic, 213
magnetization, 216
microscopic parameters, 36
nonlinear responses, 66
out of equilibrium, 255
phase transitions, 220
quasiparticle spectrum, 93
return to equilibrium, 58
scattering cross section, 232
social networks, 9
spontaneous volume fluctuations, 65
superfluid, 166
three meanings, 87
transport coefficients, 231
variance, 90

MIX
Papier aus verantwortungsvollen Quellen
Paper from responsible sources
FSC® C105338

If you have any concerns about our products,
you can contact us on
ProductSafety@springernature.com

In case Publisher is established outside the EU,
the EU authorized representative is:
Springer Nature Customer Service Center GmbH
Europaplatz 3, 69115 Heidelberg, Germany

Printed by Libri Plureos GmbH
in Hamburg, Germany